Neotropical
Rainforest Mammals
A Field Guide

Second Edition

Neotropical Rainforest Mammals

A Field Guide

Second Edition

Text by
Louise H. Emmons

Illustrations by
François Feer

The University of Chicago Press
Chicago and London

Louise H. Emmons is research associate in the Division of Mammals at the National Museum of Natural History, Smithsonian Institution. **François Feer** earned his doctorate in zoology at the University of Paris VI and the Museum of Natural History in Paris, where he is currently a researcher with the Centre National de la Recherche Scientifique, at the Laboratoire d'Ecologie of the Museum National d'Histoire Naturelle.

The University of Chicago Press, Chicago 60637
The University of Chicago Press, Ltd., London
© 1990, 1997 by The University of Chicago
All rights reserved. Published 1997
Printed in the United States of America

06 05 04 03 02 01 00 99 98 97   1 2 3 4 5

ISBN 0-226-20719-6 (cloth)
     0-226-20721-8 (paper)

Library of Congress Cataloging-in-Publication Data

Emmons, Louise
    Neotropical rainforest mammals : a field guide / text by
Louise H. Emmons ; illustrations by François Feer. — 2nd ed.
        p.   cm.
    Includes bibliographical references and index.
    ISBN 0-226-20719-6 (alk. paper). — ISBN 0-226-20721-8
(pbk. : alk. paper)
    1. Mammals—Latin America—Identification.   2. Rain
forest animals—Latin America—Identification.   I. Feer,
François.   II. Title
QL723.A1E44   1997   96-39088
599.1734'098—dc21   CIP

Nuestra responsabilidad moral es grande
frente a la extinctión de las especies
animales, sean éstas de valor económico
o no, sea su desaparición por culpa directa
de la actividad humana o no. Hay que
meditar profundamente sobre el hecho de
que el hombre no puede contemplar con
indiferencia la desaparición de una especie
a la que no podrá volver a crear, no importa
cuán intensamente lo desee.
—Marc Dourojeanni

We must assume a profound moral re-
sponsibility for the preservation of animal
species, whether they are of economic
value or not, and whether their extinction
results directly from our actions or not. One
cannot contemplate with indifference the
disappearance of a single species that can
never be recreated, no matter how deeply
we might wish to do so.

Contents

Illustrations

Maps

Only a sustained and serious commitment to research in tropical biology can provide a growing understanding and appreciation of the biological complexity of the tropics and contribute to the preservation and wise use of the diversity of life on earth. Thanks are due the Homeland Foundation for very generous support of the publication of the color plates in this book as an expression of that commitment.

Preface to the Second Edition

The seven and a half years since the manuscript of the first edition was completed have been dynamic times for Neotropical mammalogy. The number of rainforest species recognized has increased by more than 10%, from about 500 to 560–600 (depending on the primate taxonomy used). Five or more new Neotropical mammals are discovered each year: for example, one species of gray spiny mouse (genus *Scolomys*) was known in 1989, but two more have just been described, tripling the number of species in the genus. Most of the additions to the fauna were made by splitting additional species from already known ones, often by returning to earlier taxonomic views. Molecular genetics has become a standard tool for studying populations, and it is helping to clarify murky taxonomic questions.

As mammalogy dwindles in Europe and North America, it is flourishing in the tropics. Local mammalogists have snatched up the faltering torch and are adding much to our understanding of the mammals of their countries. This new edition was produced largely because the chance to publish a Spanish language translation provided an opportunity to revise the book, but also because extensive changes were needed. Almost every map has been improved, and nearly every species account has been altered. Nonetheless, many species remain poorly known, and much more fieldwork is needed. We hope we have improved the usefulness of this book, and apologize for the errors it still contains.

After the publication of the first edition, many individuals spontaneously wrote to give us their own observations of mammals and to suggest corrections. Some sent unpublished manuscripts and permitted us to incorporate their results into the new edition. Other colleagues were harassed for information by the authors, and all answered kindly. For their unsolicited time and effort or for answering specific questions, we especially thank S. Boinski, M. Carleton, J. Dalponte, G. Fonseca, L. Granjon, J. Hernández-Camacho, E. Mondolphi, J. Mead, R. H. Meade, P. Myers, N. Simmons, R. Òjeda, F. Olmos, G. Orihuela, C. Peres, A. Rylands, J.-C. Vie, M. de Vivo, and K. Zimmer. A. L. Gardner donated valuable data and information on many taxa. In an extraordinary act of generosity, T. R. Defler volunteered his unpublished maps of all primates of Colombia. T. Tarifa, L. Painter, and R. Wallace filled a large gap in the previous edition by providing Bolivian local names for mammals. Among his many contributions to almost every section of the book, we particulary thank R. Voss for his new descriptions of mouse opossums. For reading revised sections and improving the manuscript with their comments, we are grateful to C. O. Handley, Jr., E. Kalko, J. L. Patton, B. Stafford, and R. Voss. Kevin Flesher volunteered his time and skill to produce final copies of all the maps, and Erika Cuellar donated original drawings of tracks. The Smithsonian Institution once again provided the facilities and base from which to undertake this work. Its mammal collections, library, and staff were essential to this project. The usual words are inadequate to thank all for their support.

Preface

This is the first broad regional field guide to Neotropical mammals. We created it from scratch because there are no reference books for the kinds of information we needed. The knowledge of Neotropical mammals has many glaring gaps, which will take decades to fill at current rates of research. This book reflects those gaps and has many others as well, because undoubtedly we failed to unearth some sources of information. We hope that, despite its flaws, this book will fulfill its purpose of introducing the world's richest mammal fauna to its people, especially to those who live in the Neotropics. We describe the large array of mammals that live in New World lowland rainforests, what they look like, what they do, where they are, and which ones may be in danger and why. An appreciation of the mammals can come only from knowledge about them, and we hope that both knowledge and appreciation will help slow the forest destruction and overhunting that now jeopardize the future of many of the species described below.

Most of the information in this book is derived directly from specimens of mammals collected in museums. The descriptions of species, and most of the paintings in the plates, are based chiefly on material in three museums. First and foremost in its contribution to this book, and in our thanks, is the Division of Mammals of the United States National Museum of Natural History (Smithsonian Institution). Its collections were the foundation of this volume. The Smithsonian also provided many other essential facilities, including the desk at which this book was written, a base during the years of fieldwork in the Neotropics when Emmons studied its mammals, and the unflagging support of its staff. The collections of Brazilian mammals in the Museu de Zoologia, São Paulo, were of primary importance in describing geographic variation of large species, and we owe special thanks to Paulo Vanzolini for his hospitality there. The American Museum of Natural History was the third major source of specimens described and painted. We also made important use of collections in the British Museum of Natural History, the Muséum National d'Histoire Naturelle, in Paris, the Field Museum of Natural History in Chicago, and the Academy of Natural Sciences in Philadelphia. To all these museums and the many curators and staff who made us welcome and made this book possible, we express our deepest thanks. A grant from World Wildlife Fund–US supported essential museum work on the paintings.

A field guide is a product not only of its authors but of all the scientists and explorers who first discovered the species and described them, the many others who followed and collected the mass of information needed to draw each range map, and the ecologists, ethologists, and other biologists who studied the species' natural history. Many people freely shared their own superior knowledge of certain groups of mammals and donated their own unpublished observations or data to this book. J. L. Patton and P. Myers read the entire manuscript and provided much-needed criticism and additional data; A. L. Gardner generously allowed use of unpublished manuscripts and range maps; and G. Dubost contributed a large series of measurements and weights of large mammals, and critically reviewed the sketches for the paintings. All of the following read and/or added their expertise to individual sections, provided information, or helped resolve taxonomic problems: J. Cadle, M. Carleton, G. Creighton, D. Decker, J. Dietz, D. Domning, H. Greene, C. Handley, Jr., T. McCarthy, J. Mead, G. Musser, K. Redford, R. Thorington, Jr., R. Voss, D. Wilson, and P. Wright.

What Is in This Book, Where It Comes from, and How to Use It

The information in the account of each animal has been arranged under a number of headings. If a heading is missing, either no information was found (for example, we do not know what sounds some animals make) or none exists (most bats and mice have no specific local names).

Species included. This book includes the lowland rainforest mammals found in Central and South America at elevations below 1,000 meters (Maps 1, 2). These species for the most part form a discrete fauna: mammal communities within the rainforest region are very similar to one another in their composition of monkeys, opossums, sloths, bats, deer, and rodents. Communities outside the region tend to be abruptly and radically different, although there can be zones of transition into dryer or higher habitat types. About 560 species are on the current list of mammals for the region (Appendix E), but it is likely that several dozen more will be added in the future.

We treat the bats, murid rodents, and a few others to the level of genus only be-

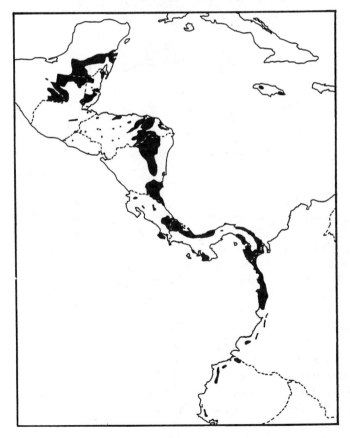

Map 1 Approximate current extent of rainforest in Central America (drawn and compiled from published sources and personal observation by Kevin Flesher).

Map 2 The four rainforest regions of South America (shaded) and major bordering habitats (after morpho-climatic domains of Ab'Saber, 1977, Univ. São Paulo Inst. Geografia, no. 52).

cause there are many similar species that are extremely difficult or impossible to identify with certainty in the hand, and because their systematics is in many cases poorly understood.

A few species known to occur only above 1,000 meters have also been included. These are members of lowland rainforest genera; by including these few extra species, all members of a genus are described (e.g., by including one montane species, all woolly monkeys and all New World monkeys are included). There are many borderline cases, and we have made some arbitrary decisions about which species to include. Our apologies if we have included our favorite borderline species, but not yours.

The way in which species are defined and arranged in this book generally follows the most recent literature. For marsupials, bats, carnivores, murid rodents, and some others, we largely follow the taxonomy in D. Wilson and D. Reeder (eds.), *Mammal Species of the World* (Washington, D.C.: Smithsonian Institution Press,

1993). The taxonomy of spiny rats and tree rats is based on Emmons's research. In the cases in which there is currently no adequate systematic view of a taxon and current classifications seem problematic, we have followed the advice of colleagues and our own intuition in wending our way through the maze of conflicting views. This book is not intended to solve systematic problems or to be a species-level systematic reference for scientists. Identification for scientific purposes, where proof of an animal's identity is needed, requires more detailed description than is possible in a field guide format, as well as examination of specimens with skulls, not just living animals. Many changes will be made in the taxonomy of Neotropical mammals as our understanding of them improves.

Common names. Many Neotropical mammals have no common names in English; some have none in any language. Because each account needed a common name, we invented many of those used ad hoc. We have tried to use descriptive or geographic names, or names derived from the scientific name. Where accounts are to genus only, as in bats, it seemed most parsimonious to simply use the scientific names for individual species rather than to invent common names that would have to be followed by the scientific name in every case. For help in understanding how animals are classified and how scientific names are composed and written, see the first section of Appendix C.

Arrangement of accounts. The accounts are arranged systematically to genus, subgenus, or group of closely related, similar species. Then species are arranged as follows: common species before rare or occasional species in a region; South American species before Central American species; and widespread species before species of limited distribution.

Measurements. Measurements and weights given in this book refer to adult, wild individuals. They give the usual or normal sizes of the animals encountered; some animals may fall outside these ranges, but they should be close to them. Most of the data were compiled directly from standard measurements taken in the field and recorded on museum specimen tags—the work of hundreds of individual collectors. In some cases published measurements from the scientific literature have been used. For rare species, the measurements are derived from few individuals, sometimes only one (single measurements in the text), so these may not accurately represent the species' size range. When a species has a geographic range much larger than the region covered in this book (e.g., Virginia opossum, northern raccoon), measurements refer only to animals from the rainforest region. Figure 1 shows how the measurements are taken. Metric units are used throughout; lengths are in millimeters unless otherwise indicated.

The following abbreviations are used to indicate measurements:

HB   Head and body length: tip of nose to inflection point of tail
T   Tail length: inflection point with body to tip of flesh
HF   Hindfoot length: heel to tip of claws, or, if followed by (su), without claws
SH   Shoulder height: point of shoulder blade to tip of toe, stretched
FA   Forearm length (bats only): elbow to wrist of folded wing

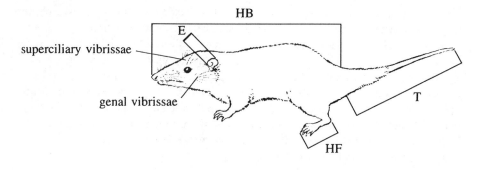

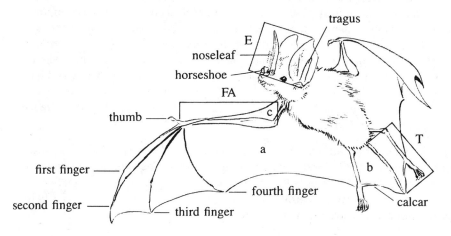

Figure 1. Measurements and parts of mammals: *(a)* wing membrane; *(b)* tail membrane or uropatagium; *(c)* antebrachial membrane or propatagium.

E   Ear length: bottom of notch to tip, on inner side
WT   Whole body weight of nonpregnant adults

The number and arrangement of the mammae often differ between genera of small rodents, so the pattern of mammae can be helpful in identification (fig. 2). The mammae are easy to find only in adult females that have nursed young.

It is important to know whether an animal is an adult. Signs of adulthood include the following: all molar teeth have erupted; body shape is that of an adult, with head and feet not disproportionately large; evidence of reproductive activity (in females, the mammae of the young are small, flat, dotlike, and unpigmented, while the mammae of parous females that have nursed young are permanently elongated and often pigmented—however, females of many mammals begin reproduction while still subadult, before full growth has been reached—e.g., humans; adult males of most species have large, descended testes at least during the breeding season, although opossums have large descended testes as juveniles); fur of young is often woollier, grayer, or duller than that of adults; in bats, the long finger bones in the wings of adults have epiphyses fused into knobs, while those of subadults are

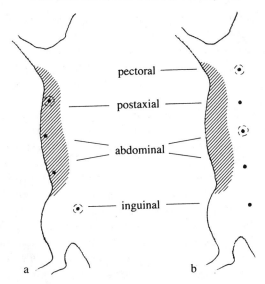

Figure 2. Mammae positions in ratlike rodents: *(a)* lateral and inguinal mammae of spiny rats and tree rats (Echimyidae); *(b)* ventral mammae of murid rodents (Muridae). Dotted lines surround mammae that are often absent; shading represents dorsal fur.

straight, with a translucent band of growing cartilage visible when the wing is held against a strong light.

Identification. The descriptions are intended to be useful both for animals seen from afar and for those in the hand. Much of the detail will not be visible from a distance. The most important distinguishing features are printed in boldface so that they can be rapidly scanned. The descriptions were taken directly from museum specimens and/or from life, with the exception of six accounts for which no specimens were available. For these we have used the original scientific description of the species, supplemented by other original published descriptions.

Variation. Our intent in this section is to give an idea of the general range and extremes of variation that can be encountered, as a help in identification. Because taxonomic problems concerning the definitions of species in this book mostly involve geographic variation in how species look, remarks on variation in color or size and problems or controversies about what to call the variants are treated together. The intricacies of disagreements between scientists over how to classify certain populations of animals will be irrelevant to most readers, but take on great importance when issues of conservation and systematics are involved. We point out where major apparent discrepancies might occur between this and other works. Appendix C contains a more detailed discussion of variation.

Similar species. This section usually includes only species found in the same geographic region as the animal in the account. Be sure to check range maps.

Natural history. We have tried to include only firsthand reliable information, both from published sources and from direct observations made by Emmons and scientific colleagues. Anecdotal information from hearsay or without a known source has been excluded; much of such folklore is correct, but much is also false, and it has an insidious way of converting itself into fact.

Geographic range. The entire known geographic range of each animal is described. The term Central America as used here includes southern Mexico and the region more correctly defined as Mesoamerica. Our knowledge of species distributions is often fragmentary, especially in the southern Amazon Basin, so some areas now blank on the maps will certainly be filled in in the future. Conversely, some areas now shaded will prove to be blank. Range maps are compiled from single points where animals have been proven to occur: for many species of the Amazon Basin, known points of occurrence are hundreds of kilometers apart, but nonetheless, the areas between them are shaded in. The effects of recent deforestation or exploitation of animal species are usually not shown, so particularly for Central America, eastern Brazil, and some sections of the Amazon Basin, the range maps are overly optimistic. Maps have been compiled from the literature and from museum specimens.

Status. Two levels of a species' status are included: (1) the 1994 official Convention on International Trade in Endangered Species (CITES) and United States Endangered Species Act (US-ESA) listings; and (2) the general situation of the species or genus, apart from official classifications, and factors that contribute to its status. For primates only, IUCN Red List categories are included, because only for primates are the listings based on the new Mace/Lande criteria.

   CITES has three classifications of species. CITES Appendix I includes species currently threatened with extinction. Special permits from both the country of export and that of import are required for importing any part of any species so listed into a country that is signatory, regardless of whether the country of export is signatory. Appendix II includes species that are not currently threatened but may become so unless trade is regulated. Appendix III includes species regulated for conservation purposes by particular countries. For species in Appendixes II and III, an export permit from the country of origin is needed for import to a signatory country. Abbreviations for countries with Appendix II and III restrictions are the same as those used to indicate country of origin of local names (see below).

   US-ESA has two categories: endangered and threatened. These are defined in the same way as CITES Appendixes I and II and are covered by similar import regulations, but they apply only to the United States. Of the species covered in this book, 33 are officially listed as endangered in at least part of their ranges, and another 37 as threatened, or a total of 70 species, 13% of all rainforest mammals.

   The general status information is derived mainly from Emmons's observations and the frequency with which the species have been recorded by scientists. "Common" animals usually occur at high densities; they are found in many places and are the species most likely to be seen. "Rare" species usually occur at low densities and/or seem to be entirely absent from many areas (patchy); they are either unlikely to be seen anywhere or may be seen only in a few localities. "Uncommon" species generally occur at low densities in many places; they are likely to be seen, but much less often than "common" species. The "rareness" of a small mammal can also be a function of how easy it is to catch, since small mammals usually cannot be identified unless caught. Some "rare" species may simply live in inaccessible places, such as canopy bats, or tend to avoid traps. If a rare monkey cannot be found after a search for it, one can be fairly certain that it is not present, but we do not yet know how, or where, to find rare small mammals in the rainforest. Therefore, we cannot evaluate the rarity of large and small mammals in the same way. We have nevertheless included statements about rarity of small mammals, both to alert people who find them to record any information they might get, and because some of these species may be truly rare and at potential risk of extinction.

Local names. Only the more widely used common names are included. The countries of use are abbreviated as follows: Argentina (Ar), Belize (Be), Bolivia (Bo), Brazil (Br), Central America (Cent Am), Colombia (Co), Costa Rica (CR), Ecuador (Ec), El Salvador (ES), French Guiana (FG), Guatemala (Gu), Guyana (Gy), Honduras (Ho), Mexico (Me), Nicaragua (Ni), Panama (Pn), Paraguay (Pa), Peru (Pe), Suriname (Su), Trinidad (Tr), Uruguay (Ur), Venezuela (Ve). Languages are abbreviated as follows: Spanish (Span), Guarani (Gua); Saramaca (Sar), Tupi-Guarani (TGua), Mayan (May), Quichua (Qui); only native languages widely used in more than one country are included. For brevity, if a name is used in many Spanish-speaking countries, we have listed it as (Span) rather than listing individual countries, even if it happens originally to be a Guarani or Quichua name, such as *anta* or *danta* (tapir).

Local names are often nonspecific and should be interpreted with caution: the same name may be used for all members of a genus or for unrelated animals of vaguely similar habits. The same name can be used for different animals in different parts of a country, or the same animal can have different names in different regions of a country. Local people who are not hunters (and even some who are) often do not know the names, or misapply them.

References. The references will help readers to find more information, but they do not necessarily represent the sources of the information used in the account, although they may do so. They sometimes represent an entirely different taxonomic view from that followed in this book. Many excellent, sometimes superior, works have been omitted; often a recent, brief paper with an up-to-date bibliography has been included rather than a far more comprehensive older work because the latter can be easily found by looking at the former, but not vice versa. We have tried to include a systematic reference and a behavior or ecology reference in each case.

Map 3 Major rivers of South America.

Species Accounts

Opossums (Marsupialia)

Didelphidae

Dental formula: I 5/4, C 1/1, P 3/3, M 4/4 = 50 teeth. All the teeth behind the canines are sharply pointed. Feet all with five toes; the first toe of the hindfoot is widely separated from the other digits, forming an opposable "thumb" used to grasp stems when climbing. Opossums are small to medium-sized mammals (10–3,000 g), with pointed snouts, short legs, long tails, and usually soft, dense fur. The tail of most species is strongly prehensile to its extreme tip, which can tightly grip an object as thin as a wire with many times the force needed to support the weight of the body. Some species will hang suspended by only the tail to reach a fruit, but this is a rare posture. Most opossums have large, delicate ears that can be curled back in pleats out of harm's way. At night they have bright, red to yellowish white eyeshine; the eyes appear small and far apart. Known diets are of insects and other invertebrates, small vertebrates, and some ripe fruit and nectar, but four species probably eat more fruit than animal matter, and one eats fish. All species give birth after a short gestation to tiny young that crawl up the mother's fur and attach by the mouth to a nipple, where they remain fastened for several weeks, until they are too large for the mother to carry them easily. The young of some species are protected within a pouch (marsupium) while they are attached to the nipples, but species in over half the genera have no pouch. Older young can be transported on the mother's back, but she rarely carries them. For several weeks after the young detach from the nipples they continue to nurse, but the mother leaves them behind in a nest while she forages. Species for which nesting habits are known make nests of dead leaves in sheltered places. The leaves are carried to the nest clutched in a coil of the tail tip (known in three genera). Recent studies suggest that opossums rarely live for more than one breeding season after they become adult.

The mouse opossums, formerly all in the genus *Marmosa,* have recently been divided into five genera, four of which occur in the rainforest region. The many species are similar externally, and much caution should be used when correct identifications are needed. Subadults look like adults, and even experts often mistake young in the hand for adults of a smaller, grayer species. A museum specimen is needed for definitive species identification, especially because new species are still being identified and the taxonomy of known forms has been undergoing frequent revision.

In most South American lowland rainforests, 1–2 species of *Marmosa, Micoureus, Monodelphis,* and *Gracilinanus* and 1–3 species of *Marmosops* can be expected to occur together, along with one each of *Didelphis, Caluromys, Metachirus,* and *Philander* (but there are zones of overlap where more than one congener of the latter co-occur). In Central America, one each of *Micoureus, Marmosa, Caluromys, Philander,* and *Metachirus* and 1–2 *Didelphis* usually coexist. Because they are difficult to inventory and identify, the complete fauna of mouse opossums is known for few localities. The opossum family is restricted to the Americas, with one species in North America and about 70 species in Central and South America.

Western Woolly Opossum

Caluromys lanatus

Plate 1, map 4

Identification. Measurements: HB = 217–295; T = 330–435; HF = 39–50; E = 31–41; WT = 310–410 g.

Upperparts red-brown to pale brown, red brightest over shoulders, forelegs, and hindlegs. Fur long, dense, and woolly. **Head gray, face with prominent dark stripe down center;** reddish brown eye rings extend in dark streak from corner of eye to nose; eye dark brown, eyeshine bright yellow-orange; ears naked, pinkish tan. **Tail thickly furred above for proximal half, and below for one-fifth; tail tip naked, whitish mottled with brown spots near base of naked area.** Feet redbrown or dark gray. Underparts yellowish white with more grayish midsection. Females seem to develop a pouch only when they are carrying young.

Variation. Animals from the Magdalena Valley of Colombia are paler, grayish brown, with gray sides, sometimes with completely gray underparts; those from Paraguay are pale brown, without mottling on tail. Rare

Map 4

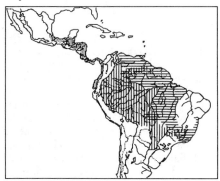

▦ Central American woolly opossum, *Caluromys*
    derbianus
▤ Bare-tailed woolly opossum, *C. philander*
▥ Western woolly opossum, *C. lanatus*

individuals have a gray patch between the
shoulders.
Similar species. Black-shouldered opossums *(Caluromysiops irrupta)* have black
shoulders and tail furred dorsally to the tip;
bushy-tailed opossums *(Glironia venusta)*
have tail furred dorsally to the tip and two
black stripes on head. Woolly mouse opossums *(Micoureus* spp.) have no dark stripe
on center of face and tail furred only near
base; bare-tailed woolly opossums *(Caluromys philander)* have tail fur ending near base.
Sounds. None usually heard in the field.
Natural history. Nocturnal; arboreal; solitary. Feeds on fruit, probably a few invertebrates, and in the dry season drinks flower
nectar. Western woolly opossums favor
dense, viny, midstory and canopy vegetation, but they are also seen in tall, open forest. Found in mature and secondary evergreen rainforest, disturbed forest, gallery
forest, and gardens and plantations.
Geographic range. South America: east of
the Cordillera Central of Colombia to western Guyana; east of the Andes in Ecuador,
Peru, Bolivia, Paraguay, Argentina, and
W Brazil. To at least 1,100 m elevation.
Status. Locally common to rare; formerly
hunted for its fur, which is no longer in
demand.
Local names. Cuica lanosa (Ar, Br);
mucura-chichica (Br); chucha lanosa, chucha rata, cinco dedos, marta (Co); mbicuré
lanoso, raposa, zorra lanuda (Ec).

Bare-tailed Woolly Opossum
Caluromys philander
Plate 1, map 4
Identification. Measurements: HB =
160–279; T = 250–405; HF = 33–40;
E = 30–40; WT = 140–390 g.
**Upperparts uniform warm reddish brown.
Head gray, face with dark brown stripe
down center,** dark brown eye rings and
streak between eye and nose; ears brown,
usually with pale yellow spot at upper and
lower corners; eyeshine bright yellow, eyes
appear small. Tail furred for first tenth of its
length; **most of tail naked, usually heavily
mottled cream and dark gray to tip, or
completely gray,** rarely tip pale and without
spots. Underparts completely orange, or
midbody grayish. Feet gray or whitish. Fur
short, dense, and soft in lowlands, longer at
higher elevations. Females have poorly developed pouch only when they are carrying
young.
Similar species. Woolly mouse opossums
(Micoureus spp.) have no central stripe on
face. Brown and gray four-eyed opossums
(Metachirus nudicaudatus, Philander spp.)
have pale spots over eyes; western woolly
opossums *(Caluromys lanatus)* have fur
reaching halfway down tail; tree rats *(Echimys* spp., *Isothrix* spp.) have small ears and
nonprehensile tails.
Sounds. None usually heard in the field.
Hisses when cornered, gives a loud distress
call when taken by a predator.
Natural history. Nocturnal; arboreal; solitary. Feeds mainly on fruit and some nectar,
with about 20% of the diet invertebrates.
Bare-tailed woolly opossums use the middle
and upper levels of the forest, where they
are sometimes seen running swiftly along a
vine or branch or sitting motionless looking
down. They seem to favor dense viny vegetation but also use the high open forest
canopy. During the day they den in a leaf
nest or tree hollow. Found in mature and
secondary rainforest, dry forest, gallery forest, and gardens and plantations.
Geographic range. South America: east
of the Andes in E Venezuela, Trinidad, the
Guianas, and northeastern and south central
Brazil, west to Santa Cruz Department, Bolivia. To at least 1,800 m elevation.
Status. Often common; thrives in disturbed
vegetation.

Local names. Mucura-chichica (Br); awari (Su).
References. Charles-Dominique, P., M. Atramentowicz, M. Charles-Dominique, H. Gérard, A. Hladik, C. M. Hladik, and M. F. Prévost. 1981. Les mamifères frugivores arboricoles nocturnes d'une forêt guyanaise: Inter-relations plantes-animaux. *Rev. Ecol.* 35:341–435.
    Charles-Dominique, P. 1983. Ecology and social adaptations in didelphid marsupials: Comparison with eutherians of similar ecology. In J. F. Eisenberg and D. G. Kleiman, eds., *Advances in the study of mammalian behavior,* 395–422. Am. Soc. Mammal. Spec. Pub., no. 7.

Central American Woolly Opossum
Caluromys derbianus
Plate 1, map 4
Identification. Measurements: HB = 225–300; T = 384–445; HF = 32–47; E = 35–40; WT = 245–370 g.
Upperparts rich reddish brown; back with pale gray patch between shoulders; hips and body behind shoulders pale gray; fur long, dense, soft, and woolly, slightly frosted. Head pale gray with dark brown stripe down center of face, brown eye rings merging to dark streak between eye and nose; ears naked and whitish or pink. Forelegs and feet creamy white; hindfeet brown. Underparts yellowish white. Tail thickly furred above for basal 30–50% of length, below for 25%, usually darker than back; tail tip naked, pale, mottled with brown spots, especially near middle. Females have a pouch only when carrying young; the pouch apparently regresses to small folds at other times. Young slightly grayer than adults.
Variation. Races from E Nicaragua and mountainous areas of the Pacific slope of the Costa Rica–Panama border are mainly or entirely pale gray, with indistinct facial markings and a variably pronounced brownish tinge over shoulders and lower back. Occasional individuals lack the gray patch between the shoulders.
Similar species. No other mammals in range have tail half furred. Gray and brown four-eyed opossums *(Metachirus nudicaudatus, Philander opossum)* have pale spots above the eyes, nonmottled tails, and no pale patch

between the shoulders. Woolly mouse opossums *(Micoureus alstoni)* are much smaller.
Sounds. None usually heard in the field.
Natural history. Nocturnal; arboreal; solitary. Feeds mostly on fruit, probably supplemented by nectar and invertebrates. Central American woolly opossums make nests of leaves and den in tree holes. Found in mature and disturbed evergreen rainforest, dry forest, and gardens and plantations.
Geographic range. Central and South America: Veracruz, Mexico, south to the Cauca Valley of Colombia and W Ecuador. To 2,500 m elevation.
Status. Seems locally common; formerly hunted for the fur trade, but skins are no longer in demand.
Local names. Chucha lanosa, chucha rata, marta (Co); zorra, raposa lanuda, comadreja baya (Ec); guazalillo dorado, zorra lanoso (Ho); zorra roja, comadreja roja (Pa); zorro de balsa (CR).
References. Bucher, J. E., and R. S. Hoffmann. 1980. *Caluromys derbianus.* Mammalian Species, no. 140.

Black-shouldered Opossum
Caluromysiops irrupta
Plate 1, map 5
Identification. Measurements: HB = 250–330; T = 310–340; HF = 47–52; E = 29–37.
Upperparts frosted grayish brown, sides paling to buffy or whitish; large black patches across shoulders extend backward in two narrow stripes beside the spine and fade to brown on rump, and forward as black inner foreleg and top of forefoot; fur long, dense, and soft. Head gray-brown, cheeks below ear pale buffy or dirty white, indistinct brown spots between eye and nose; ears bright yellow on inside, round at tips; nose pink; eyeshine bright yellow. Tail tapered, furred above to tip, below for basal 25%, naked on distal 75%, frosted dark brown above for about 60%, dirty white at tip. Underparts buffy, with orange on throat and inguinal region. Young like adults.
Similar species. Bushy-tailed opossums *(Gironia venusta)* are much smaller, with tan ears, no black on shoulders, and black stripes down the face. Lutrine opossums *(Lutreolina crassicaudata)* are plain yellow-brown with no markings. All other opossums

Map 5

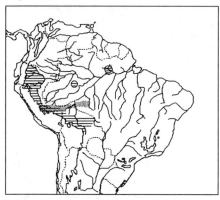

▨ Black-shouldered opossum, *Caluromysiops irrupta*
▤ Bushy-tailed opossum, *Glironia venusta*

in the region have naked tail tips. Tree rats (*Echimys, Isothrix* spp.) have small, inconspicuous ears and no black markings on shoulders. Squirrels have bushy tails.
Sounds. None heard in the field.
Natural history. Nocturnal; arboreal; solitary. In the dry season feeds on the nectar of flowers; presumably also eats fruit. Black-shouldered opossums use the upper levels of the forest and rarely seem to descend even to the middle levels. They move slowly and are most often seen sitting motionless on a branch. They will spend hours in the same flowering tree, feeding periodically on nectar, and will hang suspended by only the tail to reach flowers on the underside of a branch. Known only from mature rainforest.
Geographic range. South America: east of the Andes in SE Peru and adjacent W Brazil. Known from only three localities. Records from Leticia, Colombia, are of unknown origin as animals from Peru are sent there for export. To 700 m elevation.
Status. Generally extremely rare and patchy in distribution, but in some years can be locally common. Known from fewer than 30 specimens.
Local names. Cuíca (Br).
References. Janson, C. H., J. Terborgh, and L. H. Emmons. 1981. Non-flying mammals as pollinating agents in the Amazonian forest. *Biotropica, Repro. Bot. Supp.,* 1–6.
   Izor, R. J., and R. H. Pine. 1987. Notes on the black-shouldered opossum, *Caluromysiops irrupta. Fieldiana Zool.,* n.s., 39: 117–24.

Bushy-tailed Opossum
Glironia venusta
Plate 1, map 5
Identification. Measurements: HB = 160–205; T = 195–225; HF = 27–31; E = 22–25.
Upperparts **pale or cinnamon brown;** fur soft and dense, velvety or woolly. **Face with a broad black stripe on each side, from nose through eye to crown,** separated by a gray stripe down center of face from nose to ears; ears naked and brown or blackish. **Tail fully furred to tip** on top and sides, distal 60% below covered with short bristles, tail tip white or darker than base. Feet dirty white. Underparts gray or white tinged with orange. A female specimen has no pouch.
Similar species. Black-shouldered opossums *(Caluromysiops irrupta)* are larger, gray with black shoulders; lutrine opossums *(Lutreolina)* have no facial stripes, all other opossums have naked tail tips; brush-tailed rats *(Isothrix* spp.) have small ears and blunt muzzles.
Natural history. Nocturnal; arboreal; solitary. One seen at night in dense viny vegetation about 15 m above the ground ran around the vines quickly and with agility, jumping from one to another in a manner unlike that of other arboreal opossums. It seemed to be hunting insects but at one point spent many seconds licking the surface of a branch. Another, seen in a Bolivian dry forest, emerged from an 8 m high tree hole at nightfall and dashed up into the canopy. These rare, poorly known opossums may be completely arboreal insectivores. Known from rainforest and dry forest.
Geographic range. South America: the Amazon Basin of Ecuador, Brazil, Peru, and Bolivia. To elevations of 1,000 m. Known from only nine localities.
Status. Apparently extremely rare; known only from nine individuals, each from a different locality.
References. Marshall, L. G. 1978. *Glironia venusta.* Mammalian Species, no. 107.

Common Opossum
Didelphis marsupialis
Plate 1, map 6
Identification. Measurements: HB = 324–425; T = 336–420; HF = 51–70; E = 46–58; WT = 565–1,610 g; males larger than females.

Map 6

Common opossum, *Didelphis marsupialis*
Southeastern common opossum, *D. aurita*
White-eared opossum, *D. albiventris*

Upperparts **black or gray, fur in two layers, dense pale yellow underfur shows below long, coarse black or gray outer guard hairs.** Head dirty yellow, sometimes with indistinct black lines from nose through eye nearly to ear, and down center of crown from between ears to eye; cheeks yellow, pale orange, or dirty white, not sharply contrasting with color of muzzle; nose pink; **ears** large, naked, **black;** feet black. **Tail** usually longer than head and body, **naked, black with white tip,** black part usually shorter than white; **long hair on tail base reaches a shorter distance than stretched back length of leg and foot.** Underparts usually like back, but paler, or orange. Eyeshine bright, whitish yellow. Females have a pouch. Typical posture a crouch with back humped high and long fur bristling, sometimes in a ridge along spine, and tail coiled. Small young blackish, with pale heads, sharp facial markings, and white-tipped ears, or colored like adults.
Variation. Rainforest animals are usually black; gray animals seem more common in drier habitats, but black and gray phases occur together in the same populations. Some animals are brown from split ends on overhairs and fading with age. Rare animals have whitish instead of yellow underfur.
Similar species. Virginia opossums *(D. virginiana)* have white cheeks, sharply contrasting with the rest of the head, and the black part of the tail is usually equal to or longer than the white; white-eared opossums *(D. albiventris)* have white ear tips and sharp black-and-white facial markings, and the

furred portion of the tail is longer. See the following species. Juvenile *Didelphis* may be difficult or impossible to identify to species. Four-eyed opossums *(Philander* spp. and *Metachirus nudicaudatus)* have pale spots over eyes and short fur in only one obvious layer.
Sounds. Hisses with open mouth when cornered.
Natural history. Nocturnal; arboreal and terrestrial; solitary. Feeds chiefly on small animals—insects, worms, or small vertebrates, including snakes—with about a quarter of the diet consisting of fruit, and in the dry season sometimes nectar. Common opossums usually travel and feed on the ground, but will climb to feed on fruit or nectar in the treetops. They also climb to escape danger and to rest in the middle of the night. They travel widely during a night's activity. They den in tree hollows, leaf nests in tree forks or dense vines, or sometimes in burrows (dug by other animals). Malodorous animals, they roll in fresh dung, and they squirt vile-smelling, burning urine and defecate when handled, distributing the effluent by twirling the tail. In some areas populations explode to high levels, followed by a crash and virtual absence of the species the following year. When handled, they do not "play dead" but are generally aggressive and will threaten with open mouth, lunge, and bite when cornered or grasped. Found in humid forests and outlying gallery forests; they thrive in secondary forests and around dwellings, where they feed in garbage dumps.
Geographic range. Central and South America: Mexico to Bolivia, Paraguay, and NE Argentina, Trinidad, and the Lesser Antilles. To about 2,000 m elevation.
Status. Common, hunted for meat only where other game is scarce.
Local names. Comadreja grande (Ar); chucha común (Co); comedreja (Bo, Pa); carachupa (Bo); mucura, zorro, muca, gambá (Br, Pe); zorro chucha, gambá, jujube (Co); raposa, yalu, zorra grande, zarigüeya (Ec); pian (FG); zorra, zorro pelón, zarigüeya (Pn, CR); intuto (Pe); grote buidelrat (Su); rabipelado (Ve); mbicuré (Gua); dagoe awari (Sar); tacuazin (Cent Am); ux (May).
References. Gardner, A. L. 1973. The systematics of the genus *Didelphis* (Marsupialia: Didelphidae) in North and Middle America. Spec. Pub. Mus. Texas Tech., no. 4.

Sunquist, M. E., S. N. Austad, and F. Sunquist. 1987. Movement patterns and home range in the common opossum *(Didelphis marsupialis). J. Mammal.* 68:173–76.

Southeastern Common Opossum
Didelphis aurita
Map 6
Identification. Measurements: HB = 310–390; T = 310–370; HF = 54–60; E = 47–55; WT = 700–1,500 g.
Similar to common opossum, facial markings prominent; forehead with well-developed black center line; underfur dirty yellow, overhairs black or gray; ears black; tail with furred base as long as hindleg, black and white parts nearly of equal length, or black part longer.
Variation. This species is sometimes considered a subspecies of *D. marsupialis,* to which it is similar externally. The two forms are not known to overlap in geographic range, so site of origin can be used for preliminary identification.
Similar species. Common opossums may have a shorter black part of the tail, shorter haired portion of the same, and more indistinct facial mask. Distinguishing features between the two species have not been published, although they are said to be separable. At present, specimens are best identified by their geographic origin. White-eared opossums have white or white-tipped ears, sharp black facial markings, and whitish underfur.
Natural history. Nocturnal; arboreal and terrestrial; solitary. Diet of about equal parts fruit and insects, including beetles, ants, and butterflies. Inhabits humid tropical lowland forests on the Atlantic coastal plain, lower montane continuous forests, and *Araucaria* highlands.
Geographic range. Coastal Brazil from Bahia to Rio Grande do Sul, west to the lower Rio Paraguay. To 1,000 m elevation.
Status. May be threatened by habitat loss.
Local names. See names for common opossum.
References. Cerqueira, R. 1985. The distribution of *Didelphis* in South America (Polyprotodontia, Didelphidae). *J. Biogeography* 12:135–45.
   Leite, Y. L. R., L. P. Costa, and J. R. Stallings. 1996. Diet and vertical space use of three sympatric opossums in a Brazilian

Atlantic forest reserve. *J. Trop. Ecol.* 12:435–40.

White-eared Opossum
Didelphis albiventris
Plate 1, map 6
Identification. Measurements: HB = 305–437; T = 290–430; HF = 45–68; E = 41–60; WT = 500–2,000 g.
Similar to the common opossum except: head white with sharp black bars from ear through eye to nose, or black eye ring only; black stripe coming to a sharp point on crown of head at level of eyes; ears white, or black at base with broad white tips; throat white; tail usually slightly shorter than head and body. Overhairs often white; underfur usually white. Tail with furred basal portion as long as stretched-out hindleg and foot. Underparts white, gray, or yellowish.
Variation. There are several geographically separated populations in radically different habitats (high Andes, chaco). These could represent more than one species, in which case *D. pernigra* would apply to Andean animals, *D. imperfecta* to northern lowland and Tepui animals, and *D. albiventris* to southern dry forest opossums.
Similar species. See common opossum.
Natural history. Nocturnal. This opossum is not generally a rainforest species, but occurs in more open, arid or temperate regions adjacent to lowland humid forests, where it may rarely coexist with the common opossum. Has a wide habitat and elevational range in savannas, gallery forests, swamps, cultivated or deforested agricultural lands, cerrado, caatinga, chaco, and in humid forests at high elevations and subtropical latitudes, or near savannas.
Geographic range. South America: in and east of the Andes around the periphery of the rainforest from W Venezuela, Colombia, and the Guianas south to Ecuador, Peru, Bolivia, Paraguay, Uruguay, N Argentina, and SE Brazil. To at least 4,000 m elevation.
Status. Often common.
Local names. Saruê, cassaco, gambá (Br); comadreja mora, comadreja overa (Ar, Ur); carachupa (Bo); chucha de orejas blancas, fara, runcho (Co); zorra, raposa, zarigüeya (Ec); mbicuré (Gua). See names for common opossum.

References. Tyndale-Biscoe, C. H., and R. B. Mackenzie. 1976. Reproduction in *Didelphis marsupialis* and *D. albiventris* in Colombia. *J. Mammal.* 57:249–65.

Virginia Opossum
Didelphis virginiana
Plate 1, map 7
Identification. Measurements: HB = 370–501; T = 295–470; HF = 50–80; E = 34–60; WT = 0.5–2.3 kg; males larger than females.
Externally similar to the common opossum in the rainforest region, **except: cheek below eye white, white extending posteriorly at least to group of whiskers below and forward of ear, and contrasting sharply with color of the rest of the head; tail with black and white parts usually approximately equal, or black part longer than white part, or all black.** Young black with white head prominently marked with a black bar from ear through eye to nose, and black of body extending forward as a sharp point above eyes.
Variation. Central American Virginia opossums are smaller and darker than those in North America, which are usually pale gray.
Similar species. See common opossum.
Sounds. Hisses with mouth open when cornered.
Natural history. Nocturnal; arboreal and terrestrial; solitary. Feeds on insects and other invertebrates, small vertebrates, carrion, fruit, and other plant matter. Virginia opossums usually travel and feed on the ground, but will climb readily. By day they den in tree holes, hollow logs, brush piles, and rock outcrops. When disturbed, they may "play dead" by lying inert on the side with mouth open and salivating. Found in many types of forested and open habitats, both humid and arid, and often in towns and around garbage dumps.
Geographic range. North and Central America: southern Canada to northwestern Costa Rica. In southern part of its range to 3,000 m elevation.
Status. Common.
Local names. Zorro, zorra (Span); tlacuache (Me); ux (May).
References. Gardner, A. L. 1973. The systematics of the genus *Didelphis* (Marsupialia: Didelphidae) in North and Middle

Map 7

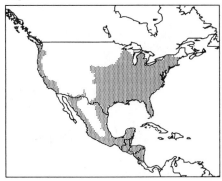

Virginia opossum, *Didelphis virginiana*

America. Spec. Pub. Mus. Texas Tech., no. 4.
———. 1982. Virginia opossum. In J. A. Chapman and G. A. Feldhamer (eds.), *Wild mammals of North America,* 3–36. Baltimore: Johns Hopkins University Press.

Common Gray Four-eyed Opossum
Philander opossum
Plate 1, map 8
Identification. Measurements: HB = 250–302; T = 253–315; HF = 35–50; E = 25–40; WT = 200–660 g.
Upperparts uniform sooty gray; fur short, dense, and soft, finely grizzled at close range by white bands on hairs. **Face with black mask around eyes and across crown, large white spots over each eye and at base of ear;** chin and lower cheeks creamy white; ears large, naked, rims black, centers pale; nose, lips above chin, and toes pink; lips at corner of mouth black. Tail dark gray at base, naked, tip white, sharply demarcated in a line, thickly furred with continuation of body fur for first 5–8 cm (about 17% of its length). Underparts yellowish white, orange, or chest and belly pale gray. The female has a pouch, stained orange if she has had young. Scrotum of males black. Young like adults.
Variation. Upperparts vary from pale to dark gray; animals from south of the Amazon in Pará, Brazil, are brownish gray; those from Paraguay are pale gray; those from west of the Andes in N Ecuador are dark gray, with gray bellies and completely dark gray tails. Molecular evidence indicates that the Atlantic forest form is distinct

Map 8

▨ Anderson's gray four-eyed opossum, *Philander andersoni*

▥ Common gray four-eyed opossum, *P. opossum*

▤ Mcilhenny's four-eyed opossum, *P. macilhennyi*

(P. frenata). The generic name of this taxon is disputed, and is sometimes given as *Metachirops.*

Similar species. Anderson's gray four-eyed opossums *(P. andersoni)* have a black mid-dorsal streak contrasting with paler sides, and ears always > 35 mm; Mcilhenny's gray four-eyed opossums *(P. mcilhennyi)* are black, with long fur and a dark belly; brown four-eyed opossums *(Metachirus nudicaudatus)* are brown with buff eyebrow spots and buff grizzling in fur, and tail without sharply demarcated white tip; water opossums *(Chironectes minimus)* have broad black bands across the back.

Sounds. No sounds usually heard in the field; hisses with open mouth when cornered; young emit sharp clicks.

Natural history. Nocturnal; arboreal and terrestrial; solitary. Feeds chiefly on invertebrates and small vertebrates, supplemented by fruit. Four-eyed opossums use the ground to middle vegetation levels of the forest; they are most common around treefalls and especially in dense undergrowth near water, where they forage like small carnivores, winding in and out of fallen brush, along fallen logs, and among low vegetation. They hunt frogs by homing in on their calls. When disturbed they often climb up into a tree, and they rest during the night on elevated perches. They swim well and voluntarily. Their nests may be 8–10 m above the forest floor, in hollow trees or in tree forks, or occasionally in hollow logs or treefalls on the

ground. Found in mature and secondary rainforest, gardens and plantations, and gallery forests.

Geographic range. Central and South America: S Mexico to Panama, west of the Andes to N Peru, east of the Andes to Paraguay and NE Argentina. To 1,500 m elevation.

Status. Often common.

Local names. Guaiki (Ar); carachupa (Bo); rajá, rabipelao, cucha gris de cuatro ojos (Co); mucura-de-cuatro-olhos, cuíca (Br); cuica común, zorro, raposa huevera (Ec); pericote (Pe); quica opossum, fo-ai awari (Su); zorro de cuatro ojos (Pa, CR); mbicuré (Gua).

References. Atramentowicz, M. 1986. Dynamique de population chez trois marsupiaux didelphidés de Guyane. *Biotropica* 18:136–49.

Anderson's Gray Four-eyed Opossum
Philander andersoni
Plate 1, map 8

Identification. Measurements: HB = 223–288; T = 267–305; HF = 36–45; E = 33–45; WT = 335–600 g.

Like the common gray four-eyed opossum except: back along midline from the neck to the furred portion of the tail black or blackish in a narrow to broad strip, with long glossy black guard hairs, distinctly darker than sides. Ears usually larger. Underparts pale gray washed with yellow, grayish buff, with or without an orange tinge, or pale gray mottled with cream. Feet black.

Variation. Color varies from pale gray or gray-brown on sides, with a narrow band of longer, glossy black hairs on midback, to almost completely black.

Similar species. See common gray four-eyed opossum *(P. opossum).*

Natural history. Nocturnal; terrestrial and arboreal; solitary. Anderson's gray four-eyed opossums favor dense viny vegetation near water. They are often confused with common gray four-eyed opossums, with which they probably share many characteristics. Found in mature and disturbed lowland rainforest.

Geographic range. South America: east of the Andes in the upper Amazon Basin of Venezuela, probably Colombia, and all of eastern Ecuador.

Status. Locally common.
Local names. Cuica común, zorro, raposa huevera (Ec); comadreja negra de cuatro ojos (Ve).

Mcilhenny's Four-eyed Opossum
Philander mcilhennyi
Map 8
Identification. Measurements: HB = 287–307; T = 265–332; HF = 36–51; E = 35–39; WT = 396–640.
Similar to the common gray four-eyed opossum except: **upperparts black or blackish, with long, glossy black guard hairs;** sides rarely paler gray. **Tail base with very long, dense, glossy black hairs extending down tail for about 25% of its length. Underparts dark gray,** tinged with yellow or orange. Feet and legs black; toes may have some white.
Variation. Animals from Brazil and Balta, Peru, are blackish overall; those from farther north in Peru seem to have paler sides.
Similar species. See common gray four-eyed opossum. Mcilhenny's opossums are blacker than Anderson's, with longer fur and darker underparts.
Natural history. Little known, probably similar to common four-eyed opossum. Found in lowland evergreen rainforests.
Geographic range. South America: Ucayali Department of Peru, and Acre and Amazonas, Brazil.
Local names. Same as common gray four-eyed opossum.
Status. Usually seems to be rarer than the common gray four-eyed opossum, with which it occurs over all of its geographic range.
References. Gardner, A. L., and J. L. Patton. 1972. New species of *Philander* (Marsupialia: Didelphidae) and *Mimon* (Chiroptera: Phyllostomidae) from Peru. *Occas. Pap. Mus. Zool. Louisiana State Univ.,* no. 43.

Water Opossum
Chironectes minimus
Plate 1, map 9
Identification. Measurements: HB = 260–298; T = 327–420; HF = 60–74; E = 25–32; WT = 590–700 g.
Upperparts frosted silver gray with four broad black or dark brown bars across back joined by a narrow black stripe down spine, **looks marbled;** fur dense and soft.

Map 9

Water opossum, *Chironectes minimus*

Head black, with U-shaped pale band from ear to ear above eyebrows, cheeks white; ears short; eyes black; chin and area around mouth almost naked; whiskers long and stiff. Tail furred for first tenth, the rest naked with large scales, black with a short white tip. **Hindfeet large, completely webbed between all toes,** black; forefeet not webbed, toes long, thin, with tips expanded into large pads; a bone of the hand (the pisiform) is enlarged into an accessory sixth "finger"; surface of palms rough like sandpaper, fingers and palms pale. Underparts creamy white or pale yellow. When swimming, appears completely black. Eyeshine brilliant yellow, eyes medium-sized and far apart. Both sexes have a pouch. Young similar to, but blacker than adults.
Variation. Some animals from drier areas are dark brown; the black parts of preserved skins of this species usually fade to brown.
Similar species. These are the only opossums with webbed feet. Gray four-eyed and common opossums (*Philander* spp., *Didelphis* spp.) have no bars across the back and have large ears; all other genera of large opossums are brown. Small carnivores have fully furred tails. See also lutrine opossum (*Lutreolina*).
Sounds. When threatened, gives sharp, screechlike spitting barks.
Natural history. Nocturnal; terrestrial and semiaquatic; solitary. Feeds on fish, crustaceans, and invertebrates that it catches in the water. The long fingers and sandpaper-like palms are used to probe for and capture prey and grasp slippery fish. Water opossums are found in and near rivers and streams.

They are active, quick-moving animals that swim and run rapidly along watercourses, swimming with just the eyes and top of the head above water. They are not wary and are easy to see where present, as a bright pair of eyes speeding along a stream. The female's pouch has a watertight seal to keep the young dry when she swims, and the male's scrotum is pulled into the pouch when the animal is in the water. They build nests of leaves and grasses, and by day they den in burrows in stream banks, under roots, or sometimes in nests on the ground surface. Found in tropical forests and cleared areas in tropical forest regions. Most records are from clear rivers, lakes, and streams in hilly areas; may be rare or absent from silt-laden lowland watercourses.

Geographic range. Central and South America: S Mexico and Belize to W Venezuela, Colombia, Ecuador, Peru, Bolivia, and W Brazil; the Guianas and mouth of the Amazon; and SE Brazil, Paraguay, and N Argentina. To elevations of 1,800 m.

Status. Seems common where it occurs but appears to be absent from many regions.

Local names. Cuica de agua (Ar, Pe); mucura do fundo, cuíca-d'agua (Br); chucha de agua, runcho de aqua (Co); yapok (Be, FG); raposa de agua, comadreja (Ec); perrito de agua (Ho); zorro de agua (Pa, CR); watrastonawari, watra-alata (Su); yapó (Gua).

References. Marshall, L. G. 1978. *Chironectes minimus.* Mammalian Species, ño. 109.

Lutrine Opossum

Lutreolina crassicaudata
Map 10, figure 3

Identification. Measurements: HB = 243–400; T = 245–358; HF = 38–53; E = 22–28; WT = 350–910 g.

Upperparts uniform pale tawny yellow to medium brown; fur dense and soft; underparts and underfur with or without infusion of glowing violaceous dark pink, which also suffuses upperparts; **no prominent facial markings,** muzzle dark brown. **Tail long, thick, and robust, fully furred;** furred like back for proximal third, **distal two-thirds dark brown,** with shorter, flat hairs; extreme tip white or brown, with small naked gripping surface below. **Ears short,** rounded, brown, naked at tip; eyes dark brown; whiskers fine, short, do not reach ears; feet and

Map 10

Lutrine opossum, *Lutreolina crassicaudata*

hands darker brown than body; claws white, long and slender. The female has a pouch. A medium-sized, **weasel-like opossum,** with a long body, short legs, and very long tail.

Variation. The striking reddish tone of the underparts and underfur of some animals is a fugitive color that vanishes from preserved specimens and skins. Underparts on skins are dirty orange. Color varies widely from yellow to brown, and size also shows great variation, even within populations. An animal from Colombia had a red muzzle. Small, dark red animals from the wet montane forests of S Bolivia and Argentina may represent a distinct species.

Similar species. This is the only plain opossum with no markings and a thick, fully furred tail. True weasels have a short tail; otters are much larger, dark brown, with tiny ears.

Sounds. Hisses when threatened. Calls with a high-pitched whistle.

Natural history. Nocturnal; terrestrial, but a good climber. Lutrine opossums are mainly carnivorous and feed on invertebrates and small vertebrates. They are closely associated with aquatic habitats, from flooded grasslands to small streams. In behavior as well as appearance lutrine opossums resemble weasels: they are agile, quick, and feisty hunters that kill their prey with a bite to the neck. They build a nest (of grass in grassland) and den amid groundcover vegetation, in tree hollows, or in old burrows of other species. These opossums are mainly known from wet savannas near forests in the northern part of the geographic range, but are found within humid forests along water-

courses in their southern range, as well as in wet grassland and savanna woodland habitats. They occupy stream courses in mossy, wet montane forest ravines in S Bolivia and N Argentina, but are not recorded from within closed Amazonian rainforest.
Geographic range. In two disjunct and locally fragmented populations: in Venezuela, Colombia, and the Guianas and in Bolivia, Paraguay, Brazil, Peru, Uruguay, and Argentina. To at least 2,000 m elevation in the Andes of S Bolivia.
Local names. Comadreja colorada (Bo); comadreja coligruesa (Ve).
References. Marshall, L. G. 1978. *Lutreolina crassicaudata.* Mammalian Species, no. 91.

Map 11

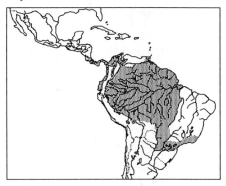

Brown four-eyed opossum, *Metachirus nudicaudatus*

Brown Four-eyed Opossum
Metachirus nudicaudatus
Plate 1, map 11
Identification. Measurements: HB = 250–280; T = 280–369; HF = 38–49; E = 28–39; WT = 300–480 g; males slightly larger than females.
Upperparts brown; center of back sometimes blackish; fur short, dense, and soft, finely grizzled with buff. **Head with a large pale yellow spot above each eye** and at the upper base of the ear; **crown black,** sometimes extending as a black stripe on neck; corner of eye to nose black; ears naked, brown; nose gray; eyes dark brown, eyeshine bright orange-yellow. **Tail naked from near its base, entirely brown or gray** or gradually paling toward tip. Underparts and lower cheeks uniform clear pale yellow or orange,

scrotum pinkish. The female has no pouch. A large, long-legged, cursorial opossum.
Variation. Some animals from Pará, Brazil, and Paraguay are grayish brown, with sides tinged pale orange-brown.
Similar species. Water opossums *(Chironectes minimus)* have black bars across back; gray four-eyed opossums *(Philander spp.)* are gray with white grizzling on fur, and have abruptly demarcated white tail tips; no other opossums have pale spots above the eyes.
Sounds. Chatters teeth rapidly when cornered, but normally silent.
Natural history. Nocturnal; terrestrial; solitary. Feeds on termites, other insects and invertebrates, and fruit. Brown four-eyed opossums seem to favor mature forest with little undergrowth, but they are also present

Figure 3. Lutrine opossum, *Lutreolina crassicaudata.*

in dense habitats. Unlike other opossums, these are extremely nervous and wary. They travel on the ground and along the tops of fallen logs, trotting or walking quickly and silently. When alarmed, they run rapidly and soundlessly away on the ground. They are difficult to observe because they run at the smallest disturbance, but seem most active and easy to see immediately following rainstorms. Their nests are in well-hidden hollows in or near the ground. Found in mature and disturbed rainforest, secondary forest, and gallery forest.

Geographic range. Central and South America: Costa Rica south to Paraguay and NE Argentina.

Status. Widespread and often common; lives at low population densities.

Local names. Cuica común (Ar); mucura de cuatro-olhos, jupati (Br); chucha mantequera (Co); raposa, zorra chica (Ec); zorricí (CR); pericote (Pe); froktu-awari, bruine opossum (Su); zorra morena (Pa); yupatí (Gua).

References. Miles, M. A., A. A. De Souza, and M. M. Póvoa. 1981. Mammal tracking and nest location in Brazilian forest with an improved spool-and-line device. *J. Zool.* (Lond.) 195:331–47.

Long-furred Woolly Mouse Opossum
Micoureus demerarae
Plate 2, map 12

Identification. Measurements: HB = 152–210; T = 195–270; HF = 25–30; E = 26–30; WT = 62–130 g; males larger than females.

Upperparts smoky gray brown; fur long, soft, and woolly, about 1 cm long on rump. Eye rings wide, black, moderately prominent, restricted to circle around eye and slight extension toward nose; cheeks to below ears pale orange; ears naked, brown; eyeshine bright yellow. Tail thickly haired for first 3–5 cm, rest naked, gray brown, usually with long whitish tip, mottled with dark spots around dark-white junction, or entirely dark, sometimes with white spots near tip. Feet yellowish, robust. Underparts orange, yellowish, or gray with orange wash; fur of underparts entirely or partially gray-based. Females have no pouch. Scrotum of males blue. Young grayer than adults, with shorter fur and base of tail less furry, but fur

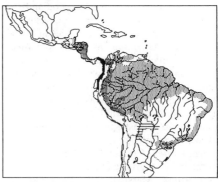

Map 12

Long-furred woolly mouse opossum, *Micoureus demerarae*

Pale-bellied woolly mouse opossum, *M. constantiae*

Alston's woolly mouse opossum, *M. alstoni*

Short-furred woolly mouse oppossum, *M. regina*

Little woolly mouse opossum, *M. phaea*

extending well onto tail. A large mouse opossum.

Variation. The genus *Micoureus* has recently been redivided from one into five species, as recognized here. There is variation among populations, however, and the distinctions between the forms are not always clear. There are probably additional species, and the geographic ranges are likely to be revised. From the accounts in this book it may not be possible to assign species names to some populations.

Similar species. Little, short-furred, and pale-bellied woolly mouse opossums *(M. phaea, M. regina, M. constantiae)* have furred portion of tail base usually shorter than 2 cm. Other rainforest mouse opossums are all smaller, have short, velvety fur, or are tiny, with no white tail tip; bare-tailed woolly opossums *(Caluromys philander)* have a dark stripe down the center of the face and no black eye rings.

Sounds. None usually heard; hisses when cornered.

Natural history. Nocturnal; arboreal; solitary. Feeds on insects and small animals, fruit, and nectar. Woolly mouse opossums are usually seen in the middle to upper levels of the forest, but they will descend to the ground when food is scarce in the dry season. They favor dense viny vegetation with many palm trees but are also found in open, high forest. They build a nest of dead leaves in the crown of a palm tree or in a vine tangle. This is a genus of aggressive species that will bite fiercely and hang on tenaciously if grasped. Found in mature and secondary evergreen rainforest, gallery forest, and gardens and plantations.

Geographic range. South America: Colombia to N Argentina, Paraguay, and E Brazil.

Status. Widespread and often common.

Local names. Raposa (Ec, Pe); muca (Pe); moismoisi-awari (Su); anguyá-guaik (Gua).

References. Charles-Dominique, P., M. Atramentowicz, M. Charles-Dominique, H. Gérard, A. Hladik, C. M. Hladik, and M. F. Prévost. 1981. Les mamifères frugivores arboricoles nocturnes d'une forêt guyanaise: Inter-relations plantes-animaux. *Rev. Ecol.* 35:341–435.

Miles, M. A., A. A. de Souza, and M. M. Póvoa. 1981. Mammal tracking and nest location in Brazilian forest with an improved spool-and-line device. *J. Zool.* (Lond.) 195:331–47.

Short-furred Woolly Mouse Opossum
Micoureus regina
Map 12

Identification. Measurements: HB = 114–151; T = 147–236; HF = 20–26; E = 22–25; WT = 35–68 g.

Upperparts brown, tinged with ochraceous yellow or pinkish; **fur short, 5 mm on rump. Eye rings narrow,** indistinct, extending to nose. Cheeks and chin orange, pinkish, or buff. Ears brown. **Tail entirely brown, occasionally with white mottling at extreme tip;** haired part of tail base shorter than 3 cm. **Underparts gray at hair base, washed with bright orange to buff.**

Variation. Quite similar to the little woolly mouse opossum, but larger and with more orange ventral tones and woollier fur. The two are sometimes considered conspecific. They are geographically separated by the Andes.

Similar species. Long furred woolly mouse opossums *(M. demerarae)* are larger and grayish, with long woolly fur on body and tail base. Slender mouse opossums *(Marmosops* spp.) have white or gray underparts without yellow or orange tones.

Natural history. Similar to long-furred woolly mouse opossum.

Geographic range. South America: the western Amazon Basin and eastern Andean slopes from Colombia to S Peru and possibly to Bolivia; Brazil at least to the Rio Juruá. To 1,634 m elevation in Andean foothills.

Local Names. Raposa (Ec).

Status. Unknown.

Little Woolly Mouse Opossum
Micoureus phaea
Map 12

Identification. Measurements: HB = 121–170; T = 153–236; HF = 20–26; E = 20–25; WT = 63 g.

Upperparts warm brown, fur short, 3–4 mm on rump; **eye rings narrow,** indistinct, extending to nose. Muzzle with short, narrow dark brown stripe on center behind nose. Cheeks and chin buff. Ears brown. **Tail entirely brown, occasionally with white mottling at extreme tip; haired part of tail base shorter than 1.5 cm. Underparts around midbody gray at hair base, washed with buff or pale orange,** chin buff or orange.

Similar species. Alston's woolly mouse opossums have long, fluffy fur stretching for 3–5 cm onto tail base. Robinson's mouse opossums are warm red-brown without any gray hair bases on underparts.

Natural history. Nocturnal; arboreal; solitary. Natural history probably like that of other woolly mouse opossums. Known from wet evergreen lowland rainforests and wet montane forests.

Geographic range. Central and South America: Panama from the Canal Zone south to N Colombia and W Ecuador.

Pale-bellied Woolly Mouse Opossum
Micoureus constantiae
Map 12

Identification. Measurements: HB = 134–150; T = 186–201; HF = 23–26; E = 20–25; WT = 68–73 g.

Upperparts medium brown, **fur** straight and **quite short, 7 mm** on rump, or longer and woolly. Eye rings brown, not extending to ears; ears brown. **Cheeks, chin, and all of underparts bright pink to clear yellow, without gray at hairbase.** Tail base thickly furred for 2 cm, naked distally; brown at base, **distal half to one-third pure white,** sharply demarcated from brown portion. Scrotum pink.

Variation. Two individuals of *M. c. budini* caught by Emmons had short fur and were bright pink on underparts. It is not clear whether this color fades to yellow on museum skins, or both colors occur in nature. Mouse opossums like those described occur in N Argentina and S Bolivia in montane forest; those from the lowlands seem to have longer fur, and perhaps pure yellow venters.
Similar species. Long-furred woolly mouse opossums have long woolly fur and gray-based fur on part or all of underparts. No other mouse opossums in range have a pure white tail tip.
Natural history. Nocturnal; arboreal. From moist forest habitats, including mossy montane forest.
Geographic range. South America: on the lower Andean foothills in SW Bolivia and NW Argentina *(M. C. budini);* to Mato Grosso, Brazil. To 2,000 m elevation.
Status. Unknown.
Local names. Catita, gaiquica (Br).

Alston's Woolly Mouse Opossum
Micoureus alstoni
Map 12
Identification: Measurements: HB = 155–180; T = 250–281; HF = 28–30; E = 25–29.
Upperparts medium brown, **fur long and woolly, 1 cm** on rump and tail base. Eye rings prominent. **Tail with large puff of fur 3–5 cm onto base,** rest naked, **proximal half of tail brown, distal half pure white,** demarcated in a line. **Underparts** washed whitish to ochraceous yellow, **fur gray at base.**
Variation. Animals from the lowlands in Costa Rica and Nicaragua have relatively short fur.
Similar species. Woolly opossums are much larger, with red shoulders and thighs and a sharp mid-facial stripe; Robinson's mouse opossums are reddish, with short fur and no gray hair base on underparts. Short-furred woolly mouse opossums *(M. regina)* are smaller, with short fur and short furred portion of tail base.
Natural history. Nocturnal; arboreal. Feeds on fruit and insects. A nest in the branches of a tree was made of green leaves. Found in lowland and montane evergreen forests.
Geographic range. Central and South America: Belize to N Colombia, but ap-

parently absent from Panama. To 1,600 m elevation.
Status. There are few specimens, which suggests that this species is uncommon or rare.
Local names. Zorra, zorricí (CR).

Murine Mouse Opossum
Marmosa murina
Plate 2, map 13
Identification. Measurements: HB = 125–150; T = 170–198; HF = 20–24; E = 20–25; WT = 43–60 g; males larger than females.
Upperparts uniform warm brown; fur short, velvety. Eye rings large, prominent, black; center of face between eye rings pale; cheeks pale yellow or orange. **Underparts delicate salmon or seashell pink, yellowish, or cream, not pure white.** Tail with short hair for about 1.5 cm at base, rest naked, tan. Feet whitish. Females have no pouch. Scrotum of males blue.
Variation. Animals from Mato Grosso, Brazil, are pale fawn. The delicate, luminous pink of the underparts of live animals fades and eventually disappears from skins.
Similar species. See woolly mouse opossums *(Micoureus* spp.); red mouse opossums *(Marmosa rubra)* are bright chestnut, with a bicolored tail; northern gracile mouse opossums *(Gracilinanus marica)* have long fur and gray-based fur on belly; other small mouse opossums in the region have white or gray bellies, except young woolly mouse opossums, which have woolly fur.
Sounds. None usually heard in the field; makes explosive clicks in threat.
Natural history. Nocturnal; arboreal; solitary. Diet consists of about two-thirds insects and other small animals, one-third fruit. Murine mouse opossums use the ground and lower levels of the forest, almost always below 5 m. They cruise the short undergrowth or walk slowly on the forest floor, exploring the undersurfaces of living and dead leaves for invertebrate prey. Especially common in disturbed zones of roadsides, swampy platanillos, extremely dense vines, thickets, river edges, and secondary forest, and usually rare in tall, open-understory, terra firme forest. Inhabits mature and secondary rainforest, plantations, gardens, camps, and houses.
Geographic range. South America: east of the Andes in Venezuela, the Guianas, the

Map 13

Map 14

▦ Murine mouse opossum, *Marmosa murina*
▤ Mexican mouse opossum, *M. mexicana*

▦ Robinson's mouse opossum, *Marmosa robinsoni*
▥ Red mouse opossum, *M. rubra*
▤ Little rufous mouse opossum, *M. lepida*

Amazon Basin of Colombia, Ecuador, Peru, Bolivia, and Brazil, and Tobago. To at least 1,365 m elevation.
Status. Widespread and often common.
Local names. Gambasinha, guaiquica, catita (Br); ratón de anteojos, ratón fara, ratón runcho (Co); raposa, zorra chica (Ec); muca, rapposa (Pe); boesi-mois-moisi (Su).
References. Charles-Dominique, P., M. Atramentowicz, M. Charles-Dominique, H. Gérard, A. Hladik, C. M. Hladik, and M. F. Prévost. 1981. Les mamifères frugivores arboricoles nocturnes d'une forêt guyanaise: Inter-relations plantes-animaux. *Rev. Ecol.* 35:341–435.

Red Mouse Opossum
Marmosa rubra
Plate 2, map 14
Identification. Measurements: HB = 121–165; T = 180–220; HF = 22–26; E = 18–23; WT = 59–67 g.
Upperparts deep chestnut-red; fur velvety. Eye rings black, prominent, especially pronounced between nose and eye; face with dark stripe on center above nose, snout appears black, cheeks yellow-orange; ears brown, relatively short for a mouse opossum. Tail with short furred base darker than back above, rufous below, rest naked, bicolored dark above, pale below. Underparts rich orange; chin, throat, inguinal region, and stripe down midventer paler orange-yellow. Outer hind- and forelegs and backs of hands blackish, toes pale. The female has no pouch.

Similar species. Murine mouse opossums *(M. murina)* are paler brown, less red, with face distinctly pale from nose to between eyes. See little rufous mouse opossum *(M. lepida).*
Natural history. Has been trapped on the ground and in the thatched roof of a house. Found in lowland rainforest and secondary forest.
Geographic range. South America: the Amazon Basin of Colombia, Ecuador, and Peru.
Status. Can be locally common, may have a patchy distribution.

Little Rufous Mouse Opossum
Marmosa lepida
Plate 2, map 14
Identification. Measurements: HB = 97–120; T = 140–150; HF = 16–19; E = 15–18; WT = 10 g.
Upperparts uniform bright chestnut or orange-red; fur long and dense. Eye rings pitch black, prominent; ear short, red brown, reaches middle of eye when laid forward. Tail very long, slender, pale brown, lightly coated with white hairs beneath. Underparts creamy white or tinged pinkish; scrotum thickly furred with cream hair. Feet cream. Females have no pouch. Young like adults. A tiny opossum.
Similar species. This is the only tiny red mouse opossum in its range. Red mouse opossums *(M. rubra)* are larger, with an orange belly and bicolored tail; murine

mouse opossums *(M. murina)* are larger and browner; Emilia's mouse opossums *(Gracilinanus emiliae)* are smaller, with shorter fur, smaller hindfeet, and browner, narrower eye rings, and are probably indistinguishable from *M. lepida* in the field, but the ranges are not known to overlap.
Natural history. Found in rainforest.
Geographic range. South America: the Amazon Basin of Colombia, Ecuador, Peru, Bolivia, and Brazil; and from Suriname.
Status. Rare but widespread; known from fewer than 10 localities and 20 specimens.

Robinson's Mouse Opossum
Marmosa robinsoni
Plate 2, map 14
Identification. Measurements: HB = 128–200; T = 170–239; HF = 20–30; E = 23–30; WT = 36–132 g; males larger than females.
Upperparts cinnamon brown; fur quite long (8–10 mm at midback). **Eye rings black, large, and prominent;** face between eye rings pale, cheeks to behind ear pale orange-yellow; ears tan, large. Tail naked, tan, relatively thick and robust. Feet large, cream or whitish above. **Underparts pale orange-yellow,** nipple area of parous females bright orange; scrotum of males blue. Females have no pouch. Young like adults.
Variation. Animals from coastal Venezuela and Ecuador are pale cinnamon-gray, with yellow-white underparts.
Similar species. Murine mouse opossums *(M. murina)* are smaller, with pinker underparts; see that species and keys for other comparisons.
Sounds. None usually heard in the field, but makes explosive clicks in threat and an insectlike twitter when excited.
Natural history. Nocturnal; arboreal and terrestrial; solitary. Feeds mainly on insects, with some fruit. Robinson's mouse opossums use the low undergrowth and ground. Found in a wide variety of habitats from evergreen or cloud forest to deciduous and thorn forest and secondary vegetation.
Geographic range. Central and South America: Belize and Guatemala south, west of the Andes from Colombia to N Peru, and east of the Andes in a narrow coastal strip from N Colombia east to the mouth of the Orinoco in N Venezuela; Roatán, Trinidad and Tobago. To 2,000 m elevation.

Status. Locally common.
Local names. Marmota (Pn).
References. O'Connell, M. A. 1983. *Marmosa robinsoni.* Mammalian Species, no. 203.

Mexican Mouse Opossum
Marmosa mexicana
Map 13
Identification. Measurements: HB = 85–144; T = 145–190; HF = 18–25; E = 19–25; WT = 40–100 g; males larger than females.
Upperparts cinnamon to warm reddish brown, **sides paler; fur short** (6–7 mm). **Eye rings prominent,** dark, face between eye rings pale, cheeks yellow or pale orange. Tail densely furred for a short section of base, rest naked, solid gray brown, relatively thick and robust. Feet whitish. **Underparts pale yellow or orange, usually mixed with gray around midbody,** nipple area of parous females stained orange. Females have no pouch.
Variation. Animals from Nicaragua are rich cinnamon; those from Mexico are browner.
Similar species. Robinson's mouse opossums *(M. robinsoni)* are difficult to distinguish from this species in the field. They have longer fur and are larger than Mexican mouse opossums. The two overlap geographically. See other mouse opossums.
Sounds. None usually heard in the field.
Natural history. Nocturnal; arboreal and terrestrial; solitary. Feeds on insects and probably fruit. Mexican mouse opossums are found on or near the ground. A leaf nest was in a burrow in the ground. Most common in moist evergreen forests, but also found in dry forests, plantations, and even arid grasslands.
Geographic range. Central America: Taumalipas, Mexico, south to W Panama. To 1,675 m elevation.
Status. Locally common.
Local names. Ratón tlacuache (Me); zorra (Pa, CR).
References. Hall, E. R., and W. W. Dalquest. 1963. The mammals of Veracruz. *Univ. Kans. Pub. Mus. Nat. Hist.* 14:165–362.

White-bellied Slender Mouse Opossum
Marmosops noctivagus
Plate 2, map 15

Map 15

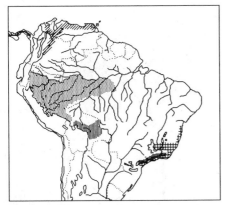

| | White-bellied slender mouse opossum, *Marmosops noctivagus* |

São Paulo slender mouse opossum, *M. paulensis*
Dorothy's slender mouse opossum, *M. dorothea*
Gray slender mouse opossum, *M. incanus*
Slaty slender mouse opossum, *M. invictus*
Gray-bellied slender mouse opossum, *M. fuscatus*

Identification. Measurements: HB = 125–160; T = 160–190; HF = 19–22; E = 18–25; WT = 55–85 g; males larger than females.

Upperparts slate brown, sides of neck and body slightly paler and with more orange tones; fur short, soft, and velvety. Eye rings black, often indistinct; eyes black; ears brown. Tail thinly haired for first 1 cm, the rest naked, uniform dark brown. Outer two hairs of triad between tail scales broad and petiolate (a unique feature of all but one *Marmosops* species). **Underparts and cheeks pure creamy white** to hair base, except nipple area orange on parous females. Feet whitish, hands brown above, toes pale. Females have no pouch. A **relatively large** mouse opossum.

Variation. Recent genetic evidence shows that an Andean form *(M. albiventris)* may belong to the same species. The species of this genus are not yet well understood. Changes can be expected in taxonomy, with more species recognized.

Similar species. See woolly mouse opossums *(Micoureus* spp.). Other mouse opossums in range are smaller if they have white underparts, or are warm brown, or have gray or partly gray underparts; short-tailed opossums *(Monodelphis* spp.) have short tails.

Kalinowski's gracile mouse opossum *(Gracilinanus kalinowskii)* is much smaller (see measurements), with relatively larger ears.
Sounds. None heard in the field. Hisses when cornered.
Natural history. Nocturnal; arboreal and terrestrial; solitary. Feeds on insects and fruit. These mouse opossums use the ground and lower vegetation levels of the forest, where they run around rapidly when active, especially on fallen trees, in dense undergrowth thickets, and up and down vines. They seem to favor dense platanillos *(Heliconia* spp.) in swamps and watersides. They build a nest of dead leaves. Found in mature, disturbed, and secondary forests.
Geographic range. South America: Amazon Basin of Ecuador, Peru, W Brazil south of the Amazon, and Bolivia.
Status. Common.
Local names. Rapposa (Pe); achocaya (Qui).
References. Creighton, G. K. 1984. Systematics and taxonomy of the marsupial family Didelphidae. Ph.D dissertation, University of Michigan, Ann Arbor.

Gray-bellied Slender Mouse Opossum
Marmosops fuscatus
Map 15
Identification. Measurements: HB = 120–155; T = 148–182; HF = 18–26; E = 22–29; WT = 41–104 g.
Upperparts gray brown; fur short and slightly stiff. Eye rings black, prominent or indistinct, face between eyes pale, **cheeks whitish. Tail robust,** gray, becoming paler toward tip, slightly bicolored pale below. Feet and lower legs whitish. **Underparts uniform gray frosted with white or yellowish;** tail base dark brown below. Females have no pouch. Young gray. Large, robust mouse opossums.
Similar species. These are the only large, grayish, short-furred, gray-bellied mouse opossums in their range. Andean slender mouse opossums *(M. impavidus)* have a white stripe or patches on midbelly, and a slender, dark tail. Robinson's *(Marmosa robinsoni)* and murine *(M. murina)* mouse opossums have pale pink, orange, or yellow bellies with fur pale at base; northern gracile mouse opossums *(Gracilinanus marica)* are smaller and have long soft fur; all are warm brown.

Natural history. Nocturnal; arboreal and
terrestrial. Probably feeds on insects and
fruit. Gray-bellied slender mouse opossums
seem to use the ground and lower vegetation
levels and are often found near water. They
mainly inhabit montane forest and cloud
forest above 1,000 m but can occur lower in
some localities. Found in evergreen forest
and clearings within it.
Geographic range. South America: northern mountains of Venezuela, Colombia, and
Trinidad. To 2,400 m elevation.
Status. Locally common.
Local names. Chucha ratón (Co).

Delicate Slender Mouse Opossum
Marmosops parvidens
Plate 2, map 16
Identification. Measurements: HB =
95–105; T = 130–160; HF = 14–17;
E = 19–23; WT = 15–27 g.
**Upperparts deep velvety smoky brown.
Dark eye rings** variable, but **often inconspicuous;** muzzle and face between eyes
pale, sometimes with a narrow dark central
stripe above nose; cheeks white or gray; ears
dark gray. Upper canine teeth small and triangular, hardly taller than tallest cheek teeth.
Tail completely naked, uniform gray brown.
**Underparts usually a mixture of gray and
white:** from gray with a white stripe down
center of belly to pure or mostly white, usually with gray spots on chest. Feet white.
Females have no pouch. **A tiny opossum
with a very long tail, long, pointed muzzle,
matchstick-thin legs, and tiny hands and
feet** with thin digits. Resembles a juvenile of
other species.
Variation. Animals from lowland Peru have
pure white underparts. A closely related
and similar form, *M. cracens,* is known only
from three specimens from the type locality
in Falcon, Venezuela. Genetic differences
show that *M. parvidens* may be a complex
of several species.
Similar species. See woolly mouse opossums (*Micoureus* spp.). *M. cracens* may not
be distinguishable in the field. Kalinowski's
gracile mouse opossum is even smaller, with
a relatively shorter tail and large ears. Other
slender mouse opossums in the region are
larger.
Sounds. None usually heard.
Natural history. Nocturnal; arboreal and
terrestrial; solitary. Feeds on insects and

Map 16

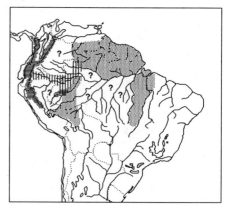

Delicate slender mouse opossum, *Marmosops
parvidens*
Andean slender mouse opossum, *M. impavidus*
Cerro Neblina slender mouse opossum, *M. neblina*

fruit. Delicate mouse opossums use the forest understory near the ground, usually in
tall, open-understory, terra firme forest. They
are slower-moving than most other mouse
opossums, and they are often found sitting
still perched on a low shrub or a branch of a
treefall. When disturbed they run a few feet
up into a sapling, where they stop and can
be caught by hand. Found in mature, closed
canopy, evergreen forest, not often in disturbed or secondary forest.
Geographic range. South America: east of
the Andes in the Amazon Basin of Venezuela, Colombia, the Guianas, Brazil, and
Peru.
Status. Usually uncommon, occasionally
locally common.
References. Pine, R. H. 1981. Reviews of
the mouse opossums *Marmosa parvidens*
Tate and *Marmosa invicta* Goldman (Mammalia: Marsupialia: Didelphidae) with a
description of a new species. *Mammalia*
45:55–70.

Andean Slender Mouse Opossum
Marmosops impavidus
Map 16
Identification. Measurements: HB =
116–152; T = 146–205; HF = 17–20;
E = 21–24; WT = 36–51.
Upperparts warm dark brown, fur 7–9 mm
on rump. **Eye rings large, black, contrast-
ing with pale brown muzzle and face**

between eye rings; cheeks pale orange to buff. Tail with short furred base 1.5–2 cm; rest naked, entirely dark brown above and below. Forefeet dusky above, toes white; hindfeet white above. Underparts with fur entirely gray-based or with complete or partial narrow, white or pale stripe down midline; whole of underparts distinctly paler than upperparts, strongly washed with orange, cream, whitish, tawny, or pinkish brown. Scrotum white-furred.
Variation. An animal from the Cordillera del Condor, Peru, has a dark face and indistinct eye rings; animals from Darién, Panama, are orange-frosted below, with orange cheeks; those from N Colombia are cream-washed below. There is uncertainty about the correct species name for these variable opossums. At least eight subspecies have been defined; some of these may be valid species.
Similar species. See Cerro Neblina slender mouse opossum.
Natural history. Nocturnal, mainly terrestrial. Chiefly known from wet cloud forest above 1,500 m; a few records are from far out into the Amazon lowlands.
Geographic range. South America: Panama on high mountains of Darién; Andes from N Colombia to Cochabamba, Bolivia; lowlands near the Rio Amazonas to W Brazil. To at least 2,400 m elevation.
Local names. Cuíca (Br); chucha ratón (Co); raposa (Pe).
References. See following species.

Cerro Neblina Slender Mouse Opossum
Marmosops neblina
Map 16
Identification. Measurements: HB = 112–142; T = 140–178; HF = 17–20; E = 19–24; WT = 37–40 g. Males larger than females.
Upperparts dark brown; fur about 7 mm long on rump. Eye rings large, indistinct against dark face; muzzle and face between eye rings only slightly paler than dorsal color; cheeks frosted white. Ears dark brown. Tail with short furred base (13 mm), naked distally, brown gradually paling toward tip, always bicolored, paler below than above, sometimes indistinctly so. Forefeet dark above, toes white; hindfeet with or without white band across meta-

tarsals, toes white. Underparts dark gray washed with brownish, with or without a complete to partial narrow, pure white stripe on midventer from chin to anus; except where white, underparts do not contrast sharply with dorsal fur. Scrotum of males white-furred.
Similar species. Andean slender mouse opossums *(M. impavidus)* are difficult to distinguish from this species; they have underparts much paler than upperparts, and a completely dark tail, not paler below. White-bellied slender mouse opossums *(M. noctivagus)* have pure white bellies; delicate slender mouse opossums *(M. parvidens)* are much smaller.
Natural history. From montane cloud forest on a Venezuelan tepui and lowland evergreen Amazonian rainforest.
Geographic range. South America: Venezuela, on Cerro Neblina from 1,400 to 2,100 m; E Ecuador; and Brazil in the lowlands along the Rio Juruá.
Status. First described in 1990, this species is known from three widely disjunct regions, but its range is likely to be much larger.
References. Gardner, A. L. 1990. Two new mammals from southern Venezuela and comments on the affinities of the highland fauna of Cerro de la Neblina. *Advances in Neotropical Mammalogy* 1990:411–24.
Mustrangi, M. A. 1995. Phylogeography of *Marmosops* (Marsupialia, Didelphidae) in the Atlantic forest of Brazil, and the phylogenetic relationships of the Atlantic forest and Amazonian species in the genus. Ph.D. dissertation, University of California at Berkeley.

Dorothy's Slender Mouse Opossum
Marmosops dorothea
Map 15
Identification. Measurements: HB = 112–136; T = 144–167; HF = 15–17; E = 21–26; WT = 28–48 g.
Upperparts medium brown to pale beige, fur soft, short to long, dark at base with pale tips; eye rings brown, prominent; muzzle and face between eyes sharply contrasting pale buff-yellow; cheeks pure cream. Tail robust, pale brown, paling gradually to completely whitish tip. Feet white, tarsus brown. Underparts completely pale yellow, sometimes with some gray-based hair around midbody. Scrotum pink. Females

have a row of mammae extending up each side of belly to axilla.

Variation. Medium-brown, shorter-haired animals seem to be from the wetter northern evergreen forests, while pale beige, long-furred animals are from dry forests. The type of this species came from montane forests of La Paz (2,300 m), and the lowland form described above may not be conspecific.

Similar species. White-bellied slender mouse opossums *(M. noctivagus)* are dark brown, have tails completely dark, white underparts, dark tops of hands, and are larger. Agile gracile mouse opossums *(Gracilinanus agilis)* have dark tails and orange cheeks, and are smaller; murine mouse opossums *(Marmosa murina)* have short fur, pink to orange underparts, and dark tails.

Natural history. Nocturnal; terrestrial and arboreal. These mouse opossums use the ground and understory of the forest, and seem to favor areas with dense vine tangles. They are found in the lowlands in the southern Amazonian evergreen rainforest and in the adjacent dry forests (Velasco formation).

Geographic range. South America: Bolivia, and Brazil south of the middle Juruá. To 2,300 m elevation.

Status. Can be extremely common.

Gray Slender Mouse Opossum
Marmosops incanus
Plate 2, map 15

Identification. Measurements: HB = 91–194; T = 110–237; HF = 15–28; E = 18–33; WT = 13–140 g. Males much larger than females: mean WT males = 63 g, females = 39 g. Because of the short life cycle (see below), size varies seasonally.

Upperparts dark gray-brown. Eye rings dark, prominent; face between eye rings pale gray; cheeks below eyes cream. Tail paling to whitish at tip, bicolored, paler below. **Underparts creamy white** or tinged with pink, **with lateral band of gray-based hairs. Legs and feet below knee and elbow white.** Scrotum well furred. Females have no pouch. A large mouse opossum.

Variation. This species has distinct juvenile and adult coats: in young the fur is long, soft, and gray-brown. In old adult females the entire coat is short, stiff, and more yellowish, while adult males show only a "collar" of short hairs around the neck and shoulders.

The pelage type of adult males was once described as a distinct species. Animals from dry inland habitats are paler gray than those from wetter coastal forests.

Similar species. São Paulo slender mouse opossums *(M. paulensis)* are smaller, with no lateral band of gray-based fur on venter, and dark lower legs. The two species overlap at some sites. Agile gracile mouse opossums *(Gracilinanus agilis)* are smaller, with orange bellies; Brazilian gracile mouse opossums *(G. microtarsus)* are smaller, redder, with grayish white bellies.

Natural history. Nocturnal, probably uses ground and low understory. This species has an interesting life cycle that may exist in other mouse opossums, but none have been sufficiently studied. Breeding occurs during three months (Sept.–Dec.), following which all adult males evidently die, so that none exist in the population for the following seven months (Feb.–Aug.). Adult females survive through May, and then die, so that none are present from June to August, when the population includes only young of the year. Males thus have a lifespan of about a year, and females of a year and a half. There may be geographic variation in the months when breeding occurs, which correspond to the rainy season. Found in humid lowland Atlantic coastal forests, montane coastal forests to 800 m, and semi-deciduous forests of the cerrado and caatinga of the Brazilian Plateau.

Geographic range. South America: Brazil, Bahia to São Paulo. To 800 m elevation on coastal mountains and to 1,300 m on the Brazilian Plateau.

References. Mustrangi, M. A. 1995. Phylogeography of *Marmosops* (Marsupialia, Didelphidae) in the Atlantic forest of Brazil, and the phylogenetic relationships of the Atlantic forest and Amazonian species in the genus. Ph.D. dissertation, University of California at Berkeley.
  Lorini, M. L., J. A. Oliveira, and V. G. Persson. 1994. Annual age structure and reproductive patterns in *Marmosa incana* (Lund 1841) (Didelphidae, Marsupialia). *Z. Säugetierk.* 59:65–73.

São Paulo Slender Mouse Opossum
Marmosops paulensis
Map 15

Identification. Measurements: HB =
94–153; T = 145–212; HF = 17–24;
E = 19–27; WT = 16–70 g. Males much
larger than females: mean WT males = 42 g,
females = 28 g.
**Upperparts brownish gray washed with
reddish;** fur length and texture varies with
age and sex as in *M. incanus.* Eye rings
prominent, contrasting with pale brown face
between them; cheeks white with reddish
tinge; ears relatively short. Tail bicolored,
dusky above, paler below, whitish at extreme
tip. Hindfeet white, **legs above wrists and
ankles dark. Underparts white, without
gray-based hairs on sides.**
Similar species. Gray slender mouse opossums *(M. incanus)* are larger and have white
legs, gray-based hairs on sides of underparts,
and lack reddish overtones. See that species
for other comparisons.
Natural history. Known only from wet montane forests and cloud forests.
Geographic range. South America: Brazil,
Atlantic montane coastal forests from S Minas Gerais to São Paulo. Above 800 m
elevation.
Status. Known from relatively few specimens and localities; habitat and geographic
ranges narrow.
References. See previous species.

Slaty Slender Mouse Opossum
Marmosops invictus
Map 15
Identification. Measurements: HB =
104–113; T = 124–147; HF = 17–19;
E = 18–22.
Upperparts dark slate gray sometimes
tinged with brown, especially on forequarters; fur short. **Face dark, eye rings inconspicuous against dark background; cheeks
gray.** Tail dark gray. Hairs in triads on tail
all slender and equal in width. **Lower legs
dusky,** feet paler. **Underparts slate gray,
frosted white.** Females have no pouch. Small,
shrewlike mouse opossums.
Similar species. These are the only dark
gray, gray-bellied mouse opossums in their
range; sepia short-tailed opossums *(Monodelphis adusta)* are dark brown, with short
ears and tail much shorter than head and
body. Apparently the only *Marmosops*
species without outer two hairs of triads on
tail broad and petiolate.

Natural history. Nocturnal; terrestrial and
arboreal. Feeds on insects and plant material. Slaty slender mouse opossums live in the
forest understory, especially around fallen
logs. They are chiefly montane cloud forest
dwellers, but can be found to below 500 m
elevation.
Geographic range. Central America: montane areas of Panama; likely to be found
in Colombia and Costa Rica. About 500–
1,300 m elevation.
Status. Rare, geographic range small.
References. See delicate slender mouse
opossum *(M. parvidens).*

Northern Gracile Mouse Opossum
Gracilinanus marica
Map 17
Identification. Measurements: HB =
96–116; T = 131–151; HF = 15–18;
E = 15–22; WT = 24–30 g.
**Upperparts pale warm brown; fur long
and soft.** Eye rings dark, prominent; face
pale brown between eyes; cheeks pale buff.
Tail slender, brown, fur of back does not
extend onto base. Feet brownish, lower legs
usually furry almost to feet. **Throat and often chest pale buff with hairs pale to base;
rest of underparts pale buff with fur dark
gray at base.** Tiny, delicate, fluffy mouse
opossums.
Variation. A similar and closely related
form, *G. dryas,* is known only from 2,200 to
4,000 m in the Mérida Andes of Venezuela.
Similar species. These are the only mouse
opossums in their range with long, pale
warm brown fur and belly hair dark gray at
base; delicate mouse opossums *(Marmosops
parvidens)* are dark gray brown; little rufous
mouse opossums *(M. lepida)* and Emilia's
gracile mouse opossums *(G. emiliae)* lack
gray on underparts; *G. dryas* is probably indistinguishable from *G. marica* in the field,
but the ranges apparently do not overlap.
Natural history. Arboreal. Found in lowland to montane rainforest, deciduous forest,
and savanna.
Geographic range. South America: northern coastal Venezuela and Colombia. To
2,100 m elevation.
Status. Rare.
References. Gardner, A. L., and G. K.
Creighton. 1989. A new generic name for
Tate's (1933) *Microtarsus* group of South

Map 17

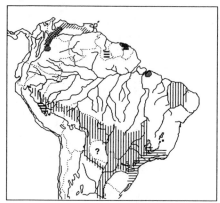

▨	Emilia's gracile mouse opossum, *Gracilinanus emiliae*
▨	Northern gracile mouse opossum, *G. marica*
▤	Kalinowski's gracile mouse opossum, *G. kalinowskii*
▤	Brazilian gracile mouse opossum, *G. microtarsus*
▥	Agile gracile mouse opossum, *G. agilis*
▤	Wood sprite gracile mouse opossum, *G. dryas*

American mouse opossums (Marsupialia, Didelphidae). *Proc. Biol. Soc. Wash.* 102:3–7.

Emilia's Gracile Mouse Opossum
Gracilinanus emiliae
Map 17
Identification. Measurements: HB = 72–87; T = 128–143; HF = 14–16; E = 16; WT = 10 g.
Upperparts dull reddish brown; fur relatively short. Eye rings narrow, black, sharply contrasting. Ears short, just covering eye when laid forward, delicate, and translucent at tips. **Tail exceptionally long, brown. Underparts pure white to cream.** Feet pale; hindfeet tiny, narrow. Scrotum unpigmented. A tiny opossum.
Variation. Some old specimens are bright orange, which is probably an artefact of bleaching with age.
Similar species. Little rufous mouse opossums *(M. lepida)* have a relatively shorter tail, larger hindfoot, longer fur, and seem to be larger, but are probably impossible to distinguish from this species in the hand; delicate mouse opossums *(Marmosops parvidens)* are slaty, with gray on underparts; Kalinowski's gracile mouse opossums

(G. kalinowskii) have relatively much shorter tail and much longer ears; other mouse opossums in range are much larger.
Natural history. Found in lowland rainforest and perhaps other habitats.
Geographic range. South America: Colombia, Suriname, French Guiana, and Brazil near Belém. To 400 m elevation.
Status. Extremely rare, known only from five individuals.
References. Voss, R. S., D. P. Lunde, and N. B. Simmons. In press. Mammals of Paracou, French Guiana: A Neotropical lowland rainforest fauna. Part 2: Nonvolant species. *Bull. Am. Mus. Nat. Hist.*

Kalinowski's Gracile Mouse Opossum
Gracilinanus kalinowskii
Map 17
Identification. Measurements: HB = 89–91; T = 110–117; HF = 15–16; E = 15–18; WT = 13–18 g.
Upperparts reddish brown. Eye rings black, wide and sharply defined, extending from base of ear to whiskers. Muzzle between eye rings pale orange; cheeks and throat white. **Ears very large, reach to base of whiskers on muzzle in front of eye when flattened forward.** Eyes large and bulging. Whiskers long and black, reaching ear tips when flattened. **Tail only slightly longer than head and body,** without fur at base. **Underparts pure white to cream, without gray hair base.** Feet orange. Scrotum blue.
Variation. This species was described in 1992 and little is known of it.
Similar species. Emilia's and little rufous mouse opossums *(G. emiliae, M. lepida)* have relatively much shorter ears and longer tails; delicate slender mouse opossums *(M. parvidens)* often have gray on underparts and relatively short ears and longer tails. Agile gracile mouse opossums *(G. agilis)* have orange cheeks and orange-tinged underparts.
Natural history. Nocturnal; probably mostly terrestrial, as some were captured in pitfall traps. From lowland and lower montane evergreen rainforests.
Geographic range. South America: E Peru, on lower Andean slopes, and Guyana and French Guiana. To 1,100 m elevation.
Status. Apparently rare, known only from five individuals.
References. See *G. emiliae.*

Brazilian Gracile Mouse Opossum

Gracilinanus microtarsus
Plate 2, map 17
Identification. Measurements: HB =
105–110; T = 145–153; HF = 16–21;
E = 18–19; WT = 19–29 g.
**Upperparts reddish brown; fur long and
soft.** Eye rings large, black, prominent.
**Forelegs and neck under ear tinged or-
ange.** Tail brown. Feet whitish. Underparts
cream, neck and chest tinged orange, hairs
gray at base. Scrotum well furred. A small,
red-brown mouse opossum.
Similar species. Gray slender mouse opossums *(Marmosops incana)* are larger and
gray. Agile gracile mouse opossums *(G. agilis)* are indistinguishable from this species
in the field; they have fur on underparts that
is white at base, and apparently occupy dryer
habitats. Emilia's gracile mouse opossums
(G. emiliae) are smaller, with a relatively
longer tail and no gray on underparts.
Natural history. Found in wet evergreen
Atlantic coastal forest.
Geographic range. South America: Brazil,
Minas Gerais to Rio Grande do Sul.

Agile Gracile Mouse Opossum

Gracilinanus agilis
Map 17
Identification. Measurements: HB =
90–106; T = 109–150; HF = 14–18;
E = 15–20; WT = 20–30 g.
**Upperparts drab gray-brown to red-
brown.** Eye rings black, prominent; face
above nose sometimes with faint dark line
down center, cheeks with pale orange not
reaching to below ear; ears light tan. Tail
relatively short, undersurface with fine silky
coat of white hairs. **Underparts pure pale
orange to cream,** or lower belly gray orange,
**base of hairs not gray on most of under-
parts.** Hands and feet whitish. Females have
no pouch.
Variation. Animals from dry areas of
Argentina and Paraguay are smaller than
those from farther north.
Similar species. See Brazilian gracile mouse
opossum *(G. microtarsus).* This species is
grayer and smaller than murine mouse opossums *(Marmosa murina);* delicate mouse
opossums *(Marmosops parvidens)* have gray
or white underparts.
Natural history. Nocturnal; arboreal. Agile
mouse opossums frequent the forest under-

story, where they use slender branches and
vines. Found in evergreen and gallery forests.
Geographic range. South America: east of
the Andes in Peru, Bolivia, E Brazil,
Paraguay, Uruguay, and Argentina.
Status. Locally common to rare. May be
present but very rare throughout W Amazonia, but common in dry forests to the south.

Red-legged Short-tailed Opossum

Monodelphis brevicaudata
Plate 3, map 18
Identification. Measurements: HB =
134–183; T = 76–105; HF = 20–26;
E = 13–20; WT = 46–150 g.
**Upperparts grizzled dark gray from nose
to tail base; sides from nose to rump red or
rusty. Ears short,** naked, tan. **Tail shorter
than head and body, densely furred for first
2–3 cm, fur tapering to fine covering on
dark tail.** Feet dark gray. Underparts glowing violaceous pink, cream, or pale orange,
grayish orange, or gray. Females have no
pouch. Young like adults, with tail less
thickly furred at base.
Variation. Highly variable in color; ranges
from almost black back with deep red sides
and legs in the Guianas and parts of Venezuela to gray with pale rusty sides and legs
south of the Amazon. Some animals from
N Venezuela and Pará are almost entirely
red, sometimes with a cream patch on the
face. Probably several species will eventually be defined from the many recognized
subspecies within this variable group. The
glowing violaceous color of the underparts
of living animals fades and eventually disappears from skins.
Similar species. Emilia's *(M. emiliae)* and
shrewish *(M. sorex)* short-tailed opossums
have red tops of rumps; Emilia's also has
a red head and a shorter tail (< 70 mm);
mouse opossums have long tails, large ears,
and black eye rings. Macconnell's rice rat
(Oryzomys macconnelli) has a long tail and
large ears.
Sounds. None usually heard in the field.
Natural history. Diurnal; terrestrial; solitary. Apparently chiefly insectivorous. Red-
legged opossums are found on the ground,
especially near or under fallen brush or rotten logs, and in hollow logs or occasionally
on them. Found in mature, disturbed, and
secondary rainforest, gardens and plantations; rarely in deciduous or dry forests.

Map 18

	Sepia short-tailed opossum, *Monodelphis adusta*
	Red-legged short-tailed opossum, *M. brevicaudata*
	Shrewish short-tailed opossum, *M. sorex*

Map 19

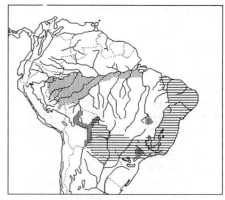

	Emilia's short-tailed opossum, *Monodelphis emiliae*
	Gray short-tailed opossum, *M. domestica*
	Pygmy short-tailed opossum, *M. kunsi*

Geographic range. South America: east of the Andes in Colombia, Venezuela, and the Guianas and in Amazonian Peru, N Bolivia, Brazil, and N Argentina. To at least 1,160 m elevation.
Status. Sometimes common, but populations show large year-to-year variation: they may be common one year and absent the next.
Local names. Colicorto (Ar); catita (Br); koort-staart opossum, moismoisi-awari (Su).
References. Charles-Dominique, P., M. Atramentowicz, M. Charles-Dominique, H. Gérard, A. Hladik, C. M. Hladik, and M. F. Prévost. 1981. Les mamifères frugivores arboricoles nocturnes d'une forêt guyanaise: Inter-relations plantes-animaux. *Rev. Ecol.* 35:341–435.

Emilia's Short-tailed Opossum
Monodelphis emiliae
Plate 3, map 19
Identification. Measurements: HB = 120–158; T = 50–70; HF = 17–24; E = 14–18; WT = 52–60 g.
Back grizzled gray; entire rump, hindlegs, and head forward of ears rufous. Snout short and broad. **Tail with rump hair extending far down length,** gradually decreasing to near tip. Forelegs and hands gray or gray-buff. Underparts in life bright glowing violaceous pink; this fades to orange on skins. Scrotum of males bluish. Females have no pouch.

Variation. The rosy wash on the belly is a "fugitive" coloration that vanishes from skins after a year or two.
Similar species. See red-legged short-tailed opossum *(M. brevicaudata).*
Natural history. Terrestrial. Found around fallen logs in lowland evergreen rainforest.
Geographic range. South America: Brazil south of the Amazon, and Peru south of the Amazon to Pando, Bolivia.
Status. Apparently rare, known from only a few individuals and localities.
Local names. Catita (Br).
References. Pine, R. H., and C. O. Handley, Jr. 1984. A review of the Amazonian short-tailed opossum *Monodelphis emiliae* (Thomas). *Mammalia* 48:239–45.

Sepia Short-tailed Opossum
Monodelphis adusta
Plate 3, map 18
Identification. Measurements: HB = 84–114; T = 51–67; HF = 11–17; E = 10–14; WT = 15–35 g.
Upperparts completely dark brown, darker on rump than foreparts; **fur very short** (≈ 3 mm). Ears naked, dark brown-gray. Tail short, **unicolored dark brown. Underparts gray or orange-gray,** sometimes with a cream streak on midline. Females have no pouch. A tiny, shrewlike opossum.
Similar species. *M. kunsi* is similar, but is smaller and has pale underparts and

bicolored tail; other species all have longer fur. Another species of short-tailed opossum, *M. osgoodi,* also occurs at higher elevations in the Andes of S Peru and Bolivia; it is gray-brown with medium-length fur and whitish underparts. Grass mice (*Akodon* spp.) and long-nosed mice (*Oxymycterus* spp.) have large ears and long fur. Shrews are smaller.

Natural history. Terrestrial. Feeds on invertebrates. Sepia short-tailed opossums have been captured in forest, grassland, rocky terrain, and in a house. In captivity they are in constant motion during activity, always investigating their surroundings. This is chiefly a montane species of wet forests at mid-elevations (1,400–2,200 m), but in some places it is found as low as 200 m in rainforest and wet grassland.

Geographic range. Central and South America: the northern Andes and their eastern slopes from E Panama to Colombia, Ecuador, and S Peru. There are two records from the lowlands disjunct from the Andes, in S Madre de Dios, Peru.

Status. Apparently rare.

Local names. Zorra de cola corta (Pa).

References. Handley, C. O. 1966. Checklist of the mammals of Panama. In R. L. Wenzel and V. J. Tipton, eds., *Ectoparasites of Panama,* 753–95. Chicago: Field Museum of Natural History.

Pygmy Short-tailed Opossum
Monodelphis kunsi
Map 19

Identification. Measurements: HB = 71–94; T = 41–42; HF = 12; E = 12; WT = 19 g.

Upperparts uniform warm brown; fur extremely short; chin, throat, and cheek buff, throat and chest by shoulder orange-tinged. Ears very short and rounded. Tail bicolored dark above, buff below, slightly hairy. Underparts buff with whitish areas, scrotum black. A tiny, shrewlike opossum with short legs and a long back.

Similar species. Sepia short-tailed opossums (*M. adusta*) are larger, with brown or gray-brown underparts, dark cheeks, and unicolored, dark tails. This is the smallest short-tailed opossum. Grass mice (*Akodon* spp.) have long tails, large ears, and long, somewhat grizzled fur.

Natural history. An individual was trapped in second-growth brush, two in a banana plantation, another in a lower montane rock ravine with dense shrubs, and one in an isolated, cerrado-like pampa in rainforest. All known habitats seem to be of disturbed vegetation in moist situations.

Geographic range. South America: Bolivia and Brazil. Lowlands to 1,500 m elevation. Likely to be found in N Argentina.

Status. Apparently rare, known from only seven specimens from six localities.

References. Anderson, S. 1982. *Monodelphis kunsi.* Mammalian Species, no. 190.

Salazar, J. A., M. L. Campbell, S. Anderson, S. L. Gardner, and J. L. Dunnum. 1994. New records of Bolivian mammals. *Mammalia* 58:125–30.

Long-nosed Short-tailed Opossum
Monodelphis scalops
Map 20

Identification. Measurements: HB = 133; T = 71; E = 8.

Head, rump, and tail bright rufous; forward part of back and shoulders grizzled olive-gray. Snout long and narrow. Forelegs and hands red. Rump hair extending only onto base of tail. Underparts grizzled olive-gray. Females have no pouch.

Similar species. Shrewish short-tailed opossums (*M. sorex*) have gray crown.

Natural history. Found in lowland Atlantic coastal forest.

Geographic range. South America: SE Brazil in Rio de Janeiro.

Status. Apparently rare.

Local names. Catita.

Shrewish Short-tailed Opossum
Monodelphis sorex
Map 18

Identification. Measurements: HB = 110–130; T = 65–85; HB = 13–16; E = 10–13; WT = 48 g.

Top of head, neck, and forequarters gray, finely grizzled with tawny yellowish; cheeks, sides of neck and body, and rump dull rusty; fur short. Tail thinly covered with fine rusty hairs. Feet reddish. Underparts pale orange, demarcated from sides, hairs gray at base on midsection only, scrotum black, furred with orange. Females have no pouch. Small, dull-colored opossums.

Map 20

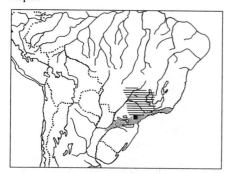

▨ Long-nosed short-tailed opossum, *Monodelphis scalops*
■ One-striped short-tailed opossum, *M. unistriata*
▤ Chestnut-striped short-tailed opossum, *M. rubida*

Similar species. Often confused with the southern short-tailed opossum *(M. dimidiata),* a grassland species, known with certainty only from Argentina and Uruguay, that has long fur and whitish or buffy feet. Gray short-tailed opossums *(M. domestica)* are gray, with large ears; red-legged short-tailed opossums *(M. brevicaudata)* have gray top of rump and tail base and are larger; long-nosed short-tailed opossums *(M. scalops)* have a red head.
Natural history. Found in Atlantic coastal rainforests.
Geographic range. South America: SE Brazil, Minas Gerais to Rio Grande do Sul.
Status. Apparently rare.
Local names. Catita.

Gray Short-tailed Opossum
Monodelphis domestica
Plate 3, map 19
Identification. Measurements: HB = 130–191; T = 70–106; HF = 17–22; E = 20–28; WT = 36–98 g.
Upperparts entirely pale gray. Face sometimes with a faint dark stripe on center; **ears large,** naked, pale gray-tan. Tail well furred for first 1–2 cm, the rest thinly haired, either completely blackish (E Brazil) or blackish above, paler below (Bolivia, W Brazil). Feet whitish. Underparts dirty white to pale gray with an orange tinge, paler than back. Females have no pouch.
Variation. An animal from Pernambuco, Brazil, has large black patches on neck and rump and black streaks from nose to eyes.

A distinct population on Isla Marajó, Brazil, is gray to gray-brown, with orange tinge on flanks; it is sometimes considered a separate species, *M. maraxina.*
Similar species. All other short-tailed opossums in range either are striped, have reddish color on body, or are tiny; mouse opossums have long tails, black eye rings, and large ears; mice and rats have four front toes, nonprehensile tails, two large incisors, and no canine teeth.
Natural history. Nocturnal and diurnal; terrestrial; solitary. Feeds mostly on insects and small animals. Gray short-tailed opossums are found in deciduous forest, savanna, and around and in houses and agricultural land. Probably not a rainforest species, but a scrub species that has followed man into cleared areas of rainforest regions.
Geographic range. South America: the southern rim of the Amazon Basin, from Bolivia and Paraguay to SE Brazil and north to Isla Marajó.
Status. Locally common.
Local names. Colicorto gris (Ar); catita (Br); mbicuré-í (Pa).

Three-striped Short-tailed Opossum
Monodelphis americana
Plate 3, map 21
Identification. Measurements: HB = 101–105; T = 45–55; HF = 16–18; E = 15; WT = 23–35 g.
Upperparts tawny-brown, faintly grizzled, rump slightly reddish; back with three prominent black stripes, a central stripe from nose to tail bordered by two lateral stripes from shoulder to tail, central stripe indistinct on face. Ears naked, brown. Tail bicolored, pale below, blackish above. Underparts pale orange-gray. Females have no pouch. Young like adults. A **very small** opossum.
Similar species. Two other three-striped species are present in SE Brazil that are much smaller, but otherwise almost indistinguishable from this species: *M. theresa,* from Rio de Janeiro, Minas Gerais, and apparently a disjunct individual from the Peruvian Andes, is tiny (HB = 92), with more whitish underparts; *M. iheringi,* from São Paulo to Rio Grande do Sul (map 21), is minuscule (HB = 77) and has the central black stripe ending between the ears. Young of *M. americana* are probably indistinguishable

Map 21

from these in the field. See other short-tailed opossums.

Natural history. Terrestrial; probably diurnal. Feeds on insects; in captivity will voraciously kill and eat small vertebrates. Found in Atlantic coastal forests, usually near water.

Geographic range. South America: Brazil, coastal forests from near Belém south to at least São Paulo.

Status. Apparently rare.

Local names. Catita.

Chestnut-striped Short-tailed Opossum
Monodelphis rubida
Map 20
Identification. Measurements: HB = 125–137; T = 56–60; HF = 15–18; E = 12; WT = 45–46 g.

Upperparts entirely warm red-brown, near chestnut; neck behind ear brighter, almost orange; back with three faint, darker brown stripes; fur short. Tail looks naked, body hair does not extend far onto base. Forefeet dusky, hindfeet yellow on inner sides, outer like back. Underparts gray, frosted yellow-white throughout, scrotum black with yellow-white hairs. Females have no pouch.

Variation. This form is sometimes considered a subspecies of the three-striped opossum *M. americana.*

Similar species. Other three-striped short-tailed opossums have distinct black stripes and are tawny brown.

Geographic range. South America: Brazil, from Goiáz, Minas Gerais, and São Paulo.

Status. Apparently uncommon or rare.

Local names. Catita.

One-striped Short-tailed Opossum
Monodelphis unistriata
Map 20
Identification. Measurements: HB = 140; T ≈ 60.

Upperparts rusty gray; midback with a single chestnut stripe. Sides, feet, and underparts yellowish orange. Tail bicolored, brown above, yellowish below.

Similar species. These are the only short-tailed opossums with a single stripe on the back. Atlantic forest mice (*Delomys* spp.) have long tails and large ears.

Geographic range. South America: Brazil, São Paulo.

Status. Apparently known only from a single individual collected prior to 1842, therefore may be endangered or extinct.

Local names. Catita.

Anteaters, Sloths, and Armadillos (Xenarthra)

This order (sometimes called Edentata) includes four families of mammals so dissimilar externally that they hardly appear related. The characteristics that unite them include simple peglike teeth (when teeth are present), with no development of milk teeth and no true incisors, canines, or premolars; extra joint surfaces between the vertebrae, and other characters of the skeleton; and features of the reproductive tract and circulatory system that are unique among mammals. They are the last living remnants of a large group of species that evolved in South America when it was an isolated island continent (see Appendix C). Most living members of this order are specialized feeders that eat mainly ants and termites or rainforest canopy leaves. A general reference is G. G. Montgomery, ed., *The evolution and ecology of armadillos, sloths, and vermilinguas* (Washington, D.C.: Smithsonian Institution Press, 1985).

Anteaters (Myrmecophagidae)

The true anteaters have no teeth; they have long, tubelike snouts and fairly small eyes and ears. The tongue can be greatly extended to reach into inaccessible crevices and is covered with sticky saliva that traps insects. The powerful curved front claws are so large that anteaters cannot walk on the soles of their forefeet; the claws are folded inward, and they walk on the outside of the hand, which has a thickened pad. The three arboreal species have strongly prehensile tails. These animals feed on social insects (ants, termites, bees), and their anatomy is specialized for opening the nests of their prey and licking up the occupants. The prehensile tail acts as a fifth paw to anchor the anteater firmly while it uses both forepaws to rip open an arboreal nest. Anteaters give birth to a single young, which rides on the mother's back, clinging tightly to her fur when she travels. Because she has no teeth or fingers, the mother cannot pick up her young; it must crawl up onto her back by itself. There are three genera and four species of true anteaters, all restricted to the New World tropics and subtropics and all found in rainforest.

Giant Anteater
Myrmecophaga tridactyla
Plate 4, map 22
Identification. Measurements: HB = 1,000–1,900; T = 640–900; HF = 150–180; E = 35–50; WT = 22–39 kg.
Upperparts grizzled gray-brown; hairs banded black, brown, and white; **shoulder to chest and neck with wide black stripe bordered with white;** midback with crest of hair increasing in length from crown to tail tip; fur thick, coarse, and dull. **Head greatly elongated, narrow, and convex;** grizzled gray, covered with short, thin, stiff hairs; nose black; ears small, rounded, gray. **Tail long,** nonprehensile, **very bushy with long, coarse, drooping plume of hair.** Chest grizzled; belly and hindlegs black. **Forelegs white, wrist crossed by black band;** elbows fringed with long hair; **forefeet with three greatly enlarged** and two smaller **claws,** hindfeet with five short claws. Young like adults. **A large, shaggy animal** that walks on its knuckles with an ambling gait.

Similar species. Tamanduas (*Tamandua* spp.) are much smaller and not shaggy; no other large terrestrial mammal has a long, shaggy tail.
Sounds. Usually silent, but adults can roar when disturbed; isolated young give a shrill whistle.
Natural history. Nocturnal and diurnal; terrestrial; solitary. Feeds mainly on ants that are licked out of nests or tunnels opened with the claws. Giant anteaters travel widely and feed from many ant colonies in a day, taking a few ants from each. Normally they walk or amble along, but they are able to gallop with surprising speed. They have a keen sense of smell, which is used to find prey. At rest they bed down on the surface of the ground in an open or sheltered spot. Giant anteaters are normally harmless, but if attacked they may rear up on the hindlegs and slash and grasp with the formidably muscled and armed forelegs; they can kill large predators. Found in a wide range of habitats from wet or dry grasslands to rainforest.

Map 22

Map 23

 Giant anteater, *Myrmecophaga tridactyla*

Northern tamandua, *Tamandua mexicana*
Southern tamandua, *T. tetradactyla*

Geographic range. Central and South America: S Belize and Guatemala to N Ecuador west of the Andes and east of the Andes to N Argentina and Uruguay.
Status. CITES Appendix II. Extirpated in parts of its range; widespread, but rare and wantonly killed or captured. Records from Central America are historic or anecdotal, and it is not known if any viable population remains there. Always rare in rainforest; more common and easy to see in grasslands with many ant mounds.
Local names. Oso hormiguero, tamandua de bandera (Bo, Ec, Pe); tamandua bandeira, tamanduá-açu, papa-formigas (Br); ant bear (Be); oso palmero; oso caballuno, oso pajizo (Co); oso caballo (CR, Ho, Pn); tamanoir (FG); yurumí, tamanduá (Gua); zam hool (May); tamanwa (Sar); reuzen-meiereneter (Su).
References. Shaw, J. H., J. Machado-Neto, and T. S Carter. 1987. Behavior of free-living giant anteaters *(Myrmecophaga tridactyla)*. Biotropica 19:255–59.

Southern Tamandua
Tamandua tetradactyla
Plate 4, figure 4, map 23
Identification. Measurements: HB = 535–880; T = 400–590; HF = 87–105; E = 47–60; WT = 3.6–8.4 kg.
Head, legs, and hindparts typically pale golden yellow with a black "vest" over the belly, lower back, and forward in a band over the shoulders and around the chest (but see variation below); fur stiff, glossy bristles. **Head long, narrow, and convexly**

curved; muzzle to level of eyes naked, blackish; eyeshine almost none, reddish. **Tail long, thick, prehensile, base furred, tip naked,** with mottled skin. **Forefeet with four long, massive claws;** hindfeet with five smaller claws. Young colored like adults.
Variation. The "vested" coloration described above is typical of animals from the eastern half of Brazil and some other populations. Those from N Brazil, S Venezuela, and the Guianas **may be pure blond,** with no vest; animals from central and western Amazonia may be vested, **partially vested,** blond, or **completely black** (fig. 4). The dark parts of the body may be either black, sooty gray, or brown; blond parts may be silvery white to red-gold. Some native peoples believe that blond and black forms are different species; they occur together in parts of Peru.
Similar species. Monkeys have short muzzles; giant anteaters *(Myrmecophaga tridactyla)* have bushy tails and are much larger.
Sounds. No calls are commonly heard, but feeding tamanduas noisily rip and tear apart insect nests and rotten wood, which rains to the forest floor if they are in a tree. Sounds of tearing wood at night almost always lead to a tamandua; by day, to a tamandua or to brown capuchin monkeys.
Natural history. Diurnal and nocturnal; arboreal and terrestrial; solitary. Feeds mainly on ants, termites, and bees extracted after ripping apart their nests with the foreclaws. Tamanduas can be seen foraging on the ground or in the canopy anywhere in the

Figure 4. Color patterns of tamanduas: *(a)* **northern tamandua,** *Tamandua mexicana (b–d)* **southern tamandua,** *Tamandua tetradactyla: (b)* Brazil, *(c)* Venezuela, *(d)* Colombia. Patterns are neither limited to those areas nor the only ones found there.

forest, but seem most common beside water-courses and in viny, epiphyte-laden habitats, where their prey may be concentrated. They move slowly and awkwardly on the ground and appear to see poorly. When alarmed, they stand upright on their hindlegs, raise the nose to sniff, then climb a tree or amble away. When attacked or cornered they rear up and slash with the foreclaws and can inflict serious wounds. By day in rainforest they are accompanied by a dense cloud of flies and mosquitoes and often brush their eyes with a forepaw. When inactive, taman-duas rest in hollow trees, burrows of other animals, or other natural shelters. Found in many habitats from mature and secondary rainforest and plantations to gallery forest and arid savannas.

Geographic range. South America: east of the Andes, Venezuela south to N Argentina and Uruguay. To about 2,000 m elevation.
Status. CITES Appendix II (SW Br); Appendix III (Gy). Widespread, always present in rainforest, but usually uncommon. Hunters often kill tamanduas, claiming that they kill dogs. They are rarely used for food, but their tail tendons are sometimes used to make ropes.
Local names. Oso colmenero, tamanduá (Co, Ec, FG, Pe); oso hormiga (Bo); oso melero (Ec, Co, Ho); tamanduá-colete, mambira, tamanduá-mirim (Br); tamandou (FG); kaguaré (Gua); shihuí (Pe); termie-teneter, boom-miereneter (Su); mirafroiti (Sar).
References. Montgomery, G. G. 1985. Movements, foraging and food habits of the four extant species of Neotropical vermilin-guas (Mammalia; Myrmecophagidae). In G. G. Montgomery, ed., *The evolution and ecology of armadillos, sloths, and vermilin-guas,* 365–77. Washington, D.C: Smithsonian Institution Press.
    Wetzel, R. M. 1982. Systematics, distribution, ecology, and conservation of South American edentates. *Pymatuning Symp. Ecol.* 6:345–75.

Northern Tamandua
Tamandua mexicana
Figure 4, map 23
Identification. Externally identical to southern tamandua; northern tamandua is distinguished from it by characters of the skull. All individuals are black vested.
Similar species. See southern tamandua.

Natural history. Similar to southern tamandua.
Geographic range. Central and South America: SE Mexico south through all of Central America, and South America west of the Andes from N Venezuela to N Peru.
Status. CITES Appendix III (Gu). Threatened by habitat destruction in much of its range.
Local names. Oso hormiguero común (Span); ant bear (Be); tamanduá, susurete, oso amarillo (Co); brazo fuerte (Me); oso mielero (CR); chab (May).
References. Montgomery, G. G., and Y. D. Lubin. 1977. Prey influences on movements of Neotropical anteaters. In R. L. Phillips and C. Jonkel, eds., *Proceedings of the 1975 Predator Symposium,* 103–31. Missoula, Mont.

Silky or Pygmy Anteater
Cyclopes didactylus
Plate 4, map 24
Identification. Measurements: HB = 154–205; T = 163–225; HF = 25–38; E = 6–15; WT = 155–275 g.
Upperparts smoky gray-gold with silver iridescence, midback usually with dark brown stripe from shoulder to rump; fur long, soft, silky, and slightly wavy. Head usually gold; ears tiny, buried in fur; button eyes round and black; nose elongated but blunt, lower jaw protruding slightly, chin prominent; nose, lips, palms, and soles pink. Tail long, tapered, prehensile, fully furred to tip dorsally, naked for last third of under-side, tip sometimes whitish. Legs gold or gray; feet with large, naked pads, hindfoot with four long claws, forefoot with one huge and one smaller claw. Underparts gray or buff, midline usually with dark brown stripe; chest sometimes dark brown or with brown spot. Young like adults; but center of back often much darker and fur longer. A beauti-ful tiny golden anteater; several of its local names translate to "little angel."
Variation. Animals from Mexico are entirely orange-gold, with shorter fur; those from the western Amazon Basin may lack the middorsal stripe; those from N Venezuela and Trinidad have brown chests.
Similar species. No other small, long-furred animals have long claws on the fore-feet. Woolly opossums (*Caluromys* spp.) have large ears and naked tail tips; black-shouldered opossums (*Caluromysiops*

Map 24

Silky or pygmy anteater, *Cyclopes didactylus*

irrupta) have large ears and black shoulders; small monkeys have flat faces and non-prehensile tails.

Sounds. Said to make a soft whistling sound.

Natural history. Nocturnal; arboreal; solitary. Feeds mainly on ants, but also on other insects. Silky anteaters travel and feed above the ground on small stems and lianas, which they grasp in a tight fold of the hindsole. They climb slowly, with two feet always in contact with the branch. To feed, they slit open an ant-filled hollow stem with the large foreclaw and lick out the ants from within. By day they rest curled in a tight ball, usu-

ally in a tangle of lianas. Young are left in a tree at night while the mother forages. They nurse on their mother's milk until they are old enough to forage for ants on their own. Adult females have large territories; a male territory includes those of several females. Pygmy anteaters defend themselves by pinching the attacker with the sharp fore-claw. Found in lowland rainforests; the distribution in different forest types is not documented.

Geographic range. Central and South America: SE Mexico south to SE Bolivia and S Brazil.

Status. Unknown, widespread. Not hunted. These animals are rarely seen but may not be uncommon.

Local names. Oso oro (Bo); tamandua-í (Br); ceibita (CR); angelito (Co); serafín (Ec, Pe); flor de balsa (Ec); lèche main (FG); perico lerdo, osito melero, perezocito (Ho); tapacara, gato balsa (Pn); wespeneter (Su); likanoe (Sar).

References. Best, R. C., and Y. Harada. 1985. Food habits of the silky anteater *(Cyclopes didactylus)* in the central Amazon. *J. Mammal.* 66:780–81.

Montgomery, G. G. 1983. *Cyclopes didactylus.* In D. H. Janzen, ed., *Costa Rican natural history,* 461–63. Chicago: University of Chicago Press.

Sloths (Bradypodidae, Megalonychidae)

Dental formula: 5 simple, peglike teeth on each side in the upper jaw and 4 in the lower = 18. No true canines or incisors: the large, bladelike "canines" of two-toed sloths are derived from other teeth. Feet with no free toes, but two or three long, curved claws that form a hook by which sloths can hang passively from a branch or clasp objects against the palm with a pincerlike grip. Sloths have long limbs, short bodies, and stumpy tails. To accommodate their upside-down lifestyle, the fur slants from the belly toward the back, and the head can rotate over 90 degrees. Sloths feed chiefly on forest canopy leaves, which they digest by bacterial fermentation in a many-chambered stomach. Their hair has microscopic grooves and notches that provide a home for greenish algae, which camouflage the fur among the canopy leaves. These algae in turn provide food for particular species of moths. Sloths move slowly and little, and they spend much time resting. The single young spends its first 6–9 months clinging to its mother, usually to her chest. Their silence, immobility, and camouflaged fur make sloths extremely difficult to see from the ground, and they are usually much more common than they seem. Eagles, however, are able to find them, and they prey on them extensively, as do jaguars in some localities. Recent studies show that the two genera of living sloths are only distantly related: the two-toed sloths belong to the family of the extinct giant ground sloths (Megalonychidae), while the three-toed sloths belong to the Bradypodidae (where the two-toed sloths were formerly placed). The two genera and five species of sloths are restricted to New World tropical rainforests.

Brown-throated Three-toed Sloth
Bradypus variegatus
Plate 4, map 25
Identification. Measurements: HB =
400–750; T = 38–90; HF = 90–180;
E = 8–22; WT = 2.3–5.5 kg.
**Upperparts usually pale brown with large
patches of dirty white** concentrated on
lower back and hindlegs; **fur long, coarse,
and wavy,** shaggy except on face, tinged
greenish. **Head small and round, ears not
visible,** face whitish or brownish, often with
darker band across brow; **eye surrounded
by a black stripe that extends back toward
ear;** mouth slightly smiling; nose black.
Throat and chest brown. Tail short and
stumpy. Limbs long; **feet each with three
long, curved, hooklike claws.** Back of
males has large patch of short orange fur
with brown stripe down middle between
shoulders (the speculum). Young like adults.
Usually seen hanging upside down by its
claws or sitting in a tree fork with head
between forelegs.
Variation. Color varies from very pale,
almost whitish in some Central American
animals, to dark red-brown.
Similar species. Pale-throated three-toed
sloths *(B. tridactylus)* have pale throat; two-
toed sloths (*Choloepus* spp.) have no pale
patches on back and no tail; all other large
arboreal mammals have longer tails.
Sounds. None usually heard, but can emit a
shrill whistle.
Natural history. Nocturnal and diurnal;
solitary; arboreal. Feeds on the leaves of
many species of trees. Three-toed sloths
usually feed high in the forest canopy, where
they are very difficult to see. They prefer
trees with crowns exposed to the sun and
may sometimes be spotted on exposed
branches as they warm themselves in the
early morning sunlight. On the ground they
are helpless and virtually unable to walk,
but they can swim and are sometimes found
crossing rivers. They have the astonishing
behavior for an arboreal animal of descending to the ground to defecate: while clinging
to a tree trunk, they dig a hole with the tail,
defecate in it, and cover it over. The necessity arises only about once a week. Found
in both evergreen and dry forests, and even
in isolated trees in pastures. Sloths occur
throughout primary and secondary forests,
but they are most easily seen in low-canopy

second-growth habitats. People often pick
them up from roadsides and later release
them in city gardens and parks, where they
can thrive.
Geographic range. Central and South
America: Honduras south, west of the Andes to S Ecuador, east of the Andes from
N Venezuela and Colombia to Bolivia and
N Argentina, except area east of the Rio
Negro and north of the Amazon. To at least
1,100 m elevation.
Status. CITES Appendix II. Widespread
and common; unlikely to be threatened except where habitat is destroyed. Sloths are
hunted widely for meat, but they are difficult
to see, which both protects them from extirpation and makes it difficult to establish their
status in any forest. When forests are cut or
flooded by dams, sloths are often found in
enormous numbers (29,000 at one dam site).
Local names. Perezoso de tres dedos (Span);
perico (Bo); preguiça-de-bentinho (Br);
perico ligero (Ec, Co); pelejo (Pe); aí aí
(Qui, Gua); q'oral (May).
References. Montgomery, G. G., and M. E.
Sunquist. 1978. Habitat selection and use
by two-toed and three-toed sloths. In
G. G. Montgomery, ed., *The ecology of
arboreal folivores,* 329–59. Washington,
D.C.: Smithsonian Institution Press.

Pale-throated Three-toed Sloth
Bradypus tridactylus
Plate 4, map 25
Identification. Measurements: HB =
445–554; T = 31–75; HF = 90–140 (cu);
E = 10–15; WT = 3.3–6 kg.
Upperparts grizzled, **smoky gray-beige or
chocolate with large cream- or orange-
tinged splotches** usually concentrated on
lower back and rump; top of head and foreparts more uniform, darker tan or brown
than hindparts; fur long, shaggy, wavy,
tinged green in wild. **Head** conspicuously
round; face flat, with short hair, ears not
visible; **forehead pale orange, yellow, or
whitish in a band from brow to behind
ear, continuous with pale cheeks and
throat;** eyes in the center of dark stripes;
mouth slightly smiling; nose black. Tail
short and stumpy. Limbs very long, **fore-
and hindfeet with three large curved
claws.** Back of males has large patch of
short orange fur with dark brown stripe
down center. Young grizzled, shaggy,

Map 25

▨ Brown-throated three-toed sloth, *Bradypus variegatus*
☰ Pale-throated three-toed sloth, *B. tridactylus*
■ Maned three-toed sloth, *B. torquatus*

gray-tan like adults. Body shape and posture like brown-throated three-toed sloth.

Variation. Color variable: dark parts from pale beige to chocolate, pale parts from dirty white to orange.

Similar species. See brown-throated three-toed sloth *(B. variegatus).*

Sounds. None usually heard, but can emit a shrill whistle.

Natural history. Probably diurnal and nocturnal; arboreal; solitary. Feeds on leaves, usually high in the forest canopy. Most knowledge of the behavior of three-toed sloths is derived from brown-throated sloths; pale-throated sloths probably have similar habits. These sloths seem most numerous in dense forests of second-growth trees or viny disturbed habitats, where there is a good food supply of young leaves and the sloth's main predator, the harpy eagle, is often absent. Found in mature and secondary lowland rainforest.

Geographic range. South America: E Venezuela, the Guianas, and NE Brazil north of the Amazon and east of the Rio Negro, and in a narrow band south of the Amazon from the Rio Negro to Belém.

Status. Widespread and sometimes common; hunted for meat; see brown-throated three-toed sloth.

Local names. Perezoso de tres dedos (Span); Preguiça-de-bentinho (Br); mouton paresseux, aï (FG); driteenluiaard (Su); sonloiri (Sar); aí (Gua, Br).

References. Wetzel, R. M., and F. de Avila-Pires. 1980. Identification and distribution of the recent sloths of Brazil (Edentata). *Rev. Brasil. Biol.* 40:831–36.

Maned Three-toed Sloth
Bradypus torquatus
Plate 4, map 25

Identification. Measurements: HB = 450–500; T = 48–50; HF = 100–115; WT = 3.6–4.2 kg.

Head and body uniform grizzled tan (a mixture of tan and cream hairs); **back of neck from nape to shoulder with pitch black "mane" of hairs to 15 cm long, projecting as plumes over shoulders;** fur long, coarse, wavy, and shaggy. Face and chin brown. Shape like other three-toed sloths. No middorsal speculum on male.

Similar species. Other sloths lack black mane; all other large arboreal mammals have long tails.

Natural history. From Atlantic coastal forests.

Geographic range. South America: Brazil, in the remaining fragments of coastal forest in Bahia, Espírito Santo, and Rio de Janeiro.

Status. US-ESA Endangered. Extirpated on much of its former range because of deforestation.

Local names. Preguiça-preta, aí-pixuna, preguiça-de-coleira (Br); aí-igapó (TGua).

References. See pale-throated three-toed sloth.

Hoffmann's Two-toed Sloth
Choloepus hoffmanni
Plate 4, map 26

Identification. Measurements: HB = 520–700; T = 14–30; HF = 100–150; E = 20–37; WT = 4.5–8.1 kg.

Upperparts completely tan, legs or forequarters sometimes brown; fur long, coarse, and wavy, slightly greenish in wild. **Throat pale, contrasting with chest.** Head round, often paler than body, muzzle protruding, brown; face pale, with short hair; ears not visible; **anteriormost teeth long and sharp, caninelike.** Limbs long; **hindfeet with three, forefeet with two long, curved claws.** Tail not visible. Young unlike adults: dark brown, with short, woolly fur. Usually seen hanging upside down by its claws.

Variation. Color varies from brown to very pale, almost whitish. Central American animals have dark brown eye rings.

Similar species. Difficult to distinguish in the field from southern two-toed sloths

Map 26

▨ Hoffmann's two-toed sloth, *Choloepus hoffmanni*
▤ Southern two-toed sloth, *C. didactylus*

(C. didactylus), which have a brown throat; three-toed sloths (*Bradypus* spp.) have large pale splotches on back, visible tail, whitish brow, three claws on front foot, and no large, bladelike front teeth. All other large arboreal animals have longer tails.
Sounds. None usually heard, but hisses in defense; low bleats in distress.
Natural history. Nocturnal; arboreal; solitary. Feeds on leaves and fruit. Two-toed sloths favor trees covered with many vines and with crowns exposed to sun. They are often high in the forest canopy and very difficult to see. At night they sometimes descend to hang in lianas of the understory, where they are more visible. Like three-toed sloths, they descend to the ground to defecate, but they do not dig a hole. These sloths do not attain such high numbers as three-toed sloths in secondary and disturbed habitats, but they are seen as often in primary Amazon Basin forest. More aggressive than three-toed sloths, they will defend themselves by slashing with the foreclaws and biting. Found in mature and secondary rainforests and deciduous forests.
Geographic range. Central and South America: Nicaragua south, west of the Andes to NW Ecuador; west of the Cordillera Oriental in Venezuela; and a disjunct population in Peru, W Brazil, and Bolivia. The range limits of the two species of *Choloepus*

in western Amazonia are not well known. To 1,800 m elevation.
Status. CITES Appendix III (CR). Status unknown, widespread, hunted for meat.
Local names. Perezoso de dos dedos (Span); preguiça real, unau (Br); perico ligero (Co, Ec); mouton paresseux (FG); cucala (Ho); pelejo (Pe); aí (TGua); intillama (Qui).
References. Wetzel, R. M. 1985. The identification and distribution of recent Xenarthra (= Edentata). In G. G. Montgomery, ed., *The evolution and ecology of armadillos, sloths, and vermilinguas,* 5–21. Washington, D.C.: Smithsonian Institution Press.
Montgomery, G. G., and M. E. Sunquist. 1978. Habitat selection and use by two-toed and three-toed sloths. In G. G. Montgomery, ed., *The ecology of arboreal folivores,* 329–59. Washington, D.C.: Smithsonian Institution Press.

Southern Two-toed Sloth
Choloepus didactylus
Plate 4, map 26
Identification. Measurements: HB = 462–860; T = 14–33; HF = 110–170; E = 20–35; WT = 4.1–8.5 kg.
Generally **the same as Hoffmann's two-toed sloth, except throat same color as upper chest; fur often brown with long, cream tips,** looks variegated. Face often same color as body. Legs often darker brown than body.
Variation. An animal from Colombia has a white throat.
Similar species. See Hoffmann's two-toed sloth *(C. hoffmanni).*
Sounds. None usually heard in the field.
Natural history. Nocturnal; arboreal; solitary. Habits probably similar to Hoffmann's two-toed sloth. Found in mature, disturbed, and secondary rainforest.
Geographic range. South America: east of the Andes, E Venezuela and the Guianas south to Ecuador and Peru and the Amazon Basin of Brazil, reaching south of, but near, the main river.
Local names. Same as Hoffmann's two-toed sloth; tweeteenluiaard (Su); skapoeloiri (Sar).
References. See Hoffmann's two-toed sloth.

Armadillos (Dasypodidae)

Dental formula: no incisors or canines; seven or more small, peglike teeth on each side of each jaw. There are three to five toes on the forefeet and five toes on the hindfeet. The dorsal surface of the body is covered with bony armor plates that shield the head, back, and sides, and sometimes the legs and tail. Around the center of the body the armor is arranged into rows or "bands" of plates separated by soft skin; this allows the animal to bend its body. The number of these bands is used to distinguish between some species. The back is smoothly rounded and the legs are short and strong, with stout claws on the toes. The belly is soft and naked. Most species have little or no hair, but one montane species has dense fur covering the armor. "Hairy" armadillos have a thin sprinkling of long, stiff bristles. Armadillos are mainly insectivorous; they feed chiefly on ants and termites, but they also eat many other kinds of small animals, carrion, and some fruit or other plant matter. With their broad, inflexible backs and short legs, armadillos trot with a rolling or scuttling gait, some like windup toys, snuffling and grubbing with their noses and forepaws and seemingly unaware of anything more than a foot or two away. They have a good sense of smell but poor eyesight, and will often run right over the feet of a person standing still in their path. All species seem to sleep and raise their young in burrows that they dig themselves; the burrow of each species has a characteristic size and shape. An armadillo burrow can be recognized by its smooth, dome-shaped roof, polished by and fitting its owner's carapace. The litter size is 1–12 young; in some long-nosed armadillos each litter is a set of four identical quadruplets formed from a single egg. There are eight genera and twenty species, all in the New World; four genera and eight species in the lowland rainforest.

Yellow Armadillo
Euphractus sexcinctus
Plate A, map 27
Identification. Measurements: HB = 401–495; T = 119–241; HF = 78–92; E = 32–47; WT = 3.2–6.5 kg.
Upperparts pale yellow, tan, or reddish tan; head and body covered with armor plates; sparsely but **prominently sprinkled with long, stiff, whitish hairs;** carapace at center of body with six or seven movable bands. **Head triangular from in front, broad between eyes and ears; ears short, not protruding above crown, widely separated by more than length of ear on top of head.** Tail moderately long, cylindrical, with two or three rings of scales near base. Feet with five toes; claws not greatly enlarged.
Similar species. Long-nosed armadillos (*Dasypus* spp.) have ears set close together and long, narrow muzzle; naked-tailed armadillos (*Cabassous* spp.) have large ears and greatly enlarged foreclaws.
Natural history. Mostly diurnal; terrestrial; solitary. Feeds on a wide range of foods including much plant material, ants, other insects, small vertebrates, and carrion. Active yellow armadillos trot rapidly about, often stopping to dig up a food item. They den in burrows with a single entrance about 19 cm high and 21 cm wide. Unlike most other armadillos, these will bite when handled. This is mainly a species of savannas, brushy grasslands, cerrado, chaco, gallery forests, and forest edges, but there are rare reports of its occurrence deep in the lowland rainforest.

Geographic range. South America: savannas of S Suriname, and Pará, Brazil, south through SE Brazil and most of the Rio Paraguay basin in Mato Grosso, Brazil, Bolivia, Paraguay, NE Argentina, and Uruguay. Distribution in rainforest poorly known.
Status. Locally common, hunted for meat.
Local names. Gualacate (Ar); peji (Bo); tatu peba, tatu peludo (Br); tatou jaune (FG); peludo (Ur); tatú podyú (Gua).
References. Redford, K. H., and R. M. Wetzel. 1985. *Euphractus sexcinctus.* Mammalian Species, no. 252.

Southern Naked-tailed Armadillo
Cabassous unicinctus
Plate A, map 28
Identification. Measurements: north of Amazon: HB = 347–445; T = 165–200; HF = 70–84; E = 30–40; WT = 2.5–3.6 kg; south of Amazon: HB = 290–345; T = 87–140; HF = 65–76; E = 25–30; WT = 1.6–4.8 kg; females larger than males.
Upperparts naked, covered with bony armor; **dark gray with sharply demarcated yellow border on lower edge of carapace;** midbody with 10–13 movable bands, not sharply demarcated; scales on body and movable bands alike, squarish and fingernail-sized. Head broad, blunt-nosed; **ears large, rounded, funnel-like,** can fold back to cover meatus, often frayed at edges, **set widely apart on crown,** armor shield extending between ears back to cover nape. **Tail long and narrow, appears naked** of armor (some small, inconspicuous scales),

Map 27

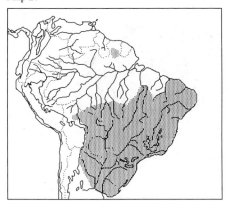

Yellow armadillo, *Euphractus sexcinctus*

Map 28

Southern naked-tailed armadillo, *Cabassous unicinctus*

Northern naked-tailed armadillo, *C. centralis*

gray, usually with a paler tip. Feet with five pale claws, **central claw of forefoot greatly enlarged.** Underparts completely naked, tan.
Variation. The race found south of the Amazon in Brazil *(C. u. squamicaudis)* is much smaller than that north of the river; the two hybridize south of the river mouth.
Similar species. Giant armadillos *(Priodontes maximus)* are much larger, with huge hindfeet; long-nosed armadillos *(Dasypus* spp.) have ears almost touching each other at base; yellow armadillos *(Euphractus sexcinctus)* have short ears, white hairs protruding between scales of back and face, and small front claws.
Sounds. Piglike grunts when handled.
Natural history. Nocturnal; terrestrial; solitary. Feeds on ants and termites. Built for heavy-duty digging, naked-tailed armadillos move slowly. Burrows have a single small, round entrance about 16 cm diameter, usually dug into flat ground in an old termite mound or into a bank. A burrow is usually used for only one night; it has a strong musky odor when the animal is in residence. Southern naked-tailed armadillos are extremely rare in rainforest, where their habits are unknown. Found in a wide range of habitats from rainforest to grassland.
Geographic range. South America: east of the Andes, N Colombia and Venezuela south to Bolivia, and Mato Grosso and Minas Gerais in Brazil.
Status. Unknown, apparently uncommon or rare everywhere, or else difficult to observe.

Local names. Peji (Bo); tatu-rabo-de-couro, cabassú, tatu-de-rabo-mole (Br); armadillo rabo de trapo (Co); lugubre, tatu iba (Ec); cabassou (FG); tatuaí (Gua).
References. Carter, T. S., and C. D. Encarnaçao. 1983. Characteristics and use of burrows by four species of armadillos in Brazil. *J. Mammal.* 64:103–8.
  Wetzel, R. M. 1980. Revision of the naked-tailed armadillos, genus *Cabassous* McMurtrie. *Ann. Carnegie Mus.* 49:323–57.

Northern Naked-tailed Armadillo
Cabassous centralis
Map 28
Identification. Measurements: HB = 305–400; T = 130–183; HF = 60–74; E = 31–37; WT = 2–3.5 kg.
Similar to southern naked-tailed armadillo, but smaller. At night, naked tail appears pale. Walks on the tips of its foreclaws, with hindfeet pigeon-toed.
Similar species. Nine-banded armadillo *(Dasypus novemcinctus)* has ears set close together on head and tail ringed with armor.
Sounds. Growls or squeals when restrained.
Natural history. Nocturnal; terrestrial; solitary. Feeds mainly on ants and termites. These armadillos move slowly; a characteristic view is the distinctive pale, naked tail disappearing under a log. When burrowing, they rotate the body from side to side, so that the forward-projecting edge of the carapace acts as an auger. Naked-tailed armadillos are strong-smelling, and if picked up they urinate copiously and sometimes defecate,

twirling the tail. Found both in rainforest and in drier habitats, often in rocky areas.
Geographic range. Central and South America: Honduras south to NE Colombia on both sides of the Andes, and NW Venezuela west of Cordillera Oriental.
Status. CITES Appendix III (CR). Appears rare everywhere.
Local names. Armado de zopilote (CR); armadillo hediondo (Co); rabo de molle, rabo de carne (Ec); tumbo armado, cusuco (Ho); armadillo rabo de puerco, morrocoy (Pa).
References. Meritt, D. A., Jr. 1985. Naked-tailed armadillos, *Cabassous* sp. In G. G. Montgomery, ed., *The evolution and ecology of armadillos, sloths, and vermilinguas,* 389–91. Washington, D.C.: Smithsonian Institution Press.

Map 29

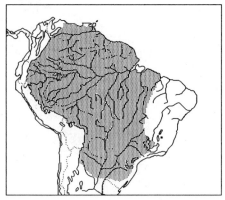

Giant armadillo, *Priodontes maximus*

Giant Armadillo
Priodontes maximus
Plate A, map 29
Identification. Measurements: HB ≈ 750–1,000; T ≈ 500–550; HF ≈ 180–190; E ≈ 47–59; Wt ≈ 30 kg.
Upperparts naked and **covered with bony armor; carapace looks several sizes too small, does not cover lower sides or legs;** carapace gray with pale yellow border on lower edge, but color often obscured by clay from digging. **Head small, ears set very wide apart,** with armor plates between; snout conical. Tail long and tapered, covered with small pentagonal scales. Underparts naked, pinkish tan. **Legs and feet,** especially hind, **enormous; forefoot with greatly enlarged central claw.**
Similar species. No other armadillo approaches this size: naked-tailed armadillos (*Cabassous* spp.) are similar, but much smaller, have carapace covering sides, and normal-sized legs; great long-nosed armadillos (*Dasypus kappleri,* often locally called "giant") have ears set close together, carapace covering sides, rings of armor on tail, and normal-sized legs.
Natural history. Nocturnal; terrestrial; solitary. Feeds mainly on ants and termites, which it obtains by digging into their nests, and other insects. Giant armadillos use their great weight to dig in firm ground: they rock back onto their outsized hindfeet and throw the head and forequarters violently forward, claws extended as if diving, and ram them with all their weight into the hole. They then

rock backward, raking the loosened dirt with the forefeet to the hindfeet, which simultaneously kick the dirt behind the body. Their foraging holes are distinctive: very large, taller than wide, with a pointed rather than rounded upper roof. They appear to hunt out the brood chambers of large termite and ant colonies, such as those of leafcutter ants (*Atta* spp.), by digging several tunnels at different points under the base of the same nest. Raided *Atta* nests are often extinct, suggesting that giant armadillos may have been important predators of these devastating ants before their own numbers were drastically reduced. When disturbed, giant armadillos raise their forequarters off the ground, sit balanced on the huge hindfeet and tail, and lift the snout to sniff the air. Their burrows have one entrance, 30 cm high by 45 cm wide, usually dug in a flat termite mound. Found in many habitats from rainforest to grassland.
Geographic range. South America: east of the Andes from N Venezuela and Colombia south to N Argentina, Paraguay, and S Amazon Basin of Brazil.
Status. US-ESA endangered, CITES Appendix I. Rare and endangered, extirpated due to overhunting for meat in parts of its range, including most savannas and areas near human settlements, where it is unable to hide.
Local names. Tatu carrera (Ar); pejichi (Bo); tatu canastra, tatu-açu (Br); ocarro, jusa trueno (Co); yungunturu, carachupa maman (Pe); tatu gigante, trueno, cutimbo

(Ec); tatou géant, cabassou (FG); tatu carreta
(Pa); reuzengordeldier, granmankapasi (Su);
tatú-guazú (Gua).
References. Clark, T. S., and C. D. Encarnaçao. 1983. Characteristics and use of burrows by four species of armadillos in Brazil.
J. Mammal. 64:103–8.

Nine-banded Long-nosed Armadillo
Dasypus novemcinctus
Plate A, map 30
Identification. Measurements: HB =
356–573; T = 245–450; HF = 75–110;
E = 35–57; WT = 2.7–6.3 kg.
**Upperparts hairless, covered with bony ar-
mor; usually nine (8–10) movable bands
around midbody;** scales on movable bands
narrow triangles, scales on rest of armor
small and roundish; upperparts usually gray,
paling gradually to yellowish sides, but
color often hidden by clay from burrowing.
**Muzzle long, narrow; ears large, set close
together, almost touching at their bases,
with no armor between. Tail long, tapered,
armored with distinct rings for first 60%,**
rings often with contrasting pale border.
Hindfeet with five claws; forefeet with four;
central claws somewhat enlarged. Underparts almost hairless, pinkish yellow.
Similar species. Great long-nosed armadillos *(D. kappleri)* are larger, with projecting
scales on knees; seven-banded long-nosed
armadillos *(D. septemcinctus)* have six or
seven bands around midbody; other armadillos have ears wide apart at base, with armor
between.
Sounds. No calls usually heard, snuffles
while foraging. One of the noisiest travelers
in the forest: treads heavily on dead leaves
and shoves through the undergrowth; a fast-
moving nocturnal crashing sound is generally a long-nosed armadillo.
Natural history. Chiefly nocturnal, sometimes diurnal; terrestrial; solitary. Feeds
mostly on ants, termites, and other insects,
but will eat many kinds of small animal
prey, carrion, and some fruit, fungus, and
other plant material. Nine-banded armadillos are found throughout the forest but are
most often seen in thickets and dense vegetation on sloping, well-drained terra firme;
rarely on flat plains subject to prolonged
flooding. Small diggings in the soil show
where an armadillo has been foraging. They
walk or trot rapidly, often using well-worn

Map 30

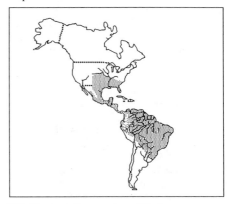

Nine-banded long-nosed armadillo, *Dasypus
novemcinctus*

pathways. They see poorly and will run into
a stationary person at night. When alarmed,
they explode crashing away through the
undergrowth for a short distance or, if surprised at close range, may make a high vertical leap into the air. They dig burrows with
several entrances of about 20 cm diameter,
usually in a bank or slope. Well-packed, narrow runways radiate from a burrow throughout the surrounding area, especially along
steep hillsides. This is the most commonly
seen armadillo. Found in a wide range of
mature and secondary habitats from deep
rainforest to grassland and dry scrub.
Geographic range. North, Central, and
South America: S USA through Central
America west of the Andes to N Peru, east
of the Andes to Uruguay and Argentina, including the islands of Grenada, Margarita,
Trinidad, and Tobago. To at least 1,500 m
elevation.
Status. Common and widespread; hunted
extensively for its excellent meat, which is
often an important food source; sometimes
scarce in areas populated with subsistence
hunters, but seems to withstand heavy hunting pressure.
Local names. Armadillo (Span); dilly (Be);
tatu-galinha (Br); tatú (Bo); cachicame (Co);
mulita, cachicambo (Ec); tatou a neuf ban-
des (FG); cusuco, pitero (Ho); tochi (Me);
carachupa (Pe); negengordelig gordeldier;
gewone kapasi, diki-diki (Su); tatueté (Gua).
References. Wetzel, R. M., and E. Mon-
dolfi. 1979. The subgenera and species of
long-nosed armadillos, genus *Dasypus* L.

In J. F. Eisenberg, ed., *Vertebrate ecology in the northern Neotropics,* 43–63. Washington, D.C.: Smithsonian Institution Press.

Redford, K. H. 1985. Food habits of armadillos (Xenarthra: Dasypodidae). In G. G. Montgomery, ed., *Evolution and ecology of sloths, armadillos, and vermilinguas,* 429–37. Washington, D.C.: Smithsonian Institution Press.

Great Long-nosed Armadillo
Dasypus kappleri
Plate A, map 31
Identification. Measurements: HB = 510–575; T = 406–456; HF = 110–135; E = 48–55; WT = 8.5–11.8 kg.
Similar to nine-banded armadillo, but much larger; seven to nine movable bands at midbody. Hind knees with two rows of long, downward-projecting, spur-like scutes, bordered by other scale-rows that do not protrude. Often covered with clay from burrowing.
Similar species. Giant armadillos *(Priodontes maximus)* have small, widely spaced ears, thick hindlegs, huge claws on forefeet, and undersized carapace; all other armadillos are smaller; this is the only species with projecting scutes on knees.
Sounds. Travels noisily, like nine-banded armadillo.
Natural history. Nocturnal; terrestrial; solitary. Feeds on a wide variety of arthropods and invertebrates, such as earthworms, centipedes, millipedes, and especially beetles. There is a record of one eating a caecelian. They forage by rooting under the forest floor litter, plowing it with the nose and digging with the foreclaws. In soft soil such as that in palm swamps, they plunge the long rostrum repeatedly into the mud up to the level of the eyes, perhaps seeking worms. When startled they release a disagreeable musky odor. These armadillos seem to favor the vicinity of streams and swamps. They den in burrows about 25 cm wide, with more than one entrance, dug into steep banks on stream edges; the floor of the entrance may be at or below water level. Networks of well-worn paths branch out from the burrow. Their behavior when seen is much like that of nine-banded armadillos. The spurs on the hindlegs may allow these animals to crawl on their knees into narrow tunnels or to wedge themselves in. Confusing local names

Map 31

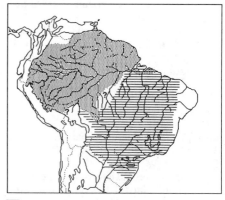

▨ Great long-nosed armadillo, *Dasypus kappleri*
▤ Seven-banded long-nosed armadillo, *D. septemcinctus*

often translate to "giant" armadillo. Found only in lowland rainforest.
Geographic range. South America: east of the Andes, Colombia and S Venezuela to Bolivia and N Pará, Brazil.
Status. Populations seem patchy, quite common in some areas, rare or absent in others; hunted for meat.
Local names. Tatú quinze kilos (Bo, Br, Pe); tatu, tatu canastra (Br); armadillo aracacho, cachicamo grande, jusachula (Co); grand tatou (FG); carachupa (Pe); maka kapasi (Sar).
References. Szeplaki O., E., J. Ochoa G., and J. Clavijo A. 1988. Stomach contents of the greater long-nosed armadillo *(Dasypus kappleri)* in Venezuela. *Mammalia* 52: 422–25.

Seven-banded Long-nosed Armadillo
Dasypus septemcinctus
Plate A, map 31
Identification. Measurements: HB = 240–305; T = 125–170; HF = 45–75; E = 30–38; WT = 1.5 kg.
Externally similar to nine-banded armadillo, except: much smaller, six to seven movable bands around midbody, more dark than pale areas on sides, and tail relatively shorter.
Similar species. See nine-banded armadillo *(D. novemcinctus).*
Sounds. Clicks, growls, grunts, and squeals have been noted in captivity.
Natural history. Most records are from dry

habitats, but a few seem to be from rainforest; most of the range is outside of rainforest regions.

Geographic range. South America: eastern Amazon Basin from the mouth of the Amazon south to N Argentina, and west to Bolivia and Mato Grosso, Brazil.

Status. Apparently rare.

Local names. Tatú chico (Bo); tatuí (Br).

References. See nine-banded armadillo.

Bats (Chiroptera)

Bats are the only mammals that can fly. Their forelimbs are true wings that are flapped in powered flight. Their wing membranes are made of remarkable skin that is strong, soft, flexible, and can heal quickly if punctured. The skin is amazingly elastic: when a bat closes its wings, the membrane does not fold, but contracts like a sheet of rubber, yet the wing opens with almost no force required to stretch it. All New World bats belong to the suborder Microchiroptera. These bats all use a kind of sonar or echolocation. High-frequency sounds are emitted through the mouth or nose, and the returning echoes inform the bat about the characteristics of nearby objects. Echolocation enables bats not only to navigate in the complete darkness of caves and the low light of the forest understory, but also to home in on flying insects. Most echolocation sounds are ultrasonic (above the human hearing range), but people with good hearing can hear a faint ticking sound as a bat flies close by. A few species that feed on large insects echolocate with audible sounds. Members of different families are often specialized for different styles of echolocation, with distinctive structures around the mouth and nose to emit the sounds, and configurations of the ears to gather them. Most echolocating bats also see quite well, especially in dim light. Despite these senses, bats flying down a familiar path may not be paying attention, or, from curiosity, may fly close to people and may brush them with their wings; a bat will not get tangled in human hair (an old superstition), and will not bite spontaneously. But like many animals, a bat will bite in its own defense if it is restrained in the hand. Relatively few species of bats in Neotropical rainforests have ever been found to carry rabies, but there is a risk, especially in populated regions with many livestock and dogs. The best advice is either not to handle live bats or to use heavy gloves when doing so. Bats are beneficial to man; they eat enormous quantities of insects, pollinate many important plants such as balsa, chicle, and calabash, and help regenerate the forest by dispersing seeds. In numbers of species, bats are the most important order of mammals in Neotropical rainforests: there are generally more species of bats in a given forest than there are of all other species of mammals combined; 39% of all mammal species in the region are bats. Worldwide, they are the second largest order of mammals, with about 950 species.

Sheath-tailed Bats (Emballonuridae)

Dental formula: I1/3, C1/1, P2/2, M3/3 = 32. The New World members of this family are all small (3–15 g). They have plain faces with no elaborate structures or warts. The nose is simple, the muzzle a little elongated and upturned, with nostrils slightly tubelike and pointing forward, the snout extending well beyond the lower lip. The ear is a simple, broad triangle, slightly pointed or rounded at the tip, often deeply ridged with parallel folds on its inner rear half and edge; the tragus is a short, simple stub. The eye is quite large and almond-shaped, coming to a gradual point in the inner corner. The tail is sheathed in the tail membrane for most of its length, but the tip sticks out freely above from the middle of the membrane. The extended tail membrane is as long as or longer than the legs. These bats all have long, fine, lax fur. In many species, either just males or both sexes have a moist glandular pocket in the wing membrane forward of the elbow (propatagium). Females are larger than males in most small species. Bats of this family are all aerial insectivores. New World species all have a slow, fluttering, butterfly-like flight and tend to forage in "beats," repeating the same flight path again and again at the same hour each night. They do not have as strong an aversion to daylight as do most bats, and many species roost in open, well-lit places. They often fly when there is some daylight at dusk or dawn. The roosting posture of bats of this family is distinctive: the bat hangs head downward from a vertical or slanted surface, with the feet spread apart and the front of the body supported against the surface by the thumbs on the tips of the folded wings, which are held out at an angle to the body (plate 5); the head is raised up and facing straight outward; the tail membrane is contracted and the tail protrudes. If disturbed, the roosting bats rock rhythmically from side to side. The family is worldwide, tropical and subtropical, and includes about 48 species, 18 of them in the New World.

Map 32

Long-nosed bat, *Rhynchonycteris naso*

Long-nosed Bat
Rhynchonycteris naso
Plates B, 5; map 32
Identification. Measurements: HB =
37–53; T = 9–19; HF = 7–9; E = 10–14;
FA = 35–41; WT = 3–7 g.
**Upperparts brown, heavily frosted or griz-
zled with dirty white,** brown base of fur
showing through gives slightly variegated
look; **lower back with two indistinct wavy
white lines; forearm with little white
tufts of hair evenly spaced along wing-
bone;** wings and tail membrane brown.
**Nose elongated, proboscis-like, extending
far beyond mouth.** Wing without sac. Tail
membrane longer than legs, calcar about
three times as long as foot; membrane thinly
haired for about length of tail on upper side.
Underparts whitish, hair brown at base.
Variation. Lines on back may be prominent
or almost invisible, color varies from tawny
brown to whitish frosted.
Similar species. This is the only small bat
with an elongated nose: see white-lined sac-
winged bats (*Saccopteryx* spp.).
Natural history. Feeds on tiny insects (about
2 mm). Long-nosed bats forage over water,
such as rivers, streams, lakes, ponds, and
marshes. At dusk they are the first bats to be
seen out foraging low over the water surface.
They are one of the most commonly seen
and locally well-known bats. They roost in
the open over or beside water, on the lower
side of a large branch, fallen tree, leaning
tree trunk, steep face of a bank, or below a
bridge. They roost in groups of 3–45, usually about 12, and hang in a characteristic

straight line, with each bat spaced 2–4 cm
from its neighbors. The frosted fur blends
perfectly with tree bark, but the silhouette of
evenly spaced little bumps along a trunk is a
giveaway. They sometimes start to rock back
and forth, with all group members swaying
like leaves in the wind. If approached too
closely, the group flies off together like a
little flock of moths. They can be seen on
almost any river trip in the rainforest region. Found in rainforest, dry forest, gallery
forest, secondary vegetation, gardens, and
pastures.
Geographic range. Central and South
America: Veracruz, Mexico, south to N Bolivia, and Mato Grosso and Minas Gerais,
Brazil.
Status. Often common; widespread.
References. Bradbury, J., and S. L. Vehrencamp. 1976. Social organization and foraging in emballonurid bats. *Behav. Ecol. Sociobiol.* 1:337–81.
   Kalko, E. K. V. 1995. Echolocation signal design, foraging habitats and guild structure in six Neotropical sheath-tailed bats
(Emballonuridae). *Symp. Zool. Soc. Lond.*
67: 259–73.

White-lined Sac-winged Bats
Saccopteryx spp.
Plates B, 5; map 33
Identification. Measurements: HB =
40–67; T = 11–20; HF = 6–13; FA =
35–50; WT = 3–12 g; females larger than
males.
**Upperparts dark, black-brown to tawny
brown, usually with two wavy white lines
from shoulder to end of rump. Wing sac
well developed in males, long, close to
forearm bone beyond elbow;** in females a
small skinfold. Tail membrane longer than
legs, calcar about twice as long as foot.
Underparts usually pale brown, fur darker
brown at base.
Species. There are four species: *S. bilineata*
is the largest (FA = 42–51; 7–12 g), almost black, with blackish wings and tail
and prominent white lines; *S. leptura* (FA =
37–42; 4–7 g) is blackish to chocolate
brown with chestnut tints, distinct whitish
or beige lines on back, black or brown
wings and tail membrane; *S. canescens*
(FA = 35–40; 3–4 g) is grizzled tawny
brown, frosted with grayish or yellowish
especially on head, indistinct white lines

Map 33

White-lined sac-winged bats, *Saccopteryx* spp.

on back, brown wings and tail membrane; *S. gymnura* (FA = 35; 3–4 g) is dark chocolate brown, with no stripes and underparts only slightly paler than back, wings and tail membrane brown.

Similar species. *S. canescens* is like the long-nosed bat *(Rhynchonycteris naso)* but has a wing sac, short muzzle, and no tufts of hair on forearm; the position of the wing sac distinguishes *S. gymnura* from doglike *(Peropteryx* spp.) or chestnut *(Cormura brevirostris)* sac-winged bats.

Sounds. *S. bilineata* often, and *S. leptura* rarely, "sing" at their roosts with a cricketlike twittering and chirping, and both species chirp loudly during encounters with conspecifics while foraging.

Natural history. White-lined sac-winged bats feed on small to tiny insects, including moths. They come out to forage during the last daylight and fly in beats, repeating the same path again and again. They forage at dusk inside the forest *(S. leptura)*, in small openings in the forest, the forest edges, around camps (where they often fly through buildings), river edges, and pastures *(S. bilineata, S. canescens)*. Later in the evening they may move higher and higher to forage above the forest canopy. They roost in small groups, each bat spaced a few centimeters from its neighbors, inside hollow trees or on the outside of a tree under an overhanging branch or knot *(S. leptura)*, or in cavelike cavities between giant buttresses of trees such as figs *(S. bilineata)*, or under bridges or the outer eaves of houses in forest. *S. bilineata* has a complex social system in which a male defends a tiny territory at the roost

site around his harem of one to eight females. Several harems may roost side by side on a tree. The male displays by waving his folded wing, with the gland open, at another bat, or sometimes he hovers in flight, singing to his mates. It is easy to see these behaviors by watching discreetly at dawn or in the late afternoon. The harem of females together defends a territory where they feed. *S. leptura* has a simpler social system; that of the other two species is unknown. Found in lowland mature and secondary rainforest, dry forest, gallery forest, plantations, gardens, and pastures.

Geographic range. Central and South America: *S. bilineata* ranges from S Mexico to SE Brazil; *S. leptura* from S Mexico to Bolivia and to Bahia, Brazil; *S. canescens* from the Amazon Basin and Colombia, Venezuela and the Guianas; *S. gymnura* is known only from Brazil near the Amazon River from the Tapajóz to Belém.

Status. All are widespread and sometimes common, except *S. gymnura,* which appears to be restricted and rare.

References. Bradbury, J. W., and L. H. Emmons. 1974. Social organization of some Trinidad bats. I. Emballonuridae. *Z. Tierpsychol.* 36:137–83.

Bradbury, J., and S. L. Vehrencamp. 1976. Social organization and foraging in emballonurid bats. *Behav. Ecol. Sociobiol.* 1:337–81.

Shaggy Bat
Centronycteris maximiliani
Plate B, map 34

Identification. Measurements: HB = 41–57; T = 20–32; HF = 7–9; E = 17–19; FA = 41.5–49; WT = 4.5–6 g.
Upperparts smoky brown with yellow-orange tint; hair very long and woolly. Head woolly with hair extending to nose and chin; ear with prominent parallel ridges on inner rear edge. **Tail long, emerging about halfway down tail membrane; tail membrane much longer than legs when extended, strongly patterned with parallel rows of dots crossed by prominent branching veins;** reddish body hair extends onto the base of the membrane. Underparts slightly more yellowish than back. No wing sac. **Bat like a small ball of fluff.**
Similar species. Long-nosed bats have a long nose and variegated fur.

Map 34

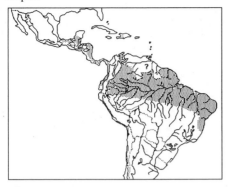

Shaggy bat, *Centronycteris maximiliani*

Map 35

Doglike sac-winged bats, *Peropteryx* spp.

Natural history. Shaggy bats forage in beats, with a slow, fluttering flight. They have been noted flying in late afternoon. They roost in tree holes. Found in mature rainforest and secondary forest.

Geographic range. Central and South America: Veracruz, Mexico, south to Ecuador and N Peru, on both sides of the Andes, and all of N South America north of the Amazon and south to coastal Bahia. Poorly known.

Status. Apparently rare everywhere in a large geographic range.

Doglike Sac-winged Bats
Peropteryx spp.
Plate B, map 35
Identification. Measurements: HB = 41–64; T = 8–17; HF = 6–12; E = 10–20; FA = 38.5–53.6; WT = 3–11 g; females larger than males.

Upperparts chocolate brown, grayish smoky brown, or yellowish smoky brown; fur long and lax. **Head with a prominent tuft on crown,** this tuft slightly wavy, **ending abruptly over naked face; mouth fringed with a narrow mustache of stiff hairs; snout short and broad; ear tip rounded,** inside of ear with deep, parallel ridges, with **fold from upper rim extending partly or completely across forehead.** Wings and tail membrane brown or wings white; **wing membrane attaches at ankle. Tail membrane** long, calcars twice as long as foot; membrane **with rows of fine dots below, undersurface covered with short, fine, stiff hairs; feels rough; above slightly hairy, with body hair extending onto membrane**

for length of tail, about one-third of length. Glandular sac on forward edge of propatagium. Underparts like back or paler.

Species. There are three species: *P. kappleri* is large (FA = 45–54; 6–11 g) and chocolate brown; *P. macrotis* is small (FA = 38–48; 3–8 g) and chocolate or smoky to tawny brown; *P. leucoptera* is intermediate (FA = 41–44), with white wings and a membrane or fold completely across forehead between ears above naked face; it is sometimes placed in a separate genus *(Peronymus).*

Similar species. Chestnut sac-winged bats *(Cormura brevirostris)* have long wing sacs that do not reach edge of membrane, no hair on top of tail membrane, and wing membrane attaches near base of toes; least sac-winged bats *(Balantiopteryx* spp.) have the wing sac in the center of the propatagium, not at its forward edge, and often a white rear edge on wing and a long snout.

Natural history. *P. macrotis* feeds mainly on tiny beetles and flies. *P. kappleri* roosts in small groups of one to six near the ground in hollow logs, moist hollow buttressed trees, or caves, where the bats hang in contact in a line, each touching with its belly the back of the one in front (they separate if disturbed). *P. macrotis* roosts in rockpiles, rock crevices, caves, hollow logs, houses, or tunnels, where *P. kappleri* also sometimes roosts. *P. leucoptera* roosts in tree holes, fallen hollow logs, and under overhanging banks. Bats of this genus fly faster than most other emballonurids. *P. kappleri* forages in open spaces around tree crowns in the rainforest canopy, and above rivers; it more rarely uses croplands. *P. macrotis* usually forages in

rainforest but also feeds in savannas, plantations, and dry forests. *P. leucoptera* is poorly
known but roosts in forest.
Geographic range. Central and South
America: *P. kappleri* and *P. macrotis* from
S Mexico to Paraguay and SE Brazil, *P. kappleri* not in the Amazon Basin. *P. leucoptera*
from the Guianas, the Amazon Basin, and
coastal Brazil to Bahia.
Status. Widespread but may be restricted by
roosting sites; generally uncommon or rare.
References. Bradbury, J., and S. L. Vehrencamp. 1976. Social organization and foraging in emballonurid bats. *Behav. Ecol. Sociobiol.* 1:337–81.

Chestnut Sac-winged Bat
Cormura brevirostris
Plate B, map 36
Identification. HB = 50–65; T = 11–16;
HF = 7–10; E = 13–16; FA = 41–50;
WT = 8–11 g; females slightly larger than
males.
**Upperparts rich chestnut red-brown or
dark black-brown;** fur medium length.
**Wings and tail membrane black; hair of
body does not extend onto base of tail
membrane. Ear tip pointed; fur of crown
gradually decreasing in length down
muzzle;** lower face naked. **Wing sac long,
spindle-shaped, a vertical slit in the center of the propatagium, not quite reaching edge, present in both sexes.** Tail short,
pierces the tail membrane at one-fourth to
one-third of its length; **wing membrane attaches high on foot near base of toe.** Calcar about twice as long as foot; tail membrane extends that much beyond toes when
stretched. Underparts slightly paler or more
chestnut than upperparts.
Variation. Blackish and chestnut colors
occur in the same populations.
Similar species. Least sac-winged bats
(*Balantiopteryx* spp.) and doglike sac-
winged bats (*Peropteryx* spp.) have short
wing sacs that reach edge of membrane, a
tuft of fur on crown that ends abruptly at
naked face, and brown wings; they have gray
tones or are chocolate brown, respectively.
Natural history. Feeds aerially on tiny insects. These bats live in the forest and forage
in small open spaces, chiefly flying in long,
slow, beats of about 20 m long between the
forest canopy and subcanopy. They roost under logs, especially logs over streams, or in-

Map 36

Chestnut sac-winged bat, *Cormura brevirostris*

side hollow logs or trees. Found in lowland
rainforest.
Geographic range. Central and South
America: central Nicaragua south through
Panama; South America east of the Andes
in the Amazon Basin from Peru to Pará; and
Colombia, Venezuela, and the Guianas. To
about 1,000 m elevation.
Status. Sometimes common; widespread.
References. Kalko, E. K. V. 1995. Echolo-
cation signal design, foraging habitats and
guild structure in six Neotropical sheath-
tailed bats (Emballonuridae). *Symp. Zool.
Soc. Lond.* 67: 259–73.

Least Sac-winged Bats
Balantiopteryx spp.
Plate B, map 37
Identification. Measurements: HB = 45–
53; T = 12–21; HF = 6–11; E = 11–16;
FA = 35–46; WT = 5–8 g; females larger
than males.
**Upperparts dark brown, smoky brown,
or gray. Crown with tuft of fur that ends
abruptly on naked face. Muzzle broad and
triangular; ear broadly triangular, round
at tip.** Wings and tail membrane brown; **tail
membrane reaching only slightly beyond
tip of foot, calcar only slightly longer than
foot; edge of wing may be white** between
leg and fingers. Wing sac short, in center of
propatagium, opening facing toward body,
well developed in males, rudimentary in
females. Underparts the same as back or
slightly paler.
Species. There are three species: *B. plicata*
(FA = 38–46) has a white rear-wing margin
and is smoky to gray; *B. io* (FA = 35–39) is

Map 37

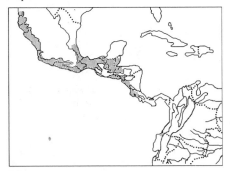

Least sac-winged bats, *Balantiopteryx* spp.

Map 38

Ghost bats, *Diclidurus* spp.

dark brown; and *B. infusca* (FA = 41) is dark chestnut brown.

Similar species. See doglike (*Peropteryx* spp.) and chestnut *(Cormura brevirostris)* sac-winged bats.

Natural history. Feeds aerially on tiny insects. Natural history well known only for *B. plicata*. Least sac-winged bats roost in small to large colonies in well-lighted parts of caves, rock crevices, culverts, or tree hollows. The roosting bats space themselves about 20 cm apart. *B. plicata* occupies dry forest and arid habitats, where it forages with slow flight in long unbroken sallies at 15–25 m above the surface of the ground or forest canopy. It may forage singly, or several may use overlapping beats. *B. io* and *B. infusca* are found in rainforest; their habits are undescribed.

Geographic range. Central and South America: *B. plicata* from the tip of Baja, and S Sonora, Mexico, to Costa Rica; *B. io* from Veracruz, Mexico, to Guatemala and Belize; and *B. infusca* only from the NW coast of Ecuador west of the Andes. To at least 1,500 m elevation.

Status. *B. plicata* is widespread and sometimes common; *B. infusca* is rare.

References. Arroyo-Cabrales, J., and J. K. Jones, Jr. 1988. *Balantiopteryx plicata.* Mammalian Species, no. 301.

Bradbury, J., and S. L. Vehrencamp. 1976. Social organization and foraging in emballonurid bats. *Behav. Ecol. Sociobiol.* 1:337–81.

Ghost Bats

Diclidurus spp.
Plates B, 5; map 38

Identification. Measurements: HB = 55–79; T = 17–26; HF = 10–14; E = 13–23; FA ≈ 50–73; WT = 12–23 g; females larger than males.

There are two color patterns: (1) **Upperparts pure white or white with gray hair base showing through;** underparts white with gray hair base showing through; **wings and tail membrane translucent,** without pigment, yellowish, bones pale brown; **or** (2) **forequarters white, lightly frosted with pale brown, hindquarters pale brown, hair white at base;** underparts white tinged with gray-brown, hairs brown at base; **wings and tail membrane pale brown, semitranslucent.** All have the following: **ear short and rounded, lower edge extending forward in a flange to below the eye,** upper edge extending as a membrane partially across the forehead; **thumb enclosed in the wing membrane, with only the claw emerging;** no wing sac, **tail membrane with prominent, brown sacs around tail near tip,** most highly developed in males; fur long and silky.

Species. There are four species: *D. scutatus* (FA = 50–57), *D. albus* (FA = 60–68), and *D. ingens* (FA = 70–73) have color pattern (1); they differ in size. *D. isabellus* (FA = 60–66) has color pattern (2) and is sometimes placed in a separate genus *(Depanycteris).*

Similar species. Honduran white bats *(Ectophylla alba)* are the only other white bats in the region; they are much smaller (5–6 g) and have a noseleaf.

Sounds. Emits a musical twittering while in flight.

Natural history. Ghost bats fly high in open spaces, such as above rivers, streams, and

lagoons, and above the forest canopy. They
have also been found flying near the water
surface of rivers. They are attracted to concentrations of insects around high floodlights
and are among the many species swooping
around such lights, even in towns. Their
roosts are not well known; they have been
found in abandoned towers, and *D. albus*
roosts between the leaves of palms. *D. isabellus* is known only from lowland rainforest habitat (over water); *D. scutatus* occurs only in the rainforest region. The other
species are found in rainforest, dry forest,
plantations, and in towns.

Geographic range. Central and South
America: only *D. albus* in Central America,
from Nayarit, Mexico, south down the coast
to N Peru, and the Amazon Basin of Brazil,
to Bahia; *D. scutatus* and *D. ingens* in the
Amazon Basin of Brazil, Venezuela, Colombia, and NE Peru, and the Guianas; *D. isabellus* is known only from the Amazon Basin
of Venezuela and a small part of adjacent
Brazil. All geographic ranges poorly known,
from few, scattered records.

Status. The high roosting and flying behavior of these bats makes detecting them extremely difficult, and it is not possible to say
whether they are common or rare.

Local names. Murciélago blanco (or albino)
(Span); jumbie bat (Tr).

References. Ceballos, G., and R. A. Medellín. 1988. *Diclidurus albus.* Mammalian
Species, no. 316.

Smoky Bat
Cyttarops alecto
Map 39

Identification. Measurements: HB = 50–
55; T = 20–25; HF = 8; E = 10–13;
FA = 45.8–47.

**Upperparts dull, dark, smoky gray, almost
black,** palest on upper back and shoulders;
**wings and tail membrane black; fur long
and silky. Ear short and rounded, with
margins as in ghost bats;** tragus with a distinct projection on lower, outer margin; nostrils tubelike and slightly diverging; **face**

Map 39

[●] Smoky bat, *Cyttarops alecto*

thickly furred down center to above nose,
almost naked on side from ear to nose. No
glandular sacs on wings or tail. Underparts
like back.

Similar species. Very like thumbless bat
(Furipterus horrens), which has tiny eye and
rudimentary thumb. White-lined sac-winged
bats *(Saccopteryx* spp.) have white lines on
back; doglike *(Peropteryx* spp.) and least
(Balantiopteryx) sac-winged bats have
longer, more pointed ears, brown wings, and
a naked face with a tuft on crown.

Natural history. Smoky sac-winged bats
have been found roosting in groups of one
to twelve under fronds of coconut palms.
They hang freely, each bat separated from
its neighbors by several cm. They leave the
roost after dark, and have been noted flying
in an oil-palm grove and in a garden.

Geographic range. Central and South
America: known only from Costa Rica,
Nicaragua, Guyana, and near Belém, Brazil.

Status. This is one of the rarest bats, known
from fewer than twenty individuals from six
lowland rainforest localities.

References. Starrett, A. 1972. *Cyttarops
alecto.* Mammalian Species, no. 13.

Reid, F. A., and C. A. Langtimm. 1993.
Distributional and natural history notes for
selected mammals from Costa Rica. *Southwestern Naturalist,* 38:299–302.

Bulldog Bats (Noctilionidae)

Dental formula: I2/1, C1/1, P1/2, M3/3 = 28. This small family derives its name from the
distinctive, bulldoglike face, with downcurved, drooping lips and jowls that expose large canine teeth. The characteristics of the family are described below. The family includes one
genus and two species restricted to the Neotropics and subtropics.

Bulldog or Fishing Bats
Noctilio spp.
Plates B, 5; map 40
Identification. Measurements: HB = 64–
109; T = 14–37; HF = 16–39; E = 22–31;
FA = 60–88; WT = 19–90 g, males larger
than females.
**Upperparts bright rust-orange, rust-red,
chestnut-brown, light yellow, dull tawny
brown, or silvery gray; midback with a
single pale stripe from neck or shoulder
to rump;** sides of neck often paler, more
yellowish; **fur extremely short, sides of
lower back naked** (wings attach high on
sides). Wings and tail membrane brown,
semitranslucent. **Head with large, forward-
pointing nosepad, no noseleaf; lips split
near nose and forming downward-
drooping, bulldoglike folds that form
cheek pouches and expose large canine
and pointed upper incisor teeth; ears long,
sharp, narrow, forward-slanting.** Tail membrane longer than hindlegs; tail shorter than
membrane, enclosed within it except for
short stub emerging from upper surface.
**Feet large, robust, toes long, claws large,
strongly curved, and sharp.** Underparts
paler than back, rich bright orange to pale
orange or pale, tawny brown in brown animals. **Strong, musty odor.**
Species. Lesser bulldog bats *(N. albiventris)*
are smaller (FA = 60–65; WT = 20–40 g)
with relatively smaller feet; they have an
extreme color form of brown frosted with
silvery or white on the neck and shoulders.
Greater bulldog or fishing bats *(N. leporinus)* are one of the largest New World bats
(FA = 77–88; WT = 45–90 g). They have
enormous feet, with scimitar-shaped, gaff-
like claws. The two species have similar,
highly variable coloration; some populations
include both brown and orange individuals,
with males often more orange and females
browner, while other populations are more
homogeneous. Ammonia in the roost atmosphere may be the cause of fur bleaching in
these and other cave- and tree hole-roosting
bats.
Similar species. All other bats as large as
bulldog bats have tall, prominent, spear-
shaped noseleaves, no tails (vampires), or
long tails extending well beyond the edge of
the tail membrane.
Natural history. Feed aerially on insects
(both species) and from the water or
ground surface on fish, aquatic inverte-

Map 40

Bulldog or fishing bats, *Noctilio* spp.

brates (shrimp), and terrestrial invertebrates,
including fiddler crabs and scorpions *(N.
leporinus).* Greater bulldog bats are specialized for fishing: with echolocation they detect fish disturbing the water surface, gaff
them with the feet, and transfer them to the
mouth, stuffing them under the loose cheeks.
Lesser bulldog bats are insectivorous; they
eat many water beetles. Bulldog bats often
forage over water, where they can be recognized from a distance by their large size, red
color, agile flight, pointed wings, and behavior of dipping their feet and skimming the
water as they fly (many species of bats come
out at nightfall and drink on the wing, dipping their jaws into the water). Both species
also forage over dry land. They come out
just at nightfall and often forage in small
flocks of several bats that circle back and
forth close to the water surface for a few
minutes in one area, then move on to another. Both species roost in large groups in
hollow trees or houses, and greater bulldog
bats also roost in caves. Usually found
around water with a still surface in rainforest, dry forest, coastal marine habitats,
marshes, and open or disturbed habitats
such as pastures, riversides, and plantations.
Geographic range. Central and South
America: Pacific coastal Mexico and Cuba
(N. leporinus) or Guatemala *(N. albiventris)*
south to N Argentina and Uruguay, and
Greater and Lesser Antilles. Lowlands only.
Status. Both species widespread and often
common.
Local names. Murciélago pescador
(Span); hazemond vleermuis (Su); mbopí
pytá (Gua).

References. Brooke, A. P. 1994. Diet of the fishing bat, *Noctilio leporinus* (Chiroptera: Noctilionidae). *J. Mammal.* 75:212–18.
Schnitzler, H. U., E. K. V. Kalko, I. Kaipf, and A. D. Grinell. 1994. Fishing and echolocation behavior in the greater bulldog bat, *Noctilio leporinus. Behav. Ecol. Sociobiol.* 35:327–45.

Leaf-chinned, Mustached, and Naked-backed Bats (Mormoopidae)

Dental formula: I2/2, C1/1, P2/3, M3/3 = 34. These are small to medium-sized bats with lips expanded into flaring plates and mouth flanked on its upper sides by a mustache of stiff, forward-pointing hairs. Two "naked-backed" species have wing membranes that are attached along the midline of the back, covering the fur beneath. The lower rim of the ear is prolonged forward as a low membrane to the corner of the mouth. The tail is enclosed in the tail membrane except for its tip, which emerges free about halfway down the membrane. The calcar is about twice as long as the foot, and the tail membrane is longer than the legs. These bats emit their echolocation calls through the mouth, with the enlarged lips acting as a megaphone. They are strong, fast flyers with long, narrow wings. All species seem to roost preferentially in dark, warm, humid caves, and their local abundance may be limited by the availability of appropriate roost sites. There are two genera and eight species in the New World tropics and subtropics.

Mustached and Naked-backed Bats
Pteronotus spp.
Plates B, 5; map 41
Identification. Measurements: HB = 38–80; T = 16–28; HF = 8–17; E = 12–20; FA = 35–56; WT = 3–28 g.
Upperparts rust-orange, gold-orange, gold-brown, smoky brown, tawny brown, or gray-brown. Back either naked, covered by wing membranes to behind shoulders, or with sides of lower back naked (wings attach high over furry sides); fur on lower back short and velvety, and fur on neck, throat, and shoulders longer, often paler, forming a paler collar. Face with lips flared out forward in meaty plates, lower lip with small round tubercles; **muzzle with stiff, forward-pointing mustaches above;** no noseleaf, **ear narrow, pointed, flared into funnel at base, pointing forward;** eye small, sometimes almost lost in fur, surrounded by ear funnel. Tail membrane longer than legs, tail tip emerging freely from middle of top surface of membrane. Underparts usually paler than back, silver, yellowish, beige, orange, or gray, fur on belly dark at base.
Species. There are six species: two have naked backs: *P. davyi* (FA = 41–49; WT = 6–11 g) and *P. gymnonotus* (FA = 49–56; WT = 12–16 g). The two furry-backed mainland species can also be separated by size: *P. personatus* (FA = 40–47; WT = 6–9 g) and *P. parnellii* (FA = 49–65; WT = 11–28 g). Two small species are found only on large Caribbean islands, one *(P. macleayii)* on Cuba and Jamaica and the other*(P. quadridens)* on Haiti and Puerto Rico as well. In all species there is much color variation both within and sometimes between populations. Some variation may be due to length of time since the last molt, fur bleaching with age, or to bleaching effects of alkaline air in caves. *Similar species.* Lesser bulldog bats *(Noctilio albiventris)* have the upper lip split from below the nose and a pale stripe on the back. *Natural history.* All mustached bats are aerial insectivores that feed on many scarab beetles, as well as other insects. *P. parnellii* forages within dense vegetation inside the forest; the other species forage over streams, in gaps, and along forest edges. They roost in warm, humid caves in colonies of up to thousands. At roost they hang free by their hindfeet, back rounded, in tight clusters of individuals in direct contact. Rarely they are found in smaller numbers in other types of roosts, such as old buildings, culverts, or tree hollows. Their scarcity in the central Amazon Basin may be due to the lack of caves and rocks. The naked-backed species are said to be agile at running and hopping quadrupedally on the ground, on feet and folded wings. Found in rainforest and dry forest, clearings, and plantations.

Map 41

Mustached or naked-backed bats, *Pteronotus* spp.

Map 42

Leaf-chinned bat, *Mormoops megalophylla*

Geographic range. Central and South America: Mexico south to Bolivia, the base of the Andes and N coast of South America through the Guianas, generally rare or absent from the central Amazon Basin. Large Caribbean islands. To 3,000 m elevation.
Status. Locally common, especially *P. parnellii.*
References. Herd, R. M. 1983. *Pteronotus parnellii.* Mammalian Species, no. 209.
 Silva, G. 1979. *Los murciélagos de Cuba.* Havana: Editorial Academia.

Leaf-chinned Bat

Mormoops megalophylla
Plate B, map 42
Identification. Measurements: HB = 61–76; T = 19–31; HF = 10–13; E = 13–17; FA = 51–59; WT = 12–20 g.
Upperparts chestnut, chocolate brown, tawny yellow-brown, pink-beige, or mouse gray; fur long; base of neck between shoulders with stiff, straight, back-slanting hair often contrasting pale yellow, pink, or pale brown; area behind ears this same color; fur on neck sometimes sparse, emphasizing shoulder crest. Face with complex skinfolds: chin with two large concave plates connected to a series of folds below and beside chin; nostrils set wide apart in short tubes, side of nose with folds; ears short, rounded, connected by a large fold across forehead that is deeply notched on midline, inside of ear with complex, deep folds; lower edge of ear extending forward in a fold to corner of mouth; eye medium-

sized, surrounded by ear funnel. Tail membrane longer than legs, tail enclosed within it, with only tip emerging. Underparts paler than back, buffy yellow, pink-brown, or reddish; fur long. Body long and slender.
Variation. Color varies greatly between populations. Another species, *M. blainvillei,* is found only in the Greater Antilles; it is smaller (FA = 43–50, WT = 6–11 g).
Similar species. Wrinkle-faced bats *(Centurio senex)* and vampires *(Desmodus, Diphylla* spp.) have simple, upstanding ears without complex folds.
Natural history. Feed on insects, chiefly moths. Leaf-chinned bats often forage over water, on forest edges, gaps, or open spaces. They roost in deep caves, where they hang singly, separated from their neighbors, in colonies of up to several thousand, but usually of only a few individuals. They seem to be nomadic. Chiefly found in arid habitats, rarely in rainforest.
Geographic range. North, Central, and South America: S Texas and Arizona south through Honduras; absent from the Isthmus of Panama and present in NE Colombia, N Venezuela, Trinidad, and NW coastal Ecuador.
Status. Generally uncommon, rarely in large numbers.
References. Barbour, R. W., and W. H. Davis. 1969. *Bats of America.* Lexington: University Press of Kentucky.
 Smith, J. D. 1972. *Systematics of the chiropteran family Mormoopidae.* Univ. Kans. Mus. Nat. Hist. Misc. Pub., no. 56.

Leaf-nosed Bats (Family Phyllostomidae)

Dental formula highly variable. In most genera, the nostrils are surrounded by a fleshy fold that has a dishlike horseshoe below and/ or beside the nostrils and a freestanding, spear-shaped leaf above and behind the nostrils. These structures act as a horn or megaphone to focus the echolocation sounds, which are emitted through the nostrils (although the bats are also able to emit them through the mouth). The point of the chin usually has a naked, V-shaped groove, often bordered by an array of warts or tubercles. The wings are broad in outline, so that the bats are able to maintain flight at slow speeds, and they can maneuver among thick vegetation. Leaf-nosed bats represent one of the largest living radiations of mammals. They have evolved a great diversity of specialized feeding habits and associated morphological adaptations and social behaviors. They give birth to single young. The family is restricted to the New World and includes about 50 genera and 140 species, mostly in the tropics, with a few species reaching temperate North America. The family is treated by subfamilies below. Members of a subfamily generally have similar food habits.

Spear-nosed Bats (Subfamily Phyllostominae)

The spear-nosed bats range from the largest New World bats to some of the smallest. They all have large, spear- or sword-shaped noseleaves, with a free horseshoe at least partially around the nostrils; narrow muzzles; long tail membranes; tail enclosed within the membrane except for the extreme tip, which may protrude on the top of the membrane; and usually large to very large ears. The colors of many species are variable, with brown, chestnut, and rust-orange individuals found in the same roosts or populations. Among greater spear-nosed bats in Trinidad, some individuals are orange when young and change to dark brown when they are several years old, while other young in the same social group, and presumably with the same father, are brown. An alkaline roost atmosphere may in some cases bleach the fur. The spear-nosed bats include all genera of leaf-nosed bats that are true carnivores and kill vertebrate prey; they also include the most strongly insectivorous genera, as well as mixed feeders that eat both insects and fruit. They mainly hunt prey that is on a substrate such as the ground, water, or vegetation, and are thought to use sounds made by the prey itself to locate it (somewhat like owls), but some species also seem able to detect prey by echolocation. Recent work shows that at least one genus can hunt insects aerially. Many species seem to have a monogamous social organization, although some have harems. The subfamily includes ten genera and about thirty-four species in the New World tropics and subtropics; nine genera and about thirty-one species are found in the lowland rainforest region, one genus is restricted to arid habitats. Recent evidence suggests that the genera in this subfamily are not all closely related and should be split into at least two groups.

Little Big-eared Bats
Micronycteris spp.
Plate C, map 43
Identification. Measurements: HB = 46–88; T = 5–16; HF = 8–20; E = 13–31; FA = 32–58; WT = 4–30.
Generally, **ears very large, usually rounded,** tragus short; noseleaf medium-sized, spear-shaped; **face and muzzle narrow; naked chin tip with single, long, smooth pads on each side angled to a V or Y shape;** eyes small; **fur long** and slightly wavy; face hairy; thumb with first joint equal to or longer than second joint; tail short, enclosed in tail membrane except for tip, which protrudes slightly on top surface; **tail membrane the same length as or shorter than legs;** calcar about one-half to one times the length of hindfoot; **size small,** with one exception. There are four pelage patterns: (1) Upperparts plain brown to chestnut or orange, **long hair on shoulders with long white base that shows prominently through parts in fur; ears often with a low membrane partly or entirely joining bases across forehead; ear hairy, with long hairs on back** or both surfaces; underparts like back or contrasting white, silvery, beige, yellowish, or orange; hair on throat pale to base. Size small (4–16 g). (2) Upperparts chestnut to pale brown

Map 43

Little big-eared bats, *Micronycteris* spp.

or gray-brown, **hairs look unicolored,** lower back sometimes with faint gray stripe; underparts slightly paler whitish or yellowish. **Ears not hairy on back** except on base. (3) Upperparts dark brown, **hairs on shoulder prominently tricolored,** dark brown at base and tip, with broad, paler tawny or silvery midsection; underparts paler tawny yellow or gray brown, hairs dark at base. **Ears not hairy.** (4) Upperparts uniform **gray-brown,** underparts paler; fur long, dense, and woolly; ears slightly hairy; **size large** (FA > 50; WT = 19–30 g).
Species. There are at least 12 species: 6–7 of them have color pattern (1); *M. nicefori* has pattern (2); *M. sylvestris* has pattern (3); *M. (Barticonycteris) davesi,* the only large member of the genus, has pattern (4) (plate C). Up to seven species can occur together at a locality.
Similar species. Short-tailed fruit bats (*Carollia* spp.) have shorter, triangular ears and chin tip V with a central, large tubercle bordered by a row of tiny tubercles; Mexican big-eared bats from arid habitats (*Macrotis,* not treated in this book) have long tails that extend all the way to the end of the tail membrane and beyond it; round-eared bats (*Tonatia* spp.) and all other similar-looking bats have rows of round tubercles on chin.
Natural history. These bats feed on large insects and occasional fruit. They forage by gleaning large insects such as cockroaches, dragonflies, and katydids from the vegetation or ground, and they carry their prey back to a feeding roost before eating. Some may also pursue flying insects. Several species

are deep rainforest inhabitants; they tend to fly along streams, gullies, and paths well shaded by the forest canopy. Big-eared bats roost in damp hollows near the ground, in small groups of a half-dozen or fewer that often seem to consist of a monogamous pair and their young. Roost sites include hollow logs or trees, cavities under roots, fallen logs, overhanging dirt banks, and culverts. The roosts are often well enough lit so that the bats can be seen without a flashlight. They hang freely suspended and spaced apart and always seem alert. They twist and turn and vibrate their large ears and will readily take flight if disturbed. Found in many habitats including mature and disturbed rainforest, gallery forest, dry forest, cloud forest, and gardens and plantations.
Geographic range. Central and South America: S Mexico south to central Bolivia and SE Brazil. All but two species are found in both Central and South America. To 2,000 m elevation.
Status. Three species are apparently rare: *M. davesi, M. pusilla,* and *M. behni,* the latter two with restricted distribution.
References. Wilson, D. E. 1971. Food habits of *Micronycteris hirsuta* (Chiroptera: Phyllostomidae). *Mammalia* 35:107–10.
  Simmons, N. B. 1996. A new species of *Micronycteris* (Chiroptera: Phyllostomidae) from Northeastern Brazil, with comments on phylogenetic relationships. *Am. Mus. Novitates,* no. 3158.

Sword-nosed Bats
Lonchorhina spp.
Plate C, map 44
Identification. Measurements: HB = 49–74; T = 46–63; HF = 9–20; E = 19–38; FA = 41–59; WT = 7–32 g.
Upperparts dark brown, chestnut, rust-red, or pale beige; neck and shoulders often paler, with hair pale at base; fur deep and plushy. **Noseleaf extremely long (18–22 mm) and narrow, about as long as ears, with smooth edges;** folds around nostrils at base complex; **ears very large, pointed, hairy inside and out on upper edge, wide at base, with lower rim extending as a fold to below eye; tragus very long and narrow, pointed, at least half the length of ear.** Legs very long, tail membrane longer than legs; **tail long, reaches rear edge of tail membrane, which forms a point at tail tip.**

Map 44

Sword-nosed bats, *Lonchorhina* spp.

Species. There are four species: *L. aurita* is widespread and geographically variable; it is large (FA = 47–54, WT = 12–22 g), and dark brown to rust. The other species have restricted distributions—*L. orinocensis* is small (FA = 41–44, E = 29–31, WT = 7–10 g) and pale brown; *L. fernandezi* is small, with shorter ears (FA = 42–43; E = 19) and stiletto-shaped processes at base of noseleaf; and *L. marinkellei* is large, with large ears (FA = 59; E = 38).
Similar species. No other bats in the region have as large a noseleaf; hairy-nosed bats (*Mimon* spp.) have a noseleaf about one-fourth to one-third shorter than the ear, and tail only half as long as the tail membrane.
Natural history. Feed on insects and some fruit. Common sword-nosed bats *(L. aurita)* roost in dense clusters of up to hundreds in caves and old tunnels, such as mine tunnels. Orinoco sword-nosed bats have been found in large rockpiles in hot grassland habitat. Three of the four species were only recently discovered and were first described between 1971 and 1982; these all have small known geographic ranges. Found in rainforest, deciduous forest, and plantations, pastures, and savanna.
Geographic range. Central and South America: *L. aurita* from S Mexico to N Bolivia and SE Brazil; *L. orinocensis* from the Colombian-Venezuelan Orinoco drainage; *L. fernandezi* from one locality in Venezuela, and *L. marinkellei* from Colombia and French Guiana.
Status. Widespread to rare. These bats may be geographically limited to areas with caves or rocks; *L. fernandezi* is currently known

from a single cave, which potentially puts it at risk.
References. Ochoa, J., and C. Ibañez. 1982. Nuevo murciélago del genero *Lonchorhina* (Chiroptera: Phyllostomidae). *Mem. Soc. Cien. Nat. La Salle* 42 (118):145–59.

Long-legged Bat
Macrophyllum macrophyllum
Plate C, map 45
Identification. Measurements: HB = 40–53; T = 38–49; HF = 13–16; E = 17–20; FA = 34–40; WT = 7–11 g.
Upperparts plain medium brown; wings and tail membrane brown; fur medium length. **Ears quite large, pointed, slightly hairy on upper rim,** lower rim not prolonged downward and forward as a fold on head; **noseleaf a large, broad spear with prominent central rib, horseshoe at base of noseleaf broad, with expanded lobe behind nostrils;** eyes small. **Legs and tail membrane long, membrane with a distinctive row of about seven vertical lines of dots, each leading to a bump on the straight rear edge of the tail membrane; tail long, enclosed in membrane and reaching its edge, but not forming a longer point;** calcar about two times as long as foot. **Foot with elongated toes and claws.** Underparts about the same color as back.
Similar species. This is the only bat with parallel rows of dots on the edge of the tail membrane. Little big-eared bats (*Micronycteris* spp.), short-tailed and little fruit bats (*Carollia, Rhinophylla* spp.), and hairy-nosed bats (*Mimon* spp.) have tails much shorter than membrane, or no tail; sword-nosed bats *(Lonchorhina)* have greatly elongated noseleaf and tragus. This is one of the smallest phyllostomines.
Natural history. Long-legged bats feed on insects, including water striders. They fly with a fluttering, butterfly-like flight and may use their long toes and tail membranes to skim insects from on or near the water surface. These bats are usually found near water pools. They roost singly or in small groups, in wet tunnels, culverts, under bridges, and in caves and abandoned buildings. Chiefly found in rainforest and rainforest clearings, sometimes in deciduous forest.
Geographic range. Central and South America: extreme S Mexico south to NE Bolivia, Paraguay, and SE Brazil.

Map 45

Long-legged bat, *Macrophyllum macrophyllum*

Map 46

Round-eared bats, *Tonatia* spp.

Status. Uncommon, but very widespread.
References. Harrison, D. L. 1975. *Macrophyllum macrophyllum.* Mammalian Species, no. 62.

Round-eared Bats
Tonatia spp.
Plate C, map 46
Identification. Measurements: HB =
54–95; T = 7–25; HF = 10–19; E =
19–38; FA = 33–59; WT = 8–40 g.
**Upperparts usually gray-brown, faintly
frosted with silver, or warm brown,** rarely
rusty or dark, ashy gray frosted with white;
**fur on neck pale at base; fur long and soft,
not very dense. Ears very large, rounded,**
usually hairy on base of back and upper edge;
in all but one species **ears when touched
flatten back with tips curled down;** noseleaf large, broad, spear-shaped; **naked tip of
chin with U of tiny round tubercles.** Thumb
often hairy, with second, claw-bearing phalange longer than first phalange. Tail membrane longer than legs; tail less than half the
length of membrane, within it except for extreme tip, which protrudes above. Feet and
claws robust, not elongated. **Underparts
often washed with silver, often with pale
collar on neck.** Small to large size.
Species. There are 6–11 species; up to five
can occur together in South America. The
more distinctive species include: *T. bidens*
and *T. saurophila,* large, with relatively short
ears (FA = 52–61; E = 29–32) and hairy
face, feet, thumbs, forearms, and ears (the
two were recently separated as species and
are distinguished by skull characters and
geography); *T. silvicola,* large, with large

ears (FA = 50–59; E = 36–38), face naked,
throat usually white or gray, curls its ears back
against the crown when they are touched;
T. carrikeri, large with pure white underparts; *T. schulzi,* medium-sized, with dark
head, white venter, and distinctive warty
skin on forearms, fingers, ears, and noseleaf;
and *T. brasiliensis,* small (FA = 33–40),
warm brown. The latter is divided by some
into as many as five species.
Similar species. The small species are very
similar to little big-eared bats (*Micronycteris*
spp.), which have single chin pads in a V, not
round tubercles in a U on point of chin. Hairy-
nosed bats (*Mimon* spp.) and spear-nosed bats
(*Phyllostomus* spp.) have pointed ear tips.
Natural history. Round-eared bats feed on
insects and perhaps occasional fruit. *T. bidens*
is reported to feed intensively on small birds
(4–24 g), which are taken to a feeding roost
for consumption. Round-eared bats seem to
fly everywhere within the forest. They forage by hanging from a perch and scanning
for sounds made by prey, then sallying out to
capture it. Several species, including *T. silvicola, T. carrikeri,* and the small *T. brasiliensis,* roost in small groups of up to ten in holes
in arboreal termite nests. The folding ears of
these species may be an accommodation to
these specialized roosts. *T. saurophila* roosts
in tree hollows and seems to lack folding
ears. *T. silvicola* lives in small home ranges
and has a social system including harems.
Found in mature and secondary rainforest, or
more rarely deciduous forest and plantations.
Geographic range. Central and South
America: extreme S Mexico south to N Argentina and Paraguay.

Status. Common to rare.

References. Williams, S. L., M. R. Willig, and F. A. Reid. 1995. Review of the *Tonatia bidens* complex (Mammalia: Chiroptera), with descriptions of two new subspecies. *J. Mammal.* 76:612–26.

Martuscelli, P. 1995. Avian predation by the round-eared bat (*Tonatia bidens,* Phyllostomidae) in the Brazilian Atlantic forest. *J. Tropical Ecol.* 11:461–4.

Hairy-nosed Bats
Mimon spp.
Plates C, 5; map 47
Identification. Measurements: HB = 52–75; T = 14–29; HF = 9–18; E = 24–38; FA = 46–58; WT = 10–15 g.
Ears large and pointed, thickly haired with long, pale hair on back of base and with long hairs on both sides reaching at least halfway up inner rim; tragus narrow and pointed; noseleaf very long, narrow, spear-shaped, with simple basal horseshoe; tail reaches about halfway down tail membrane, tip just perforates top side; tail membrane longer than legs, calcar longer than foot; foot with long toes and claws; medium-sized. There are two forms: (1) **upperparts dark, blackish brown with a pale stripe down midback from crown to rump,** wings and tail membrane blackish; **sides of noseleaf serrated, noseleaf thinly haired to tip; chin with a U-shaped row of tubercles; underparts frosted silvery buff,** washed orange around throat; (2) **pale, warm brown, hairs pale at base, fur long and dense, wings and tail membrane brown; noseleaf not hairy, smooth on sides; chin with a single large, triangular central wart; underparts about the same as back,** slightly paler on neck (FA = 53–58; E = 35–38).
Species. There are three or four species: *M. crenulatum* has color pattern (1); it is placed by some in a separate genus, *Anthorhina. M. bennettii* (E = 35–38) and a possibly conspecific form, *M. cozumelae,* have color pattern (2). Another species occurs at higher elevations in the Andes.
Similar species. There is no other bat with a long, hairy noseleaf and a striped back. Only one species of little big-eared bat (*Micronycteris daviesi*) is as large as *M. bennettii,* and it is dark gray-brown; round-eared bats (*Tonatia* spp.) have round ears and quite short noseleaves. Spear-nosed bats, *Phyllostomus*

Map 47

▨ Hairy-nosed bats, *Mimon* spp.

elongatus, have no long hairs on base and side of ears, and ears shorter than 31 mm.
Natural history. These bats feed on insects, spiders, and small lizards. *M. bennettii* and *M. cozumelae* roost in small groups in wet caves, and *M. bennettii* in tree hollows. *M. crenulatum* roosts in tree hollows and crevices in rotten stumps and under bark. These roosts have a very strong batty odor. *M. crenulatum* is often mist-netted in male-female pairs, suggesting that pairs forage together. Found in rainforest, gallery forest, and elfin forest, rarely in deciduous forest, and in secondary forest and plantations.
Geographic range. Central and South America: S Mexico south to central Bolivia and SE Brazil. *M. bennettii* is found only along the coast of South America from Colombia to SE Brazil.
Status. *M. crenulatum* is uncommon and *M. bennettii* is rare; both are widespread.

Spear-nosed Bats
Phyllostomus spp.
Plate C, map 48
Identification. Measurements: HB = 76–150; T = 7–30; HF = 12–27; E = 21–36; FA = 55–94; WT = 33–140 g.
Males larger than females.
Upperparts dark blackish brown, rich chestnut red-brown, rusty, or gray-brown; fur short, velvety, flat and smooth on back, usually with a silvery sheen; head, neck, and shoulders often paler than back, more grayish. **Ears broadly triangular, medium length; noseleaf a simple, broad, medium-length spear; horseshoe at base forms a**

Map 48

Spear-nosed bats, *Phyllostomus* spp.

complete semicircular fold free of lip under nostrils; head and muzzle broad and powerful; naked tip of chin with V of small tubercles; eyes large; noseleaf and lips black or dark brown. **Tail short, about one-third the length of tail membrane;** calcar slightly shorter to longer than foot. Underparts slightly to distinctly frosted paler than back.

Species. There are four species: *P. hastatus* is very large (FA = 79–94, WT = 92–140 g); *P. elongatus* is large (FA = 61–68, WT = 38–57 g) and usually has extreme wing-tips pure white, a dark belly, and large ears (E = 28–31); *P. discolor* is slightly smaller (FA = 55–69, WT = 33–45 g) and has short ears (E = 21–26) and distinctive pale grayish underparts; *P. latifolius* is the smallest (FA = 56–60) and has dark gray underparts, short ears (E = 27), and short legs.

Similar species. See pale-faced spear-nosed bat *(Phylloderma stenops).* Only one other leaf-nosed bat is larger than *Phyllostomus hastatus:* the false vampire bat *(Vampyrum spectrum)* is huge (FA > 100 mm), with large, round-tipped ears, cup-shaped horseshoe, and no tail; all other large and medium-sized spear-nosed bats (Phyllostominae) have long fur and long ears; large spear-nosed fruit-eating bats *(Artibeus)* have no tails. Other very large bats have no noseleaf.

Sounds. These bats are social and very noisy; they squabble at the roost and often call with loud chattering squawks while they feed or fly in the forest at night.

Natural history. Spear-nosed bats feed on fruit, large insects, and pollen and nectar.

P. hastatus is noteworthy for specialized feeding on the hanging fruit of a vine *(Gurania)* as well as large beetles, cicadas, and katydids. *P. elongatus* and *P. hastatus* also hunt aerially above the forest canopy. *P. latifolius* is little known. *P. elongatus* roosts in small groups in tree holes, or large groups in caves, while the other two species roost in large congregations of tight clusters in tree holes or caves. These roosting congregations are divided up into smaller permanent social units consisting of harems of a single male and a few *(P. discolor)* to as many as several dozen females and their young *(P. hastatus),* and groups of nonbreeding bachelors. In *P. hastatus,* females of a harem have fixed feeding areas adjacent to each other, but separate from those of other harems. The bats emerge from their roosts at complete darkness and forage for about two hours before returning to the roost. Several *P. discolor* fly single file to forage. Found in mature and secondary rainforest, gardens and plantations, gallery forest, dry forest, and cloud forest.

Geographic range. Central and South America: S Mexico to central Bolivia and SE Brazil. *P. elongatus* and *P. latifolius* only in South America east of the Andes. To 1,300 m elevation.

Status. *P. hastatus* is locally common and widespread in the rainforest. The other species are less common or rare; *P. latifolius* is known from only a few sites and individuals.

References. McCracken, G. F., and J. W. Bradbury. 1981. Social organization and kinship in the polygynous bat *Phyllostomus hastatus. Behav. Ecol. Sociobiol.* 8:11–34.

Bradbury, J. W. 1977. Social organization and communication. In W. A. Wimsatt, ed., *The Biology of Bats,* 3:1–72. New York: Academic Press.

Pale-faced Spear-nosed Bat
Phylloderma stenops
Plate C, map 49

Identification. Measurements: HB = 82–115; T = 12–24; HF = 20–23; E = 25–32; FA = 66–73; WT = 41–65 g.

Upperparts faded brown to slightly reddish brown, often with faint frost or sheen; sides of neck and front of shoulders gray-frosted, continuous with grayish underparts; fur very short, often sparse, or with sparse patches. Wings and tail membrane

Map 49

Pale-faced bat, *Phylloderma stenops*

Map 50

Fringe-lipped bat, *Trachops cirrhosus*

blackish brown, **wingtips white.** Wings attach high on sides of body. Ears large; muzzle narrow; **noseleaf a large, broad spear, horseshoe merging with upper lip at center, without any free fold at center below nostrils; lips, noseleaf, and skin in general pale pinkish mottled with tan.** Tail membrane long, calcar about the same length as foot. **Large, long-bodied bats; adults always seem to look scruffy with faded, sparse hair and mottled skin on face and wing membranes.**
Similar species. Very similar to the larger spear-nosed bats (*Phyllostomus* spp.), which all have horseshoe with continuous free flange or fold around entire base over lip, and blackish noseleaf, lips, and wings; round-eared (*Tonatia* spp.) and woolly false vampire bats *(Chrotopterus auritus)* have long fur; false vampires *(Vampyrum spectrum)* are larger, with a cup-shaped noseleaf and no tail.
Natural history. Has been recorded feeding on the larvae and pupae from a wasp nest, and on fruits, especially those of curcurbit vines. One was collected while it was attacking a large rat, apparently a *Proechimys*. These bats are found around streams and swamps or marshes. The roost is undescribed. Found in rainforest, dry forest, scrub, and plantations.
Geographic range. Central and South America: S Mexico south to N Bolivia and SE Brazil. To 2,600 m elevation.
Status. Uncommon to rare, but widespread.
References. Jeanne, R. L. 1970. Note on a bat *(Phylloderma stenops)* preying upon the brood of a social wasp. *J. Mammal.* 51: 624–25.

Fringe-lipped Bat
Trachops cirrhosus
Plates C, 5; map 50
Identification. Measurements: HB = 71–92; T = 13–21; HF = 16–22; E = 30–40; FA = 57–64; WT = 28–45 g.
Upperparts smoky brown to tawny brown, faintly frosted with paler brown; fur long, pale at base on neck; wings and tail membrane brown, **wingtips often white. Ears large, rounded, prominently hairy on both sides of upper edge; chin, lips, and muzzle around nose studded with long, projecting tubercles; noseleaf moderately large, spear-shaped, edges of both basal horseshoe and spear serrated with tubercles on edges;** muzzle narrow. Tail membrane slightly shorter than legs, tail short, about one-third as long as membrane. Toes long, claws robust and strongly hooked, calcar shorter than foot. Underparts frosted silvery.
Similar species. This is the only bat with large projecting tubercles on the muzzle.
Natural history. Feeds on frogs and lizards, with some insects and occasional small mammals and birds. Fringe-lipped bats are carnivores that appear to feed largely on frogs. Experiments have shown that they home in on the calls of frogs and learn to avoid poisonous species by recognizing their calls. The tubercles on the chin may be chemical sensors; the bat touches a frog's skin (where poisons are found) with the tubercles as it is about to grasp it. They hunt insects by hanging from a perch and sallying out to capture insects whose sounds they detect. Fringe-lipped bats fly low in the forest understory and are especially common around streams, swamps, and ponds. They

roost in small groups, usually a dozen or fewer, in hollow trees, caves, culverts, under bridges, or in houses in the forest. Found in rainforest, gallery forest, deciduous forest, and plantations and secondary vegetation.
Geographic range. Central and South America: S Mexico south to Bolivia and SE Brazil.
Status. Widespread and common.
References. Tuttle, M. D., and M. J. Ryan. 1981. Bat predation and the evolution of frog vocalizations in the Neotropics. *Science* 214:677–78.

Woolly False Vampire Bat
Chrotopterus auritus
Plate C, map 51
Identification. Measurements: HB = 94–114; T = 0–15; HF = 23–28; E = 40–49; FA = 77–87; WT = 61–94 g.
Upperparts gray-brown, often faintly frosted paler; fur very long, dense, and woolly; wings and tail membrane dark brown, **wingtips almost always pure white. Ears very large, rounded, hairy on back of inner rim; eyes large; muzzle long and narrow; noseleaf** moderately large, sword-shaped with prominent central rib, **horseshoe and spear with continuous edge forming a hollow cup around nostrils;** lips, ears, and noseleaf pale pink-brown. Tail membrane long, calcar slightly shorter than foot, **tail a tiny or short stub** within membrane, **sometimes absent.** Underparts pale gray, fur brown at base. **A very large woolly bat.**
Similar species. False vampires *(Vampyrum spectrum)* are similar, but larger, with shorter fur, often a stripe down back, and wingtips usually dark; no other bats have noseleaf and horseshoe that form a hollow cup around nostrils.
Natural history. Feeds on small vertebrates such as small mammals, birds, and lizards, and large insects. Woolly false vampires take prey of up to 70 g weight, but mostly feed on smaller vertebrates of 10–35 g, especially mice. Prey is sometimes brought back to the roosts, which are in caves and hollow logs. These bats roost in small groups of two or three that can consist of a male-female adult pair and a young. Found in primary and secondary rainforest and plantations, and in dry forest and chaco.
Geographic range. Central and South America: S Mexico south to Bolivia, Paraguay, and SE Brazil.

Map 51

Woolly false vampire bat, *Chrotopterus auritus*

Status. Uncommon, but seems ubiquitous in rainforest; very widespread.
References. Medellín, R. 1988. Prey of *Chrotopterus auritus,* with notes on feeding behavior. *J. Mammal.* 69:841–44.
    McCarthy, T. J. 1987. Additional mammalian prey of the carnivorous bats, *Chrotopterus auritus* and *Vampyrum spectrum. Bat Res. News* 28:1–3.

False Vampire Bat
Vampyrum spectrum
Plate C, map 52
Identification. Measurements: HB = 135–158; T = 0; HF = 29–38; E = 39–49; FA = 98–110; WT = 126–190 g.
Upperparts dark brown, chestnut brown, or rust-orange; midback from crown to rump with a broad, pale stripe, often indistinct; fur medium length, woolly, often sparse; wings and tail membrane blackish brown; wingtips usually dark, sometimes partly bleached, but not pure white. **Ears very long, rounded; eyes large; muzzle long and narrow; canines large; noseleaf medium-sized, lance-shaped, horseshoe and spear with continuous rim raised to form hollow cup around nostrils.** Tail membrane long, calcar longer than foot; tail absent. Feet large, robust, with huge curved claws. Underparts pale, dirty gray-brown to yellow-brown; fur much shorter than on back. **A huge bat with a wingspan of about 80–90 cm.**
Similar species. No other New World bats attain this size; see woolly false vampire *(Chrotopterus auritus).*

Map 52

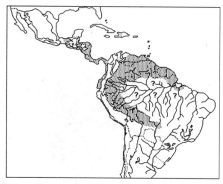

False vampire bat, *Vampyrum spectrum*

Natural history. Feeds on birds, bats, and rodents, possibly rarely takes insects or fruit. Birds brought by false vampires back to a roost were most often large species that weigh 20–150 g, such as cuckoos, anis, parakeets, trogons, and orioles. False vampires roost in small, tight groups in hollow trees. Groups seem to consist of a monogamous adult pair and one to three of their nonbreeding young. A member of the group that catches a large prey will sometimes bring it back to the roost and may share it. One member of the group seems to stay behind to guard the young while the other hunts. These bats often hunt in open areas such as over swamps, forest edges, and secondary vegetation, but they are also found deep in the forest.

Geographic range. Central and South America: S Mexico south to N Bolivia and the southern Amazon Basin. To 1,650 m elevation.

Status. Widespread. False vampires are the top predators of their order, and, like other top predators such as harpy eagles and jaguars, they are always relatively rare.

References. Vehrencamp, S. L., F. G. Stiles, and J. W. Bradbury. 1977. Observations on the foraging behavior and avian prey of the Neotropical carnivorous bat, *Vampyrum spectrum. J. Mammal.* 58:469–78.

Navarro, D., and D. E. Wilson. 1982. *Vampyrum spectrum.* Mammalian Species, no. 184.

Nectar-feeding or Long-tongued Bats (Subfamilies Glossophaginae, Lonchophyllinae)

These subfamilies include bats that all feed on flower nectar. Most are small (7–20 g). They have elongated muzzles, often with the tip of the lower jaw overshot, projecting farther than the upper jaw; and a narrow tongue that can be greatly extended and usually has a brush of hairlike papillae at its tip. The lower incisor teeth are absent or reduced in size, and the lower lip has a concave pad or deep, V-shaped notch; these features provide a channel for the tongue. The noseleaf is a short triangle; there are fine whiskers, generally at least as long as the noseleaf, growing from around the nose and forward from the chin; the ears are relatively short and set low on the sides of the head. The tail is short, and the tail membrane is shorter than the legs. This is one of the most poorly known groups of bats; almost nothing has been recorded of the natural history of many species. These little bats are important or essential pollinators of many plants on whose nectar they feed. Such plants often have large, pale flowers that open at night, and musky odors. Bat-pollinated New World plants include balsa trees, ceiba (silk-cotton) trees, and jícaro (calabash), as well as desert plants such as saguaro cactus and some agaves. Most species of long-tongued bats also eat some insects and fruit. There are thirteen genera in two subfamilies, eight in the rainforest region. Five genera of tiny species (5–10 g) are distinguished by differences in their teeth and skulls, but are very similar externally and are difficult to identify with certainty in the hand. Four of the genera not described herein occupy arid New World habitats, and one is found on Caribbean islands.

Chestnut Long-tongued Bat
Lionycteris spurrelli
Plate D, map 53
Identification. Measurements: HB = 50–60; T = 6–9; HF = 9–13; E = 12–15; FA = 32–37; WT = 7–11 g.

Upperparts rich, uniform chocolate to chestnut brown; fur unicolored to base,

Map 53

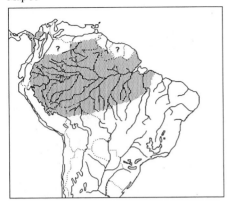

Map 54

Spear-nosed long-tongued bats, *Lonchophylla* spp.

Chestnut long-tongued bat, *Lionycteris spurrelli*

short to medium length, **velvety;** area between shoulder blades sometimes with some gray hairs; wings and tail membrane blackish brown. **Muzzle slightly, not greatly, elongated,** lower jaw about as long as upper; lower incisors present, **tongue with deep groove on side; chin with a single large, concave V-shaped pad with smooth sides, not a deep groove;** noseleaf small, without horseshoe; ear short, tip rounded. Tail short, half the length of membrane; tail membrane short, about half the length of leg at center; calcar about half as long as foot. **Underparts about the same as back or only faintly frosted slightly paler.** Young like adults.

Variation. Color varies individually, within populations, from chestnut to brown.

Similar species. Common long-tongued bats (*Glossophaga* spp.) have fur distinctly pale at base and a deep groove on chin with tubercles on edges; see common long-tongued bat for other comparisons.

Natural history. Roosts in caves and crevices and is found in rainforest, gardens, and plantations.

Geographic range. Central and South America: barely enters Panama near the Colombian border, south to N Bolivia and the southern Amazon Basin of Brazil, east to Belém. To 1,400 m elevation.

Status. Uncommon over large geographic range.

References. Williams, S. L., and H. H. Genoways. 1980. Results of the Alcoa-Foundation-Suriname expeditions. II. Additional records of bats (Mammalia: Chiroptera) from Suriname. *Ann. Carnegie Mus.* 49:213–36.

Spear-nosed Long-tongued Bats
Lonchophylla spp.
Plate D, map 54

Identification. Measurements: HB = 46–78; T = 5–11; HF = 8–14; E = 12–18; FA = 30–45; WT = 5–21 g.

Upperparts pale orange, pale brown, smoky brown, or dark gray; fur bicolored with distinctly pale base, neck and shoulders often paler than back. Muzzle moderately, not extremely, long and narrow, lower jaw longer than upper; lower incisors present; central upper incisors much larger than outer incisors, higher than wide, **outer incisors surrounded by gaps, tongue with deep groove on side; noseleaf a well-developed spear;** no free horseshoe flange around base; chin tip a broad concave pad with a few tubercles on sides near lip, smooth sides farther down. Tail short; tail membrane short to quite long, but shorter than leg; calcar shorter than foot. Underparts the same as back or paler, grayish.

Species. Six of the approximately seven species occupy lowland rainforest: *L. thomasi* is small (FA = 31–33; 5–6 g), dark smoky brown with fur distinctly pale at base; *L. mordax* is small (FA = 32–35; 9 g), with fur barely paler at base than at tip, usually with a dark belly washed with grayish (but one from Bahia has completely white underparts); *L. hesperia* is medium-sized (FA = 39) and pale brown with pale, gray-brown underparts; *L. bokermani* (endemic to Atlantic forest of Brazil) is large (FA = 38–41), has dorsal fur with pale base, and ventral fur paler than dorsum, with dark-

based hairs; *L. robusta* is large (FA = 39–45; 14–17 g) and pale orange; *L. handleyi* (FA = 46–47) is similar to *L. robusta*.
Similar species. The noseleaves on these bats are larger than on any similar species: common long-tongued bats (*Glossophaga* spp.) have no groove on tongue, central upper incisors not much larger than outer incisors, and all incisors crowded together; chestnut long-tongued bats (*Lionycteris spurrelli*) have groove on tongue, unicolored fur; long-nosed (*Choeroniscus* spp.), Underwood's (*Hylonycteris underwoodi*), and Ega (*Scleronycteris ega*) long-tongued bats have extremely long, narrow, tubelike muzzles, and no lower incisors; dark long-tongued bats (*Lichonycteris* spp.) have tricolored fur; hairy-legged long-tongued bats (*Anoura* spp.) have no tail.
Natural history. Feed on nectar, insects, and probably fruit and pollen. *L. thomasi* roosts in small groups in hollow logs and trees and in small caves; *L. robusta* roosts in small groups in caves. These are the most common long-tongued bats in parts of the Amazon Basin. Found in rainforest, gardens, and plantations; some species occupy cloud forest and deciduous forest.
Geographic range. Central and South America: Nicaragua south to Bolivia and SE Brazil; *L. thomasi, L. mordax,* and *L. robusta* in Central and South America; *L. hesperia,* W coast of Peru and Ecuador; *L. handleyi* is from Ecuador and Peru; *L. bokermani* is from SE Brazil.
Status. Common to rare: *L. hesperia* is known from only five individuals.
References. Hill, J. E. 1980. A note on *Lonchophylla* (Chiroptera: Phyllostamatidae) from Ecuador and Peru, with the description of a new species. *Bull. Brit. Mus. Nat. Hist. Zool.* 38:233–36.

Common Long-tongued Bats
Glossophaga spp.
Plates D, 6; map 55
Identification. Measurements: HB = 53–65; T = 3–10; HF = 9–13; E = 12–16; FA = 32–40; WT = 10–13 g.
Upperparts pale gray to pale brown, fur bicolored pale at base, dark at tip, pale showing through on neck and shoulders; wings and tail membrane brown. **Muzzle slightly, not greatly, elongated, lower jaw only slightly longer than upper;** lower incisors present; central upper incisors not

Map 55

Common long-tongued bats, *Glossophaga* spp.

much larger than outer incisors, crown wider than high, incisors not separated by gaps; **tongue without deep groove on side.** Noseleaf a small spear, reaches close to corner of eye when laid back; **horseshoe around nostrils with no free flange on sides or bottom; chin tip a deep V with sides a ripple of tiny tubercles, no tubercles in center;** ears relatively short, slightly hairy on back of inner edge, tips rounded. **Tail very short; tail membrane shorter than legs;** calcar about half as long as foot. Young dark gray-brown.
Species. The five or six species are externally similar: two, *G. soricina* and *G. commissarisi,* occur mainly in rainforest, and three occur more often in drier habitats, but two or three species may occur together in Central America and N South America. There is individual variation in color within a species: bats from dry habitats, and young bats, are gray.
Similar species. Chestnut long-tongued bats (*Lionycteris spurrelli*) have fur uniformly dark to base; these and spear-nosed long-tongued bats (*Lonchophylla* spp.) have a deep groove on side of tongue; other long-tongued bats lack lower incisors; short-tailed fruit bats (*Carollia* spp.) have a free flange on the side of the horseshoe and a tubercle in the center of the chin notch; little fruit bats (*Rhinophylla*) have no tail and a flanged horseshoe.
Natural history. Feeds on nectar, pollen, fruit, and insects. Common long-tongued bats are the least specialized of the long-tongued bats, and they have a broad diet. Like hummingbirds, these bats will either defend temporary territories around nectar

plants or exhibit "trapline" foraging, depending on the richness of the nectar source. They roost in caves, buildings, tunnels, hollow trees, culverts, drains, or under bridges. They hang spaced apart in roosting colonies of a half-dozen to many hundreds of individuals. Found throughout the rainforest and in secondary forest, plantations, deciduous forest, cerrado, chaco, and savanna.

Geographic range. Central and South America: N Mexico south to SE Brazil, Paraguay, and N Argentina. *G. soricina,* from Mexico to Argentina, is the only species in the Amazon and Rio Paraguay basins. *G. longirostris,* from Colombia, Venezuela, and Guyana; *G. alticola* on the Pacific coast of Central America from Mexico to El Salvador; *G. commissarisi* on the Pacific coast from Mexico to Peru.

Status. Common and widespread.

References. Webster, W. D., and J. K. Jones, Jr. 1980. Taxonomic and nomenclatorial notes on bats of the genus *Glossophaga* in North America, with description of a new species. *Occas. Pap. Mus. Texas Tech. Univ.* 7:1–12.

Lempke, T. O. 1984. Foraging ecology of the long-nosed bat, *Glossophaga soricina,* with respect to resource availability. *Ecology* 65:538–48.

Hairy-legged Long-tongued Bats
Anoura spp.
Plate D, map 56

Identification. Measurements: HB = 56–78; T = 0–7; HF = 10–14; E = 12–17; FA = 35–45; WT = 10–20 g.
Upperparts medium to dark brown, pale grayish brown, or blackish brown; fur quite short, bicolored, pale at base, often faintly frosted silver or pale; wings and tail membrane black or blackish brown. Muzzle long and narrow, lower jaw longer than upper; lower incisors absent, upper incisors minute; noseleaf a small triangle, whiskers around muzzle sometimes prominent; ears short, rounded, triangular. Tail absent or a tiny, inconspicuous stub; tail membrane virtually absent, reduced to a narrow band down inside of hindleg to rump, this band hairy along edge; calcar less than half of length of foot. Underparts lightly frosted paler than back. Young like adults.

Species. The four species are all found in rainforest: *A. caudifera* is small (FA = 35–39; 10–13 g), brown, with a tiny tail, and a

Map 56

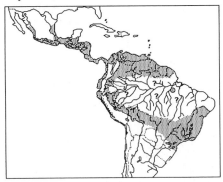

Hairy-legged long-tongued bats, *Anoura* spp.

narrow, semicircular remnant tail membrane; *A. cultrata* is large (FA = 38–45; 14–23 g), blackish, often with a tiny tail; *A. geoffroyi* is large (13–19 g; FA = 41–45), smoky brown, with no tail; *A. latidens* is large (FA = 42–46), pale grayish brown, with pale throat and chest. Three or four more species are awaiting description.

Similar species. Short-haired long-tongued bats (*Leptonycteris*) of arid habitats (not described in this book) have extremely short fur and are much larger; all other long-tongued bats have well-developed tail membranes; other bats with noseleaves and reduced or no tail membranes have short, broad muzzles.

Natural history. Feed on nectar, insects, fruit, and pollen. *A. cultrata* is restricted to montane areas and is only known to roost in caves and tunnels. The other species roost in small groups in caves or tree hollows and are found in lowland rainforest, deciduous forest, gardens, and plantations. Although they will use other roosts, these bats seem common only where there are caves or rock crevices, and they are rare or perhaps absent from lowland Amazonian forests lacking high ground, rocks, and caves. They reach their highest species diversity in the N Andes.

Geographic range. Central and South America: Central Mexico (*A. geoffroyi*); or Costa Rica (*A. cultrata*) south to Peru, Bolivia, and SE Brazil. *A. caudifera* in South America east of the Andes only; *A. latidens* from Venezuela. To about 3,000 m elevation.

Status. Common to rare, restricted to widespread.

References. Tamsitt, J. R., and D. Nagorsen. 1982. *Anoura cultrata.* Mammalian Species, no. 179.

Handley, C. O., Jr. 1984. New species of mammals from northern South America: A long-tongued bat, genus *Anoura* Gray. *Proc. Biol. Soc. Wash.* 97:513–21.

Dark Long-tongued Bats
Lichonycteris spp.
Plate D, map 57
Identification. Measurements: HB = 49–55; T = 6–10; HF = 8–11; E = 11–13; FA = 31–34; WT = 6–8 g.
Upperparts pale to dark brown; fur tricolored, dark at base and tip, pale brown in middle, dark base indistinct, looks like a shadow; elbows thinly furred above. Muzzle elongated, robust, not narrow and tubelike; lower jaw longer than upper; lower incisors absent; **deep chin notch extends under chin as a slit without rippled border of tubercles.** Noseleaf small; whiskers prominent, chin whiskers as long as muzzle whiskers; ear very short, rounded. Tail short; **tail membrane long; calcar almost as long as foot. Underparts brown, paler than back,** about same color as central pale band of dorsal **fur, bicolored faintly dark at base, paler at tip.**
Species. Both species occupy rainforest: *L. obscura* is dark brown; *L. degener* is pale brown; some think the latter is a subspecies of the former.
Similar species. These are the only long-tongued bats with tricolored fur in South America; Underwood's long-tongued bats (*Hylonycteris underwoodi,* Central America) are similar but have a longer, narrower, tubelike muzzle and thickly furred elbows.
Natural history. Diet and roosts apparently unrecorded. Found in lowland and lower montane evergreen rainforest and plantations.
Geographic range. Central and South America: Guatemala south to Peru, NE Bolivia, and the Amazon Basin of Brazil. To about 1,000 m elevation.
Status. Rare, but widespread.

Ega Long-tongued Bat
Scleronycteris ega
Map 58
Identification. Measurements: HB ≈ 57; T ≈ 6; FA = 35.
Upperparts blackish brown; fur bicolored, paler brown at base, long; wings and tail membrane blackish brown. Muzzle greatly

Map 57

Dark long-tongued bats, *Lichonycteris* spp.

elongated, lower jaw projecting prominently forward and downward; a very long whisker grows from below front of ear. Distal thumb joint (with claw) distinctly long and thin, longer than first joint (in wing membrane). Tail short, tail membrane long, calcar about three-fourths length of foot; hindfoot tiny. Underparts brown, slightly paler than back, throat like back.
Similar species. Difficult to distinguish from long-nosed long-tongued bats (*Choeroniscus* spp.), which lack a long whisker in front of the ear; the distal thumb joint is robust and about the same length as the first joint, and fur of medium length; dark long-tongued bats (*Lichonycteris* spp.) have tricolored fur; chestnut long-tongued bats (*Lionycteris spurrelli*) have unicolored fur.
Natural history. From lowland rainforest.
Geographic range. South America: central Amazon and upper Orinoco drainages of Brazil and Venezuela. Known from only four localities.
Status. Rare; known from only four individuals.

Long-nosed Long-tongued Bats
Choeroniscus spp.
Plate D, map 59
Identification. Measurements: HB = 51–66; T = 6–11; HF = 7–11; E = 11–13; FA = 31–41; WT = 5–8 g.
Upperparts dark brown, fur bicolored, paler at base, sometimes indistinctly so; fur sometimes slightly frosted, medium length; elbows thinly furred; wings and tail membrane blackish brown. **Muzzle greatly**

Map 58

Map 59

Underwood's long-tongued bat, *Hylonycteris underwoodi*

Ega long-tongued bat, *Scleronycteris ega*

Long-nosed long-tongued bats, *Choeroniscus* spp.

elongated into a long, narrow tube; lower jaw longer than upper, upper incisors minute, lower incisors absent; **noseleaf small, triangular; line of nostrils close to horizontal; whiskers prominent, those from chin as long as those around nose; ears short, triangular, barely protruding above crown. Border of deep chin notch rippled with tubercles on edge throughout length. Two thumb joints of about equal length, or second only slightly longer than first.** Tail very short; tail membrane long, shorter than legs; calcar about three-fourths as long as foot. Underparts the same as back or slightly paler. **Tiny, very long-nosed bats.**
Species. The three to five species all occupy rainforest: *C. godmani* has fur pale at base; *C. minor* has fur almost unicolored; *C. periosus* is large (FA = 41); *C. intermedius* has medium brown fur, pale at base, and is small (FA = 33–35).
Similar species. Ega (*Scleronycteris ega*) and dark (*Lichonycteris* spp.) long-tongued bats are similar; all have the second thumb joint (with claw) distinctly longer than the first (in wing membrane); dark long-tongued bats have tricolored fur and nostrils set at about 45 degrees; Ega long-tongued bats have a long whisker growing from near ear and long fur.
Natural history. A roosting group of eight was under a fallen log over a stream, and a group of six *C. intermedius* was beneath the buttress of a fallen tree. Found in rainforest and openings in the forest such as marshes and plantations, and in montane forests.

Geographic range. Central and South America: Mexico to N Bolivia and the Amazon Basin of Brazil.
Status. Uncommon to rare but widespread.

Underwood's Long-tongued Bat
Hylonycteris underwoodi
Map 58
Identification. Measurements: HB = 48–60; T = 4–10; HF = 7–11; E = 10–13; FA = 31–34; WT = 6–9 g.
Upperparts uniform dark brown; fur tricolored, dark at base and tip with paler brown center, dark band at base indistinct, looks like a dark shadow; wings and tail membrane blackish brown; elbows thickly furred above and below. **Muzzle elongated, narrow, tubelike, lower jaw longer than upper, no lower incisors; border of deep chin notch with long, smooth pads near lip that extend under chin as deep slit with edges of rippled tubercles.** Ear short, rounded, narrow across base; noseleaf short. Tail short; tail membrane long; calcar more than half as long as foot. **Underparts brown, about the same as back** or with more orange tints, fur bicolored, dark at base. Tiny, dark, narrow-muzzled bats.
Variation. Some Underwood's bats from drier areas of Mexico have bicolored fur (some of these have tricolored fur on throat) and are paler; those from Costa Rica and Panama are very dark, almost blackish, with dark band at base of fur distinct.
Similar species. Dark long-tongued bats (*Lichonycteris* spp.) are the only other long-tongued bats with tricolored fur; they have robust muzzles and thinly haired elbows.

Long-nosed long-tongued bats (*Choeroniscus* spp.) are difficult to distinguish from bicolor-furred individuals; Underwood's bats generally have hairier elbows and a longer second thumb joint, but probably cannot be distinguished externally with certainty.

Natural history. Feeds on pollen, fruit, insects, and probably nectar. Underwood's bats roost in small groups of two to eight in caves, hollow logs, tunnels, and under bridges. They may be chiefly montane.

Found in rainforest, deciduous forest, and cloud forest.

Geographic range. Central America: SW Mexico to W Panama. To about 2,600 m elevation.

Status. Rare, geographic range patchy and much deforested.

References. Jones, J. K., Jr., and J. A. Homan. 1974. *Hylonycteris underwoodi.* Mammalian Species, no. 32.

Little Spear-nosed and Short-tailed Fruit Bats (Carolliinae)

This subfamily includes plain little bats without any striking distinguishing features. They have medium-sized muzzles, well-developed, medium-sized noseleaves and ears; short or absent tails; and moderately to greatly reduced tail membranes. They all seem to live in the rainforest understory, where they feed on small fruits, but there is no natural history information on several species. The two genera and seven species are all found in the Neotropical rainforest region.

Short-tailed Fruit Bats
Carollia spp.
Plates E, 6; map 60
Identification. Measurements: HB = 44–74; T = 6–12; HF = 9–18; E = 15–23; FA = 33–45; WT = 9–25 g.
Upperparts gray to warm brown, rarely rusty; **fur moderately long and soft, parts easily to show strongly to indistinctly three- to four-banded hairs, dark at base;** wings and tail membrane gray-brown. **Muzzle short and narrow, noseleaf quite large, reaches to eye when flattened, spear-shaped, horseshoe around nostrils with free flange on side; chin with a V-shaped line of small round warts surrounding a large central tubercle; ears medium, triangular, pointed, reach about to nose when flattened forward.** Tail short, tip sticks up through tail membrane; tail membrane long, but shorter than legs, calcar shorter than foot. Underparts uniform, about the same color as back, or frosted. Young gray.
Species. The four species all occupy lowland rainforest: *C. perspicillata, C. brevicauda,* and *C. subrufa* are all usually gray-brown, with distinctly banded variegated-looking fur. *C. perspicillata* is large (FA = 40–45; 10–23 g), and in some populations there are rusty-tinted individuals (not seen in the other gray-brown species). *C. castanea* is small (FA = 33–38; 10–13 g) and uniform chestnut, with bands on fur in-

distinct. *C brevicauda* is intermediate in size (FA = 37–42), with bright bands on fur, and can be difficult to distinguish from *C. perspicillata.* Two to three species occur together in most of the rainforest region.
Similar species. Common long-tongued bats (*Glossophaga* spp.) have fur pale at base and no tubercle in center of chin V; little fruit bats (*Rhinophylla* spp.) have no tail.
Natural history. Feed on fruit and insects, supplemented by nectar in the dry season. These are often the most numerous bats in lowland rainforests, and they seem most common in disturbed habitats. They use the understory vegetation levels, where they concentrate their feeding on the fruits of shrubs and treelets, especially the slender, green, candlelike fruits of plants of the genus *Piper.* Because of their high numbers, short-tailed fruit bats are probably one of the most important seed dispersers for *Piper* and many other plants with small fruits. They have been shown to be instrumental in carrying seeds into cleared areas and thus stimulating forest regeneration after a small clear-cut. They usually fly about a kilometer from their roost to feed. Short-tailed fruit bats roost in small groups, with individuals separated, in tree hollows, caves, overhanging banks, tunnels, culverts, or abandoned buildings. Roosting groups consist of small harems of females with one male, or bachelor males. Found in mature and disturbed

Map 60

Short-tailed fruit bats, *Carollia* spp.

Map 61

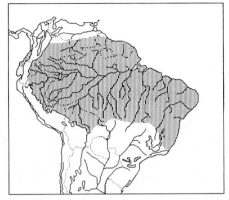

Little fruit bats, *Rhinophylla* spp.

rainforest, gardens and plantations, deciduous forest, and gallery forest.
Geographic range. Central and South America: S Mexico to Bolivia, Paraguay, and SE Brazil. To at least 2,400 m elevation.
Status. Common and widespread.
References. Fleming, T. H., and E. R. Heithaus. 1977. An experimental analysis of the food location behavior of frugivorous bats. *Ecology* 58:619–27.
De Foresta, H., P. Charles-Dominique, C. Erard, and M. F. Prévost. 1984. Zoochorie et premiers stades de la régénération naturelle après coupe en forêt guyanaise. *Rev. Ecol.* 39:369–400.

Little Fruit Bats
Rhinophylla spp.
Plate E, map 61
Identification. Measurements: HB = 43–68; T = 0; HF = 8–12; E = 13–17; FA = 31–38; WT = 7–13 g.
Upperparts pale gray, pale brown, or blackish brown; fur unicolored gray or brown to base, except tips of hairs slightly darker, no distinct pale basal bands on hair; fur dense. Muzzle short, medium width; noseleaf large, spear-shaped, when flattened reaches well behind eye to center of forehead, horseshoe well developed, with free flange on sides; chin with a large, smooth, sculptured pad, no tiny warts; ears triangular, medium-sized, reach to front of eye when flattened forward. Tail absent; tail membrane rudimentary, either a deeply notched V or reduced to a narrow band on inside of leg from calcar to

rump; this band with or without hairy fringe; calcar half the length of foot, or less. Underparts about the same as back or frosted, fur unicolored gray or brown to base. Young like adults.
Species. The three species all occupy rainforest: *R. pumilio* is brown or gray, with tail membrane V-shaped and almost hairless, with a band across rump; *R. fischerae* is beige, with tail membrane a hairy band down leg, not forming a band between legs; *R. alethina* is blackish, with tail membrane a band down leg only, legs and feet very hairy.
Similar species. Short-tailed fruit bats (*Carollia* spp.) have tails; long-legged long-tongued bats (*Anoura* spp.) have long muzzles, short noseleaf and ears, and pale hair base; yellow-shouldered fruit bats (*Sturnira* spp.) are larger (> 15 g), and have broad muzzles and a short, broad noseleaf that does not reach behind eye; fruit-eating bats (*Artibeus* spp.) have broad muzzles and often stripes on face.
Natural history. Feed on fruits of understory shrubs. *R. pumilio* roosts under tents made from palm and aroid leaves (see following section for descriptions of bat tents). Found only in lowland rainforest or gardens and plantations in forest; rare in deciduous forest.
Geographic range. South America: *R. pumilio* in the Amazon Basin and the Guianas; *R. fischerae* along the main Amazon and its upper western tributaries in Colombia, Ecuador, and Peru; *R. alethina* is known only from the west coasts of Colombia and Ecuador. To 1,400 m elevation.

Status. *R. alethina* apparently rare; *R. fisch-erae* and *R. pumilio* locally common to uncommon.

References. Tuttle, M. 1970. Distribution and zoogeography of Peruvian bats, with comments on natural history. *Univ. Kans. Sci. Bull.* 49:45–86.

Neotropical Fruit Bats (Subfamily Stenodermatinae)

Dental formulas are highly variable. Stenodermatines are small to large; they have short, broad muzzles, large eyes, medium-length plain triangular ears; and all but two species have medium-sized, fleshy, spear-shaped noseleaves. They have no tails; tail membranes are much shorter than the legs or sometimes absent. Many species have striking pale stripes; the basic stripe pattern is a single stripe down the middle of the back from neck to rump and four stripes on the face, one above and one below each eye. The development of each of these stripes varies from species to species, from bright to faint to none. These bats all feed on fruit, supplemented by flower nectar or sometimes leaves in the dry season when fruit is scarce. To feed, they pick a fruit from a tree and fly away, carrying it in the mouth to a feeding roost, which is usually a thin horizontal stem beneath the shelter of foliage. A favored spot is under a horizontal palm leaf. There the bat chews up the fruit and sucks the juices from the indigestible fibers, which are spit out on the ground in small wads, along with any large seeds. A bat will use the same roost many times, and one can find these throughout the forest: a pile of seeds, feces, and wads of fibers beneath a roost support (which may be meters above the ground). Many stenodermatines roost by day among foliage, where their white stripes probably act as camouflage. Bats of six genera in this subfamily make "tents" of large leaves, usually palms or palmettos, by biting the sides and ribs of a leaf in a characteristic pattern (forming a row of jagged holes, unlike the rows of round holes drilled by insect larvae through rolled leaves), so that the sides of the leaf drop down to form a shelter. Tent-making species can be found during the day by quietly approaching a strangely drooping leaf with bitten sides and looking under it without touching it (the bats will fly off if too much noise is made or the roost is touched; beware of wasps, which often shelter their own nests in an old bat tent). Neotropical fruit bats are the main seed dispersers for many plants. They are particularly important in carrying seeds for early secondary or successional growth into gaps in the forest caused by natural disturbances or cutting by man, thus restoring new forest, and they are important for maintaining the plant species richness of the forest. This is the largest subfamily of leaf-nosed bats, with about sixteen genera.

Yellow-shouldered or Hairy-legged Fruit Bats
Sturnira spp.
Plate E, map 62
Identification. Measurements: HB = 54–101; T = 0; HF = 7–21; E = 15–22; FA = 39–62; WT = 13–67 g.
Upperparts gray, pale brown, or yellow brown; fur conspicuously tricolored, frosted with dark hair tips overlying silver or pale beige, hair base dark; shoulders and sometimes neck of adults stained yellow to orange with secretion from glands, hair sometimes stuck together with thick, dark secretion; yellow shoulders usually most prominent on large males. **Muzzle short, broad;** noseleaf a short, broad spear, not reaching beyond eye when flattened against muzzle; ear short, triangular; eye quite large. **Tail absent; tail membrane virtually absent, reduced to a narrow band; feet and inner side of legs very hairy;** thumb long and robust. Underparts gray or gray-brown, like pale part of dorsal fur. **Medium to large robust, roly-poly, woolly-looking bats with large heads, thick necks, burly bodies, and short, strong, hairy legs.**
Species. There are about thirteen species; six occupy lowland and montane rainforest; the others are mostly montane in the N Andes. Species are distinguished by features of the teeth and are difficult to identify in the hand. New species are awaiting description.
Similar species. Vampires (*Desmodus* spp., *Diphylla*) do not have spear-shaped noseleaves; other hairy-legged bats with no tail membranes have facial stripes or are much smaller.

Map 62

Yellow-shouldered fruit bats, *Sturnira* spp.

Map 63

Tent-making bats, *Uroderma* spp.

Natural history. Feed on fruit and nectar; especially fond of the nectar of banana flowers and the small round fruits of members of the tomato family (Solanaceae). These bats are often common, but little is known about them. They seem most numerous in open spaces in the forest such as light gaps, gardens, clearings, streams, and edges, where *Solanum* fruits are common. Their normal roosts are unknown and are probably somewhere high in the forest canopy (or they would have been discovered). Rarely they have been found in caves, hollow trees, tunnels, or culverts. The advantage of lack of a tail membrane is seen in a bat in the hand, which can twist and turn its legs and body like a contortionist. It seems likely that yellow-shouldered bats are adapted for clambering quadrupedally among flowers and fruits. Found in lowland and montane rainforest, dry forest, cloud forest, gardens, and plantations.

Geographic range. Central and South America: central Mexico south to N Argentina. To at least 3,500 m elevation.

Status. Common to very rare *(S. nana)*; widespread *(S. lilium)* to restricted *(S. aratathomasi, S. nana)*.

References. Tamsitt, J. R., and C. Haüser. 1985. *Sturnira magna.* Mammalian Species, no. 240.

Soriano, P. J., and J. Molinari. 1987. *Sturnira aratathomasi.* Mammalian Species, no. 284.

Molinari, J., and P. J. Soriano. 1987. *Sturnira bidens.* Mammalian Species, no. 276.

Tent-making Bats
Uroderma spp.
Plates E, 6; map 63

Identification. Measurements: HB = 55–68; T = 0; HF = 10–15; E = 14–20; FA = 39–47; WT = 14–21.

Upperparts dark to pale gray or gray-brown or beige; fur bicolored, pale at base, midback with prominent white stripe that does not extend onto head. Face with faint to prominent white stripes above and below eye; muzzle short and broad; upper incisors bilobed and only slightly longer than outer incisors; noseleaf a large fleshy spear, reaching to eye when pressed back, **horseshoe well developed, with a large free flange all the way around base,** pale-colored on sides where facial stripe ends; ears triangular, cream-colored along side of rim by facial stripes, tragus pale. **Tail membrane naked, a deeply notched V, extends well beyond body fur in center.** Underparts gray or gray-brown, slightly paler or more gray than back.

Species. There are two species, both found in rainforest: *U. bilobatum* is dark gray-brown, with bright facial stripes; *U. magnirostrum* is medium brown, with faint facial stripes, head, neck, and shoulders paler than back, and edges of wings translucent. The tongue is black in some populations.

Similar species. White-lined fruit bats (*Platyrrhinus* spp.) and great stripe-faced (*Vampyrodes* spp.) bats have a hairy fringe on the tail membrane; yellow-eared bats (*Vampyressa* spp.) are smaller (FA = 30–37); big-eyed bats (*Chiroderma* spp.) have fur with a dark band at base and upper incisors with single points; fruit-eating bats (*Artibeus*) have no back stripe.

Natural history. Feed on fruit, nectar, and insects. These bats roost under "tents," which they make by cutting a line down each side of the midrib of a large leaf, such as banana, so that the sides droop downward. A small (2–52) group of bats hang in a tight cluster from the midrib of the leaf. Found in mature and disturbed rainforest, gardens and plantations, and deciduous forest. *U. bilobatum* is abundant in rainforest, and *U. magnirostrum* in southern dry forests.

Geographic range. Central and South America: S Mexico south to central Bolivia and SE Brazil. To about 1,000 m elevation.

Status. Common and widespread.

References. Baker, R. J., and C. L. Clark. 1987. *Uroderma bilobatum.* Mammalian Species, no. 279.

White-lined Fruit Bats
Platyrrhinus spp.
Plate E, map 64

Identification. Measurements: HB = 53–103; T = 0; HF = 10–19; E = 15–25; FA = 35–63; WT = 13–63 g.

Upperparts blackish brown to warm brown; midback with bright white stripe that usually extends onto head; shoulders but not top of neck often paler than back; elbows often densely furred; fur without dark band at base. Muzzle short; central upper incisors bilobed, parallel at tips, and much longer than outer incisors; **facial stripes above eyes bright to indistinct, often fading out forward of rear edge of ear; stripes below eye less distinct, often narrow or fainter;** noseleaf a large fleshy spear, horseshoe at base broad and with free flange on sides and usually front, sides of horseshoe usually dark, sometimes cream, not bright yellow; ear sometimes with pale rim, usually not. **Tail membrane a deeply notched V, very short at center, barely extending beyond fur of body; always fringed with hair;** calcar less than half the length of foot. Underparts the same as back, or slightly frosted paler gray or brown. Young like adults. **Small to large, robust bats.**

Species. There are about ten species; most are found in lowland rainforest. The most common and widespread species in the region is the smallest, *P. helleri* (FA = 35–

Map 64

White-lined fruit bats, *Platyrrhinus* spp.

40; WT = 13–16 g), which is pale brown. Up to six species can occur in the same area; some are mostly montane. The name *Vampyrops* was formerly used for this genus.

Similar species. Great stripe-faced bats (*Vampyrodes* spp.; FA = 45–56) are warm brown, have yellowish sides of horseshoe and lower ears, central incisors without lobes, all facial stripes broad and bright, upper stripes sharp to behind ear, and elbows thinly haired. Most yellow-eared bats (*Vampyressa* spp.) are smaller, with no fringe of hairs on tail membrane. Big-eyed bats (*Chiroderma* spp.) have a depression behind noseleaf, single pointed incisors, and a dark band at base of fur. Tent-making bats (*Uroderma* spp.) have a naked tail membrane.

Natural history. Feed on fruit, nectar, leaves, and insects; diets of some species are unknown. These bats roost in small groups of 6–20, probably harems, in caves and in tents they make from palm leaves. Found in mature and disturbed lowland rainforest, gardens, clearings and plantations, cloud forest, deciduous forest, chaco, and caatinga.

Geographic range. Central and South America: S Mexico south to Bolivia, Paraguay, Uruguay, and SE Brazil.

Status. CITES Appendix III for *P. lineatus* (Ur). Common to uncommon, widespread. To about 2,500 m elevation.

References. Willig, M. R., and R. H. Hollander. 1987. *Vampyrops lineatus.* Mammalian Species, no. 275.

Gardner, A. L., and D. C. Carter. 1972. A review of the Peruvian species of *Vampyrops*

Map 65

Great stripe-faced bat, *Vampyrodes* spp.

(Chiroptera, Phyllostomatidae). *J. Mammal.*
53:72–84.

Great Stripe-faced Bats
Vampyrodes spp.
Plate E, map 65
Identification. Measurements: HB =
73–80; T = 0; HF = 12–17; E = 20–25;
FA = 45–57; WT = 28–38 g.
**Upperparts warm cinnamon brown, neck
and shoulders paler than back; midback
with sharp, bright white line from between ears to end of rump;** elbows thinly
furred; fur pale at base. Muzzle very broad,
U-shaped; noseleaf a large fleshy spear;
**white stripes above and below eye broad
and distinct,** expanding posteriorly to almost coalesce behind ear, stripes reach rear
edge of ear; **ear rim, tragus, and side of
noseleaf pale to bright yellow;** lips and
tongue pink. **Tail membrane a deeply
notched V, very narrow at center, fringed
with hair;** calcar less than half the length
of foot. Underparts gray-brown. Young like
adults. **Large, brightly striped, brown bats.**
Species. Two species are usually recognized, *V. caraccioli* from South America and
V. major from Central America; they are
similar and perhaps conspecific.
Similar species. White-lined fruit bats
(*Platyrrhinus* spp.), if brown, are usually
smaller, with stripes less distinct, rims of ear
and horseshoe whitish or gray, elbows often
thickly furred; yellow-eared bats (*Vampyressa* spp.) are all much smaller; big-eyed
bats (*Chiroderma* spp.) have dark band at
base of fur and usually indistinct stripes;

tent-making bats (*Uroderma* spp.) have no
hairy fringe on tail membrane; other Neotropical fruit bats do not have back stripe.
Natural history. Feeds on fruit, especially
figs. Great stripe-faced bats roost in small
groups of two to four in foliage and under
palm leaves; the leaves are not modified to
form a tent, but few roosts have been reported. Found in lowland rainforest and in
gardens and plantations.
Geographic range. Central and South
America: S Mexico south to N Bolivia and
the S Amazon Basin of Brazil. To about
1,000 m elevation.
Status. Sometimes common, widespread.

Big-eyed Bats
Chiroderma spp.
Plate E, map 66
Identification. Measurements: HB =
53–82; T = 0; HF = 10–15; E = 16–21;
FA = 38–52; WT = 13–26 g.
**Upperparts brown to gray-brown; fur with
distinct dark band at base, especially on
lower back; midback with sharp, faint,
or no white line. Muzzle short and very
broad, noseleaf with wide, flared-out sides
in leaf shape,** horseshoe with free flange
around all of base; **skull lacks nasal bones,
in live bats a depression can be felt behind
the noseleaf; middle upper incisor teeth
long, narrow, with simple points;** eyes
large; ears with rims or base and sides of
noseleaf yellow or brown; **facial stripes
strong to indistinct,** most ending forward
of ear, rarely to middle of ear; tongue pink.
**Tail membrane slightly hairy above, with-
out hairy fringe, quite long at center, calcar about three-fourths of length of foot.**
Underparts gray-brown. Young like adults.
Species. Three of the five species are found
in the rainforest region; two or three can
occur together. *C. trinitatum* is the smallest
(WT = 13–15 g, FA = 38–43) and has a
middorsal stripe; *C. salvini* is large (FA =
46–52) and has large, bright facial stripes;
C. villosum is large (FA = 44–50) and has
indistinct facial stripes; *C. doriae* occupies
deciduous forest in SE Brazil.
Similar species. Yellow-shouldered bats
have no tail membrane and no stripes; no
other striped bats of this size have single-
pointed incisors, a depression behind noseleaf, and a dark band at base of hair (see
yellow-eared bats).

Map 66

Big-eyed bats, *Chiroderma* spp.

Map 67

Yellow-eared bats, *Vampyressa* spp.

Natural history. Feed on fruit, especially figs. *C. villosum* roosts in tree hollows; roosts of the other species are apparently unrecorded. Found in mature and disturbed rainforest, gardens, and plantations.
Geographic range. Central and South America: S Mexico south to Bolivia and the S Amazon Basin of Brazil. *C. trinitatum* and *C. villosum* from Central America to South America east of the Andes, *C. salvini* to west of the Andes. To 2,200 m elevation.
Status. All species common to uncommon and widespread except *C. doriae* and *C. improvisum* (Guadeloupe), which are geographically restricted and rare.

Yellow-eared Bats
Vampyressa spp.
Map 67
Identification. Measurements: HB = 45–62; T = 0; HF = 7–12; E = 11–18; FA = 29–39; WT = 7–16 g.
Upperparts gray-brown to tawny or cinnamon brown, sometimes slightly frosted; neck and shoulders paler; **fur with faint (like a shadow) to distinct dark band at base; midback with or without white stripe; wings and tail membrane brown,** usually pale on forward edge. Muzzle short and broad; noseleaf a large fleshy spear, horseshoe with complete free flange around base; **rims and base of ears, tragus, and sides of noseleaf bright yellow; white facial stripes distinct, stripes above eyes broad, sometimes coalescing above nose,** often extending to rear of ear, stripes below eyes fainter; tongue pink; central upper incisors bilobed and converging at tips, much

longer than outer incisors. **Tail membrane deeply notched, lightly haired above near body, without hairy fringe in bats with middorsal stripe, with slight hairy fringe in center in bats without stripe;** calcar less than half the length of foot. Underparts gray-brown, a little paler than back. Young gray or gray-brown. Small, chunky bats.
Species. Four of the five species occupy lowland rainforest; the other is montane. Up to three species can occur together. *V. pusilla* (FA = 29–33, 7–8 g), the smallest, is the only lowland species with no back stripe, and is tawny brown. *V. bidens* has two lower incisors (the other species have four). *V. brocki* (FA = 33) has brown ears; *V. nymphaea* (FA = 36–39) has a stripe only on lower back; *V. melissa* (montane) has a hairy interfemoral membrane and no back stripe.
Similar species. White-lined fruit bats (*Platyrrhinus* spp.) have a bright middorsal stripe, hairy fringe on tail membrane, and fur with no dark band at base; tent-making bats (*Uroderma* spp.) have fur with no dark band at base; big-eyed bats (*Chiroderma* spp.) of this size have facial stripes ending at base of ear; Macconnell's bat (*Mesophylla macconnelli*) has no facial stripes; dwarf fruit-eating bats (*Artibeus* spp.) have no middorsal stripe and short facial stripes that end at base or middle of ear.
Natural history. Feed on fruit, especially figs; the diet of most species is unrecorded. *V. pusilla* have been found roosting in tents of *Heliconia* leaves, and singly in a tent made from a *Philodendron* leaf. Found in mature and disturbed rainforest, gardens, and plantations.

Geographic range. Central and South America: S Mexico south to central Bolivia, the S Amazon Basin of Brazil, Paraguay, and SE Brazil.
Status. Most species widespread, uncommon.
References. Peterson, R. L. 1968. A new bat of the genus *Vampyressa* from Guyana, South America, with a brief systematic review of the genus. Roy. Ontario Mus. Life Sci. Contrib., no. 73.
Lewis, S. E., and D. E. Wilson. 1987. *Vampyressa pusilla.* Mammalian Species, no. 292.

Macconnell's Bat
Mesophylla macconnelli
Map 68
Identification. Measurements: HB = 44–50; T = 0; HF = 9–11; E = 12–15; FA = 29–33; WT = 7–9 g.
Head and shoulders pale whitish beige, back pale beige; no white stripes on head or body. Wings gray-brown, **wrist, distal forearm, and thumb yellow; skin on forearm with tiny pale speckles. Noseleaf yellow,** a narrow spear, horseshoe with complete free flange at base; **ears all yellow** or pale beige toward tip; eye large, sometimes with dark smudge in front of corner. **Tail membrane short, semitranslucent, naked,** calcar less than half the length of foot. Underparts pale whitish beige, palest on throat. Tiny, delicate bats.
Variation. This species is close to and may belong in the genus *Vampyressa;* it is sometimes placed in the genus *Ectophylla,* to which it does not seem closely related.
Similar species. Little yellow-eared bats *(Vampyressa pusilla)* are similar but have facial stripes, gray wrists and thumbs, and a slight fringe of hair in center edge of tail membrane; dwarf fruit-eating bats *(Artibeus [Dermanura])* usually have facial stripes and gray thumb, and are usually uniform gray or brown.
Natural history. Macconnell's bats have been found roosting in a hollow tree, and a group of three was roosting under a tent made from a single broad palm leaf about 150 cm high. The leaf had been cut in a broad V from its opposite edges to the midrib. Found in mature and disturbed lowland rainforest.
Geographic range. Central and South America: Costa Rica south to N Bolivia

Map 68

░ Macconnell's bat, *Mesophylla macconnelli*

and the S Amazon basin of Brazil. To about 1,000 m elevation.
Status. Uncommon and widespread.
References. Gardner, A. L. 1977. Chromosomal variation in *Vampyressa* and a review of chromosomal evolution of the Phyllostomidae (Chiroptera). *Syst. Zool.* 26:300–318.

Honduran White Bat
Ectophylla alba
Plate 6, map 69
Identification. Measurements: HB = 37–47; T = 0; HF = 9–10; E = 14–15; FA = 26–30; WT = 4–6 g.
Head and shoulders white, grading to gray-white on midback and pale brownish white on lower back; wings blackish brown except forearms and thumbs yellow, propatagium translucent. Face with dark smudge from corner of eye to ear; **noseleaf bright yellow,** a narrow, large spear, horseshoe with free flange entirely around base; **ears bright yellow;** eyes large, blackish. Tail membrane naked, semitranslucent, short; calcar less than half as long as foot. Throat white; belly pale gray or gray-brown. **Tiny, delicate bats.**
Similar species. Ghost bats *(Diclidurus* spp.) have white ears and no noseleaf, and are larger; there are no other white bats in the region.
Natural history. Roosts in groups of one to six under tents made from large, horizontal, banana-like leaves of platanillos *(Heliconia* spp). The bats bite the leaf along each side of the midrib until the sides fold down to form a shelter. Found in mature, disturbed, and secondary rainforest and in plantations of trees.

Map 69

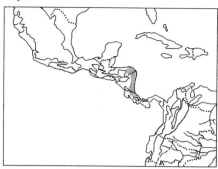

▨ Honduran white bat, *Ectophylla alba*

Map 70

▨ Fruit-eating bats, *Artibeus* spp.

Geographic range. Central America: Honduras and Nicaragua south to Panama, mostly in the Caribbean lowlands.

Status. Unknown; rare and potentially threatened by deforestation.

References. Timm, R. M. 1982. *Ectophylla alba.* Mammalian Species, no. 166.

Timm, R. M., and J. Mortimer. 1976. Selection of roost sites by Honduran white bats, *Ectophylla alba* (Chiroptera: Phyllostomatidae). *Ecology* 57:385–89.

Large Fruit-eating Bats
Artibeus spp.
Plates E, 6; map 70

Identification. Measurements: HB = 70–103; T = 0; HF = 12–21; E = 18–28; FA = 48–75; WT = 25–86 g.
Upperparts almost black to gray or brown; fur pale at base or unicolored; wings and tail membrane black in gray or black animals, brown in brown; no white stripe on midback. Head large and powerful; four pale facial stripes usually present, sometimes absent, **stripes below eye often indistinct or absent, upper stripes end at level of ear; muzzle short, broad; upper tooth rows horseshoe-shaped;** tongue often black; **noseleaf a broad fleshy spear, reaches to rear corner of eye or beyond when pressed back;** horseshoe either with free flange all around base or only at sides, usually blackish, with pale edges in brown species; eyes large, dark brown; ears medium-sized, triangular, tragus and lower rims pale or whole ear dark gray; chin with V-shaped row of small tubercles with pad in center; **central incisors short, bilobed.** Tail membrane

usually short to very short at center, U- or V-shaped. **Thickset, large, muscular bats** that bite fiercely when handled.

Species. There are about nine species, all but one in lowland rainforest (the dozen small species are described in the following account). Up to four species can occur together. The most common include: *A. jamaicensis,* large (50–60 g) dark brown to dark gray, fur slightly frosted, with distinct pale band at base, facial stripes distinct but faint; *A. lituratus,* very large (50–86 g), brown, wings brown, facial stripes broad and bright (these two species are among the most common bats throughout the Neotropical rainforest); *A. obscurus,* large (40 g), rich blackish to dark chocolate, fur long, only slightly paler at base or unicolored, facial stripes faint; *A. hirsutus,* medium-sized (24–30 g), pale gray with long fur, faint facial stripes, hairy tail membrane. There is disagreement about how many species are represented in the group of forms near *A. jamaicencis.*

Similar species. Big-eyed bats (*Chiroderma* spp.) have a distinct dark band at base of fur, long, narrow, single-lobed middle upper incisors, and usually a middorsal stripe; spearnosed bats (*Phyllostomus* spp.) have tails.

Natural history. Feed mainly on fruit, especially figs, nectar, and insects. Large fruiteating bats are some of the most important dispersers of seeds into early successional forest. Jamaican fruit bats (*A. jamaicensis, A. planirostris*) are often the most common bats in the forest, along with common shorttailed fruit bats (*Carollia* spp.). They fly low in the understory and use trails as flyways.

When encountered as they fly to a feeding roost carrying a fruit, they may drop the fruit in surprise, seeming to throw it toward the observer as they veer sharply to avoid collision. Fruit-eating bats roost in small groups in hollow trees and branches, among dense foliage, under palm leaves, in caves, or occasionally in buildings, and at least one species sometimes uses tents made from leaves. *A. jamaicensis* groups consist of harems of one male and several females that prefer to roost in tiny tree hollows, and groups of bachelor males that roost in foliage or tents. In caves, many harems may roost close together. Found in mature and secondary rainforest and deciduous forest, gardens, and plantations.

Geographic range. Central and South America: central Mexico south to Paraguay and N Argentina.

Status. Widespread and common.

References. Handley, C. O., Jr., D. E. Wilson, and A. L. Gardner. 1991. Demography and natural history of the common fruit bat, *Artibeus jamaicensis,* on Barro Colorado Island, Panama. Smithsonian Contrib. Zool., no. 511.

Morrison, D. W. 1978. Foraging ecology and energetics of the frugivorous bat *Artibeus jamaicensis. Ecology* 59:716–23.

Dwarf Fruit-eating Bats
Artibeus (Dermanura) spp.
Map 70
Identification. Measurements: HB = 47–69; T = 0; HF = 9–14; E = 16–20; FA = 34–52; WT = 10–22 g.
Upperparts usually gray or gray-brown, sometimes pale brown or dark chocolate brown, fur either pale or faintly dark at base; wings and tail membrane blackish. Facial stripes present or absent, usually end at upper edge of ear, rarely at middle of ear; no midback stripe. Noseleaf spear-shaped, often pale on sides of base; ears triangular, rims sometimes cream or yellow. Central upper incisors bilobed and less than twice as long as outer incisors, and parallel at tips. Muzzle broad and powerful, upper tooth row horseshoe-shaped. Tail membrane with or without fringe of hairs on edge, V-shaped, often short; if very short, then edge with hairy fringe; if longer, without hairy fringe.
Species. There are about a dozen species; all probably occur in lowland rainforest.

Most cannot reliably be distinguished in the hand. Species include: *A. hartii* (formerly *Enchisthenes*), medium-sized (13–17 g), dark chocolate brown, blackish on head, tail membrane short and hairy, facial stripes narrow, pale brown; *A. glaucus,* small (12 g), uniform smoky brown or gray, fur uniformly dark to base, facial stripes bright, rims of ears, tragus, and sides of noseleaf pale yellow, wings black; *A. watsoni,* small (13 g), pale beige to gray, stripes prominent, tail membrane naked. Some place this group in a separate genus.

Similar species. Yellow-eared bats *(Vampyressa)* have central upper incisors more than twice as high as outer incisors, and converging at tips, facial stripes to middle or rear of ear; Macconnell's bats *(Mesophylla macconnelli)* have no stripes, yellow thumb and wrist, semitranslucent tail membrane.

Natural history. Feed on small fruits and insects. At least two species roost in tents made from palm leaves, and another species has been found roosting under a banana leaf. Found in mature and disturbed lowland and montane rainforest, plantations and gardens, and cloud forest and deciduous forest.

Geographic range. Central and South America: central Mexico south to central Bolivia and the S Amazon Basin of Brazil. To at least 2,400 m elevation.

Status. Common to rare; most species widespread.

References. Handley, C. O., Jr. 1987. New species from northern South America: Fruit-eating bats, genus *Artibeus* Leach. *Fieldiana Zool.* 39:163–72.

Ipanema or Double-lipped Bat
Pygoderma bilabiatum
Plate E, map 71
Identification. Measurements: HB = 60–85; T = 0; HF = 12–15; E = 17–21; FA = 36–41; WT = 15–22 g; females strikingly larger than males.
Upperparts warm brown; forequarters paler than rump; fur long, tricolored with dark bands at base and tip; shoulder at junction of wing with pure white spot; wings and tail membrane brown. Head broad and short; mouth wide, upper lip with a fold from base of noseleaf to corner of mouth forming a "double" lip; insides of upper and lower lips at corner of mouth with dense fringe of long papillae; noseleaf

Map 71

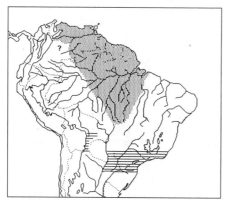

▨ Little white-shouldered bat, *Ametrida centurio*
☰ Ipanema or double-lipped bat, *Pygoderma bilabiatum*

a large, narrow, fleshy spear; **chin bald, without tubercles;** ears triangular, broad at base, with a partial membrane across head from back of upper side, tragus yellow; eyes large, chocolate brown. Tail membrane short, V-shaped, well haired, with hairy fringe. **Chest thinly haired,** almost naked in males, hairier in females; belly medium brown, frosted.
Similar species. No similar bat has a double lip; little white-shouldered bats *(Ametrida)* are smaller and have a white spot below the ear; the two other bats with shoulder spots do not have large spear-shaped noseleaves.
Natural history. Apparently feeds on soft, pulpy fruits. Found in mature and secondary forests and plantations with fruit trees, in rainforest and deciduous forest. Seems most abundant in dry forests.
Geographic range. South America: known from two disjunct areas: south-central Bolivia and adjacent Argentina; and E Paraguay to SE Brazil.
Status. Locally common but patchy, rare in many areas; widespread.
References. Webster, W. D., and R. D. Owen. 1984. *Pygoderma bilabiatum.* Mammalian Species, no. 220.
  Myers, P. 1981. Observations on *Pygoderma bilabiatum* (Wagner). *Z. Saügetierk.* 46:146–51.

Little White-shouldered Bat
Ametrida centurio
Map 71
Identification. Measurements: HB = 38–55; T = 0; HF = 10–11; E = 13–16; FA = 25–33; WT = 8–12 g; females larger than males.

Upperparts warm pale brown to cinnamon, males more dusky, with grayer tints than females; forequarters paler than hindquarters; **fur tricolored, dark at base and tip; shoulder at angle of wing with pure white spot; neck below ear with smaller, fainter white spot;** wings and tail membrane brown. **Face short and broad; mouth wide and monkeylike, muzzle naked,** lips with fringe of papillae inside; **noseleaf a short, broad spear flattened back against base of eyes; eyes large and bulging, males with a swollen pad below.** Ears triangular, broad at base, brown with yellow base and tragus. Tail membrane short, V-shaped, hairy, with hairy fringe.
Similar species. See Ipanema bat.
Natural history. Found in lowland rainforest and in clearings and edges derived from it.
Geographic range. Central and South America: one individual is known from the Canal Zone of Panama; all others are from Venezuela, the Guianas, and the eastern Amazon Basin of Brazil.
Status. Rare in a large geographic range.

Visored Bat
Sphaeronycteris toxophyllum
Plate E, map 72
Identification. Measurements: HB = 52–63; T = 0; HF = 11–13; E = 13–17; FA = 37–42; WT = 18 g; females larger than males.

Upperparts cinnamon brown, males sometimes tinted gray; fur tricolored, dark band at base and tip; **shoulder at angle of wing and neck below ear with pure white spots;** wings and tail membrane brown. **Face short and broad, mouth wide, monkeylike, smiling; forehead with large, fleshy horizontal fold (visor), much larger in males than females, males appear to be wearing a baseball cap; noseleaf an inverted U shape, pushed forward by bottom of visor; muzzle and chin naked; eyes large, bulging, metallic golden brown;** ears triangular, base yellow, tip brown, tragus yellow; **chin with fold of skin that can be pulled up over face, fold has translucent spots over eyes when pulled up.** Tail membrane short, V-shaped, hairy. Underparts like back.
Similar species. No other bats have a face like this (see Ipanema bat [*Pygoderma bilabiatum*]); wrinkle-faced bats *(Centurio senex)* have no forward-projecting visor or noseleaf.

Map 72

▨ Wrinkle-faced bat, *Centurio senex*
▤ Visored bat, *Sphaeronycteris toxophyllum*

Natural history. Found in many habitats, including mature and secondary rainforest, cloud forest, and deciduous forest, and in gardens, plantations, and pastures.
Geographic range. South America: Venezuela and the western Amazon Basin in Colombia, Peru, Brazil, and N Bolivia. To 2,200 m elevation.
Status. Rare to uncommon in a large geographic range.

Wrinkle-faced Bat
Centurio senex
Plate E, map 72
Identification. Measurements: HB = 53–67; T = 0; HF = 10–15; E = 13–19; FA = 42–47; WT = 16–25 g.
Upperparts warm pale brown to beige; fur tricolored, dark at base and tip; shoulder at angle of wing with pure white spot; wing brown, membranes between third-fourth and fourth-fifth fingers (nearest to body) with striking ladderlike pattern of alternating translucent and opaque bars, most pronounced in males; forearms very hairy. Head and face a round ball, face naked, with deep, complex skinfolds and channels between eyes and around nostrils; mouth very wide; eyes large and bulging; no free noseleaf; chin with loose fold of skin that can be pulled up over eyes, trans-

parent spots where fold covers eyes, this mask well developed in males, less so in females; only males have additional large folds or lappets on the chin; ears long, narrow, thin, naked, rounded, brown distally, yellow basally, upper edge of ears with large, distinctive horizontal lobes covering part of brow. Tail membrane narrow, V-shaped, hairy. Throat whitish, belly colored like back. Neck very short, head appears stuck onto large, square, squat body. Males have a musky odor.
Variation. Males have more extremely developed facial folds than females.
Similar species. No other bats have a face like this or laddered wings; visored bats *(Sphaeronycteris toxophyllum)* have a noseleaf and projecting visor on forehead.
Natural history. Feeds on fruits, from which it extracts the juice. Wrinkle-faced bats roost among foliage. Males roost singly or in groups of two or three; a number may be dispersed around the same tree. Females roost in dense clusters. While roosting, the mask of skin is pulled up tightly completely over the face and forehead and covers the horizontal flaps of the ears; a ridge across the crown holds it in place and/or forms an air channel for breathing. Wrinkle-faced bats fly rapidly, like big, heavy beetles, with a wobbly motion, sometimes with the body vertical to the ground. Found in mature and secondary rainforest, dry forest, gallery forest, plantations, and gardens. Apparently adapts well to extremely disturbed habitats and can live in city parks and scrubby forest near cane fields.
Geographic range. Central and South America: S coastal Mexico south through Panama, coastal N Colombia, Venezuela, and Trinidad. To about 1,400 m elevation.
Status. Rare, but widespread and with a wide habitat range.
References. Snow, J. L., J. K. Jones, Jr., and W. D. Webster. 1980. *Centurio senex.* Mammalian Species, no. 138.
    Goodwin, G. G., and A. M. Greenhall. 1961. A review of the bats of Trinidad and Tobago. *Bull. Am. Mus. Nat. Hist.* 122:187–302.

Vampire Bats (Desmodontinae)

Dental formula: I1/2 or 2/2; C1/1; P1/2; M1/1, 2/1, or 2/2 = 20 or 26. These are the only true vampire bats in the world. They are highly specialized for feeding on the blood of birds and mammals. They have small folds above the nose and no spear-shaped structures; long, razor-sharp, caninelike, forward-pointing upper incisor teeth; and large, bladelike canine teeth. There is no tail, and the tail membrane is reduced. The legs and forearms are long and the body is slender. The thumbs are long and thickened or have pads, and are used as feet to walk, hop, or climb quadrupedally along the ground, or on branches, to approach prey quietly. These bats never suck blood from their prey, but lick up the drops flowing from a small scoop they bite in the skin. The razor-sharp teeth make a quick and painless wound, without waking sleeping prey, and the saliva has an anticoagulant that keeps the blood from clotting. Vampires are usually feared by local people, who often have never seen one at close range and thus do not discriminate between vampires and other species of bats. There is a tendency to believe that all large bats are vampires. People therefore kill bats at random (much as they behave toward snakes)—a practice to be discouraged because bats are beneficial. Vampires are an extraordinary example of complex evolutionary adaptation and should be objects of wonder rather than fear. They are related to the leaf-nosed bats, but they are sometimes placed in a separate family of their own. The three genera and three species are restricted to the New World tropics and subtropics.

Common Vampire Bat
Desmodus rotundus
Plates F, 6; map 73
Identification. Measurements: HB = 70–93; T = 0; HF = 13–22; E = 16–21; FA = 49–63; WT = 26–42 g.
Upperparts rich dark brown, rarely orange; fur short, stiff and straight, with a strong sheen, often silvery; hair on neck and shoulders sometimes whitish, with a prominent pale base; **fur on back often sparse,** especially on forequarters; wings blackish. **Muzzle short, nostrils in small M-shaped fold, with another, inverted U-shaped fold behind this, no free noseleaf, no horseshoe under nostrils;** ears triangular with pointed tips, tragus hairy; chin tip a deeply notched V, undersurface of chin naked, with dot of a gland at base of throat; **middle upper incisors larger than canines, triangular and forward-pointing, canines triangular, very sharp. Thumbs long, with two pads under joints;** legs long, feet with long toes and claws. **Tail membrane brown, a narrow U-shaped band, hairy;** tiny calcar present. Underparts sometimes sharply demarcated from sides, silvery gray or white with sheen, brown hair base showing through.
Variation. Some animals from Argentina have whitish wingtips.
Similar species. White-winged vampires *(Diaemus youngi)* have white wingtips; hairy-legged vampires *(Diphylla ecaudata)* have long, dense fur and no tail membrane; wrinkle-faced *(Centurio senex)* and visored *(Sphaeronycteris toxophyllum)* bats have white shoulder spots; all other tailless bats have noseleaves (except some bats restricted to Caribbean islands and not described in this book).
Sounds. May make a bizarre, machinelike, pulsed, rasping, whining sound, probably defensive, when held in the hand.
Natural history. Feeds on blood. Common vampires feed on the blood of mammals, including man, nowadays often on domestic ungulates, especially cattle and horses. More rarely they feed on birds, and captives even on reptiles and amphibians. They feed by silently approaching an animal (these bats fly extremely quietly), landing near it or on it, making a small scoop in the skin, and lapping up flowing blood with the tongue. Vampires are agile quadrupedal runners and will readily run and hop on all fours, using the thickened thumb as a foot. They roost in caves and hollow trees. Common vampires have a complex social organization in which females form stable groups of 8–12 individuals that roost together, feed in the same home range, and cooperate by sharing food with each other by regurgitation. Males form more temporary associations and compete for dominance of a group of females. These bats seem naturally rare in deep rainforest, but large concentrations of prey in the form of livestock introduced by man have caused them to become common in some areas, where they can transmit rabies, cause wounds susceptible to screwfly invasion, and bite humans. This species should be protected in forest, and only populations that feed on livestock should be controlled. Found in rainforest, deciduous forest, secondary vegetation, gallery forest, cerrado, chaco, pastures, and gardens.

Map 73

Common vampire bat, *Desmodus rotundus*

Map 74

White-winged vampire bat, *Diaemus youngi*

Geographic range. Central and South America: N coastal Mexico, S Mexico south along both sides of the Andes to central Chile and S Argentina and Uruguay. To 1,500 m elevation.

Status. Common and widespread.

Local names. Vampiro (Span, Br); vampir (Su).

References. Greenhall, A. M., and U. Schmidt. 1988. *Natural history of vampire bats.* Boca Raton, Fla.: CRC Press.

Wilkinson, G. S. 1985. The social organization of the common vampire bat. *Behav. Ecol. Sociobiol.* 17:111–21.

White-winged Vampire Bat
Diaemus youngi
Map 74
Identification. HB = 80–93; T = 0; HF = 16–22; E = 16–21; WT = 32–48 g. Upperparts brown; fur generally sparse, smooth, shiny, and silky, without pale tips, hair white at base, this evident on shoulders; wingtips pure white, sharply demarcated, leading edge of wing also white. Similar to common vampire, except: no secondary fold behind that containing nostrils; mouth with large glands inside corners that can emit a jet of nauseating fluid in defense; thumb with a single long, thickened pad under joints; calcar absent; feet hairy, with hairs extending beyond claws.

Variation. Similar to the common vampire and sometimes placed in the same genus.

Similar species. Common vampire *(Desmodus rotundus)* has two thickened pads under thumb and black wingtips; hairy-legged vampire *(Diphylla ecaudata)* has dense fur, especially on legs, and no tail membrane.

Sounds. Similar to common vampire.

Natural history. Feeds on the blood of birds, including domestic poultry, and sometimes domestic mammals. The behavior in the wild is poorly known, but captives kill most wild bird species during a single feeding. White-winged vampires roost in caves and tree hollows. One cave colony of about 30 bats roosted in male-female pairs, each pair spaced 20–40 cm from its neighbors, suggesting a monogamous social system unlike that of common vampires. Mainly from humid lowland evergreen forests.

Geographic range. Central and South America: from Taumalipas, Mexico, south to N Argentina and SE Brazil; Trinidad and Margarita.

Status. Rare in a large geographic range.

References. Foglietta, L. M., and O. J. Linares. 1992. Contribución a la biologia general del murciélago vampiro de alas blancas *(Diaemus youngi)* en Venezuela. *XIX Congreso Brasileiro de Zoologia y XII Congreso Latino-Americano de Zoologia, 1992,* Resumos.

Greenhall, A. M., and U. Schmidt. 1988. *Natural history of vampire bats.* Boca Raton, Fla.: CRC Press.

Hairy-legged Vampire Bat
Diphylla ecaudata
Plate F, map 75
Identification. Measurements: HB = 72–86; T = 0; HF = 13–19; E = 15–19; FA = 50–56; WT = 23–33 g.

Map 75

Hairy-legged vampire bat, *Diphylla ecaudata*

Upperparts rich brown, pale to almost blackish, or near chestnut; fur soft and dense, with strong sheen, hair on shoulders white at base; forearms hairy. **Noseleaf very short, consisting of two inconspicuous inverted U-shaped folds, one behind the other; face attractive, mouth with a permanent pleasant smile; eyes large; ears short and rounded, lower edge prolonged forward in a small fold, forms a line under eye to nose.** Incisors as in common vampires. **Tail membrane almost absent, a narrow band down side of leg, hairy above and below; calcar forms a spur with free tip.** Thumb long and thick. Chin naked; underparts gray-brown with sheen.

Variation. Paler animals seem to come from drier habitats.
Similar species. Other vampires *(Desmodus, Diaemus)* have short fur and tail membrane that forms a band between legs; see common vampire for other comparisons.
Natural history. Feeds on blood, chiefly or exclusively of birds. Hairy-legged vampires roost singly or in small colonies in caves, tunnels, and hollow trees. These vampires do not seem to be as agile at walking quadrupedally on flat surfaces as are common vampires; they approach their prey by crawling upside down along the bottom of a branch. They feed like common vampires, by biting a small incision and licking up the drops of blood that flow from the wound. They sometimes attack chickens or other domestic poultry, but do not usually seem to do them much harm. Found in rainforest, deciduous forest, dry forest, caatinga, and gardens and plantations.
Geographic range. North, Central, and South America: Texas (one record) south along the Caribbean coast of Mexico, all countries of Central America, south to Bolivia and SE Brazil. To about 1,500 m elevation.
Status. Rare or uncommon, but extremely widespread.
References. Greenhall, A. M., U. Schmidt, and G. Joermann. 1984. *Diphylla ecaudata*. Mammalian Species, no. 227.

Funnel-eared, Thumbless, and Sucker-footed Bats (Natalidae, Furipteridae, Thyropteridae)

Dental formula: I2/3, C1/1, P3/3, M3/3 = 38. These are three small families of tiny insectivorous bats. The funnel-eared bats (Natalidae) include one genus and five species. The thumbless bats (Furipteridae) include two genera, each with one species. The sucker-footed bats (Thyropteridae) include one genus with three species. The ecology of all species is poorly known; they all have a fluttering style of flight and may be specialized for hunting insects within dense vegetation. All are restricted to the New World tropics and subtropics.

Funnel-eared Bats
Natalus spp.
Plates F, 7; map 76
Identification. Measurements: HB = 45–64; T = 45–60; HF = 7–11; E = 12–16; FA = 36–42; WT = 4–7 g.
Upperparts pale beige, tawny yellowbrown, gold, or bright rust-orange; fur

long, lax, slightly wavy; wings and tail membrane pale brown. **Muzzle slightly elongated and narrow, forehead rising abruptly above; nose and chin simple, with no leaf or folds, mouth somewhat funnel-shaped; ears triangular, broad at base, funnel-shaped, pointing forward; eyes small,** inconspicuous. **Tail very long,**

Map 76

Funnel-eared bats, *Natalus* spp.

Map 77

Thumbless bat, *Furipterus horrens*

about equal to head and body, entirely enclosed within membrane; **tail membrane
slightly longer than legs, comes to a point
at tail tip;** calcar longer than foot; **hindlegs
very long, leg to tip of claws about the
same length as head and body;** thumbs
tiny. Underparts the same color as back or
slightly paler. **Tiny bats with a large ex-
panse of delicate, crinkled membrane.**
Species. Two of the four species are found
in lowland continental rainforest; the other
two occupy Caribbean islands. *N. stramineus* is slightly smaller (HB = 45–48), with
the tail usually slightly longer than the head
and body; *N. tumidirostris* is larger (HB =
50–64), with the tail shorter than the head
and body; color varies individually from
beige to orange in both species.
Similar species. Thumbless bats *(Furipterus
horrens)* have a tail much shorter than the
tail membrane; in all vespertilionid bats the
tail and tail membrane are much longer than
the legs, and held curled up forward in a coil
at rest. No other bats with no noseleaf have
a long tail entirely within the tail membrane
that forms a point at its edge.
Natural history. Feed on insects. Funnel-
eared bats roost in the dark recesses of humid
caves, where they hang singly or in groups
of well-separated individuals, sometimes
in colonies of thousands. They may be restricted to regions with caves (which are
absent in much of the Amazon Basin).
They fly with a fluttering flight low over the
ground, and use the large tail membrane to
catch insects. Found in rainforest, but more
often in more arid habitats of deciduous or

dry forest, and in gardens and plantations.
Geographic range. Central and South
America: both coasts of Mexico almost to
the United States border, south through
Central America and along the north and
east coast of South America to SE Brazil,
W Bolivia, and the lower Amazon Basin.
Status. Widespread but generally patchy in
distribution, numerous in a few caves.

Thumbless Bat
Furipterus horrens
Plate F, map 77
Identification. Measurements: HB =
33–43; T = 20–27; HF = 6–9; E = 9–12;
FA = 30–40; WT = 3 g.
Upperparts dark smoky gray, slate blue-
gray, or dark gray-brown; **fur long, dense,
and lax,** wings and tail membrane dark
brown. Muzzle short, narrow, **nose and chin
plain,** without folds or leaf, **forehead large,
rising abruptly at right angle to muzzle;**
ear short, broad at base; eye tiny, lost in fur.
**Tail about half as long as tail membrane,
entirely enclosed within it; tail membrane
longer than leg, entirely filled with pattern
of close-set, fine, parallel lines at right
angles to tail; thumb a rudimentary stub,**
with almost no claw; legs long, almost as
long as head and body. Underparts slate gray;
nipples abdominal. **Tiny bats like long-
legged balls of gray fluff.**
Similar species. Superficially very like
smoky bat *(Cyttarops alecto),* which has a
large eye and thumb. All other bats in rainforest have well-developed thumbs with
claws.

Natural history. Feeds on insects. Thumbless bats roost in small clusters in colonies
of up to at least 60 in caves, horizontal fallen
logs, and deep cracks between rocks; one
such roost was among large boulders in a
riverbed, which were exposed during the dry
season. Sometimes they are found singly,
roosting under logs. They forage near the
forest floor, with a slow, fluttering, mothlike
flight. Found in lowland rainforest.
Geographic range. Central and South
America: Costa Rica south to the southern
Amazon Basin of Peru and Brazil; and to SE
coastal Brazil.
Status. Rare, but widespread.
References. LaVal, R. 1977. Notes on some
Costa Rican bats. *Brenesia* 10–11:77–83.

Map 78

▓ Sucker-footed bats, *Thyroptera* spp.

Sucker-footed Bats
Thyroptera spp.
Plates F, 7; map 78
Identification. Measurements: HB =
37–57; T = 24–37; HF = 4–7; E = 11–14;
FA = 31–41; WT = 3–5 g.
**Upperparts rich chocolate brown to chest-
nut; fur long and soft or shorter and plushy.**
Face short, muzzle narrow, without noseleaf; ears triangular, broad at base, sharply
pointed; eyes tiny. **Ankle and thumb bases
with moist, fleshy sucker disks. Tail membrane pointed, longer than legs; tail longer
than membrane,** within it to posterior edge,
**tip protrudes free of membrane for a short
distance beyond it. Hindfoot tiny, with
third and fourth digits fused into one thick
digit.** Underparts white or brown. Tiny, delicate bats.
Species. The three species are found in rainforest: *T. tricolor* (FA = 35.5–38) has long,
lax fur, sharply contrasting white or cream
underparts, and free tail projecting 3.5–
10 mm beyond edge of membrane; *T. discifera* (FA = 32–35.5) has long, lax fur, underparts the same as back or only slightly
paler, and free tail of 1–3 mm; and *T. lavali*
(FA = 38.8–40.7) is larger, has shorter,
plushy fur, brown underparts, and free tail
of 5.3–7.2 mm.
Similar species. No other bats in region
have suckers on feet.
Natural history. Feed on insects. *T. tricolor* roosts in small groups of 1–9 inside
the rolled new leaves of platanillos (*Heliconia* spp.) and bananas, at the stage when the

leaves are just unrolling to open, or sometimes in curled, dead leaves of the same
plants. Bats move from leaf to leaf every few
days. The bats in a group stay together when
they change roosts. Platanillos tend to grow
on relatively good soil in forest openings,
such as treefall gaps, and on riversides, and
that is where these bats can be found, but
they choose roosts in the shade. When the
females have young, they may move to "ma-
ternity" roosts in hollow logs on the forest
floor. *T. discifera* is poorly known, but it has
been found roosting beneath open, dead banana leaves. Unlike most other bats, sucker-
footed bats roost head upward, attached to
the flat surface of a leaf by their suckers.
Their local distribution seems to be patchy;
they are apparently limited to areas where
they find numerous appropriate roosts. Found
only in lowland rainforest and gardens and
plantations derived from it.
Geographic range. Central and South
America: SE Mexico south through Central
America to the southern Amazon Basin and
the Atlantic forests of SE Brazil.
Status. *T. discifera* is rare; *T. tricolor* is
common to uncommon and apparently
patchy in a huge geographic range; *T. lavali*
may be known only from the type series of
four individuals from Peru.
References. Pine, R. H. 1993. A new species
of *Thyroptera* Spix (Mammalia: Chiroptera:
Thyropteridae) from the Amazon Basin of
northeastern Perú. *Mammalia* 57:213–25.
    Wilson, D. E., and J. S. Findley. 1977.
Thyroptera tricolor. Mammalian Species,
no. 71.

Vespertilionid Bats (Vespertilionidae)

Dental formula: I1/3 or 2/3; C1/1; P1/2, 2/2, or 3/3; M3/3. This is a large family of small to medium-sized bats that feed on insects. They have plain faces with no noseleaves, small eyes, and long, wide tail membranes that come to a point at the tip of the long tail. At rest the tail and membrane are curled forward under the body. During flight at the moment of prey capture, the large tail membrane is brought forward and used as a scoop to help trap flying insects. Most species are agile and rapid flyers. In northern regions, most species hibernate in caves during the winter, or both migrate and hibernate, or migrate far enough south to be able to remain active. Vespertilionids have one or, occasionally, two to four young, which are left behind at the roost while the mother forages. The natural history and social behavior of Neotropical species are poorly known. This is the most geographically widespread family of bats; members are found on all continents, with about 37 genera and 318 species worldwide.

Little Brown Bats

Myotis spp.
Plate F, map 79
Identification. Measurements: about HB = 36–54; T = 30–42; HF = 7–10; E = 11–17; FA = 29–44; WT = 4–7 g.
Upperparts blackish brown, smoky brown, chestnut, reddish, or orangish; fur short or medium length, **unicolored entirely dark to base or with tips frosted;** wings and tail membrane usually blackish, sometimes brown. **Head triangular, broad between ears, coming to point at nose, without noseleaf** or other folds; eyes tiny; ears triangular, pointed; **tragus pointed; first tooth behind canine tiny and peglike, creating a large gap between canine and first large tooth behind it. Tail membrane much longer than legs, coming to a point at tip of tail, tail entirely within membrane,** usually curled forward under body when not in flight; **calcar shorter than foot.** Thumb with long distal joint. Underparts paler than back with hairs frosted pale at tips, dark at base, rear of abdomen usually paler than chest.
Species. This is the largest and most widespread genus of bats in the world, with about 88 species. About 8 of the 35 New World species are found in rainforest; 3 or 4 can occur together at one locality. They are often quite similar. Some of the most common and widespread in rainforest include: *M. nigricans,* almost black, fur smooth; *M. riparius,* smoky gray or brown, fur waved; *M. albescens,* fur usually frosted, with pale tips; *M. simus,* fur short and orange-red.
Similar species. Most big brown bats (*Eptesicus* spp.) are larger (10 g); they have even fur that is very dark at the base, often blackish, evenly tipped over the entire upperparts with slightly to distinctly paler brown, which

gives the appearance of a rich, smooth sheen; the first tooth behind the canine is large, leaving no gap between the canine and the first large tooth behind it. Little yellow bats (*Rhogeessa* spp.) have fur pale at base; sheathtailed bats (Emballonuridae) have a short tail whose tip sticks out above the tail membrane partway along its length; thumbless bats (*Furipterus* spp.) have a short tail; funnel-eared (*Natalus* spp.) bats are pale, with funnel-shaped ears and mouth.
Natural history. Feed on tiny insects. Little brown bats seem to favor flying in open areas or gaps in the forest, such as those formed by treefalls, streams, trails, roads, swamps or pools, and gardens, or in clearings around houses. They fly swiftly, with rapid wingbeats, making frequent dives and turns to catch their prey. They may be high or low in the forest. Forest species roost in tree hollows, rock crevices, or buildings. Most species of this genus roost in close contact in tight clusters. Found in all lowland habitats.
Geographic range. Worldwide; subarctic to equatorial; found in the entire rainforest region. To at least 2,400 m elevation.
Status. Rainforest species are common to rare; most appear widespread.
References. LaVal, R. K. 1973. A revision of the Neotropical bats of the genus *Myotis. Los Angeles County Nat. Hist. Mus. Sci. Bull.* 15.
  Wilson, D. E., and R. K. LaVal. 1974. *Myotis nigricans.* Mammalian Species, no. 39.

Big Brown Bats

Eptesicus spp.
Plates F, 7; map 80
Identification. Measurements: HB = 48–84; T = 33–49; HF = 8–13; E = 11–19; FA = 37–54; WT = 5–16 g.

Map 79

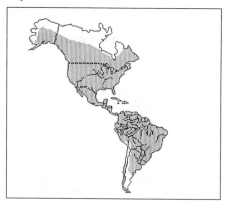

Little brown bats, *Myotis* spp.

Map 80

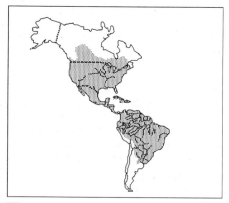

Big brown bats, *Eptesicus* spp.

Upperparts black-brown to tawny yellow-ocher; fur deep blackish at base with tip always paler, from slightly paler, rich brown to much paler tawny, on dark animals paler tips impart a sheen; **fur medium-long, of even length, often silky; wings and tail membrane usually black,** sometimes dark brown, strong and leathery. **Face plain, pointed, without noseleaf; ears triangular, pointed; tragus rounded;** eyes small; **first tooth behind upper canine large, leaving little gap between canine and first large tooth. Tail shorter than head and body length (40–90%), entirely within tail membrane, membrane comes to a point at tail tip, tail held curled forward under body at rest;** calcar longer than foot. Underparts yellowish, paler than back, hair blackish at base, usually tawny to gold at tips.
Species. There are about 30 species worldwide; 4 of the approximately 6 New World species can occupy lowland rainforest habitats: *E. fuscus* is large (FA = 48–54; WT ≈ 15 g) and tawny yellow-ocher; *E. brasiliensis* is dark brown to reddish and smaller (FA = 40–46; WT = 9–12 g); *E. furinalis* is dark brown to orange-tinged brown to dark tawny (FA = 37–43; WT = 7–10 g); and *E. andinus* is blackish (FA = 42–48; WT = 11 g) and largely montane. There are probably more species in the region than are currently recognized.
Similar species. Little brown bats (*Myotis* spp.) have a pointed tragus and tiny first tooth behind the canine, leaving an apparent gap, and usually have unicolored fur, except that *M. albescens,* which is colored very

similarly to *E. furinalis,* but smaller (FA = 34–37), has underparts frosted with more whitish or grayish, whereas in *E. furinalis* they are yellow or tawny.
Natural history. Feed aerially on insects. Big brown bats roost in tree hollows, holes in snags in lagoons, buildings, and caves. *E. fuscus* flies in a steady, straight course, with occasional dives to catch prey, and it favors unencumbered areas. These bats are often in and around buildings and country towns. Found in rainforest, cloud forest, gallery forest, deserts, gardens, plantations, and fields.
Geographic range. Worldwide, including the Americas from Canada to Argentina and Uruguay.
Status. Rare in rainforest; common in other more open habitats.
References. Hill, J. E., and D. L. Harrison. 1987. The baculum in the Vespertilioninae (Chiroptera: Vespertilionidae) with a systematic review, a synopsis of *Pipistrellus* and *Eptesicus* and description of a new genus and subgenus. *Bull. Brit. Mus. Nat. Hist.,* zool. ser. 52, no. 7.

Black-winged Little Yellow Bat
Rhogeessa tumida
Plate F, map 81
Identification. Measurements: HB = 36–49; T = 26–31; HF = 6–8; E = 12–14; FA = 25–33; WT = 3–5 g.
Upperparts pale yellow, lightly to heavily frosted or overlaid with brown; fur bicolored pale yellow at base with brown tips; wings blackish or brown. Face triangular,

Map 81

Black-winged little yellow bat, *Rhogeessa tumida*

Map 82

Big-eared brown bats, *Histiotus* spp.

nose pointed, plain, no noseleaf; eyes tiny; **ears triangular,** narrow at base, pale at base with dark tips; tragus a long, narrow blade. Tail membrane much longer than legs, coming to point at tip of tail, tail entirely within it; tail shorter than head and body. **Underparts pale yellow or gold.** Minuscule bats.

Variation. Varies geographically from pale in Central America to dark brown-frosted in South America. Five other species in the genus occupy dry habitats.

Similar species. Little brown bats (*Myotis* spp.) have fur dark at base; funnel-eared bats (*Natalus* spp.) have pale membranes, funnel-shaped ears, tail as long as head and body.

Natural history. Feeds on insects. Little yellow bats roost in colonies in hollow trees. They start to forage at dusk, sallying out from the shelter of trees with a more rapid flight than sheath-tailed bats. This genus includes the smallest New World bats. Found in a wide range of habitats from rainforest to dry forest.

Geographic range. Central and South America: E Mexico south through all of Central America to W and N Colombia and Ecuador, Venezuela, the Guianas, Trinidad, N Bolivia, and the Amazon Basin of Brazil to coastal Bahia.

Status. Apparently rare but very widespread.

References. LaVal, R. K. 1973. Systematics of the genus *Rhogeessa* (Chiroptera: Vespertilionidae). *Occas. Pap. Mus. Nat. Hist. Univ. Kans.* 19:1–47.

Big-eared Brown Bats
Histiotus spp.
Plate F, map 82

Identification. Measurements: HB = 54–68; T = 47–58; HF = 8–10; E = 27–35; FA = 45–51; WT = 14–16 g.

Upperparts dark brown, tawny brown, or beige; fur long and lax, very dark brown at base with paler tips. Face triangular, nose plain and sharply pointed, without noseleaf; ears enormous, broad and long, inner edges almost meet over brow; tragus long, lance-shaped. Tail membrane much longer than legs, tail within it to edge, then tip extends free beyond it for a short distance. Underparts frosted white or beige, hair dark brown at base.

Species. There are about five species; they are mostly montane or temperate, but *H. velatus* may be found in Atlantic coastal rainforests, and the cloud forest species *H. montanus* may occasionally be found below 1,000 m.

Similar species. There are no other small bats with enormous ears and no noseleaf on the South American continent.

Natural history. Feed on insects. *H. velatus* has been found roosting in small clusters in the roofs and attics of buildings. Found in mature and disturbed premontane and montane humid forests and in temperate forests.

Geographic range. South America: the entire Andean chain and subtropical and temperate South America, mountains of Venezuela, greater Rio São Francisco drainage of E Brazil, and Paraguay. To above 2,000 m elevation.

Status. All species apparently rare, widespread to restricted in range.
References. Mumford, R. E., and D. M. Knudson. 1978. Ecology of bats at Vicosa, Brazil. *Proc. 4th Int. Bat Res. Conf.,* 287–95.

Hoary or Hairy-tailed Bats
Lasiurus spp.
Plates F, 7; map 83
Identification. Measurements: HB = 52–90; T = 41–61; HF = 7–14; E = 10–20; FA = 36–57; WT = 6–21 g.
Upperparts rust-red, yellow, brown frosted with white, or black; often frosted or variegated; fur dense, hairs with several color bands, dark band at base; forearm hairy below and with patches of hair near thumbs; **tail membrane long, wide, and thick, entirely filling space between legs, thickly furred above for half or more of its length;** wing membranes with pale patches surrounding bones. **Head short, muzzle blunt, nostrils often widely separated, no noseleaf; ears thick, short, broad, rounded, scarcely protruding above crown;** tragus a narrow panel displaced on a stalk in front of ear. Tail long, entirely within membrane, **at rest tail is bent forward and upward so that hairy tail membrane covers lower belly; posture at rest characteristic, curled forward in a tuck.** Underparts dirty yellow or red, hair dark at base. Medium-sized, often beautifully patterned and brightly colored bats.
Species. Eight of the 12 species can be found in lowland rainforest: *L. cinereus* is large (FA = 50–57), either brown (Peru, Central America) or pale yellow (Venezuela), strongly variegated or frosted with white; *L. blossevillii* is small (FA = 36–42), bright rust-orange on lower back, forequarters frosted with red over yellow, underparts yellowish; *L. ega* is medium-sized (FA = 47–51), entirely dirty yellow, without frosting; *L. egregius* is large (FA = 50), rust-red over yellow, with red underparts; *L. castaneus* (Central America) is small (FA = 44.8), chestnut over yellow above and yellow over black below; *L. ebenus* (FA = 45.7) is entirely black. The newly described *L. atratus,* from the Guiana region, is medium-sized (FA = 45–47), red dorsally with contrasting black and white chest, and black face and wings.
Similar species. No other bats in region have the combination of short, blunt head and

Map 83

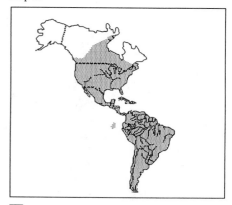

Hoary or hairy-tailed bats, *Lasiurus* spp.

ears, no noseleaf, and long, thickly furred tail membrane.
Natural history. Feed on insects, especially moths and other large kinds, and have been reported preying on tiny species of bats. Hoary bats often fly high, with a strong, fast flight and slow wingbeats. They sometimes forage around streetlights in towns. They roost singly among foliage, especially along forest edges, in places well sheltered from above, and their variegated colors probably serve as camouflage against the vegetation. They can have litters of up to four young. Hoary bats include the most widespread species of New World bats. In North America they are seasonally migratory, but their behavior in the tropics is poorly known. Found in a wide variety of habitats from deep undisturbed Amazon rainforest to many kinds of disturbed forest, dry forest, and temperate and boreal forest.
Geographic range. North, Central, and South America: Canada to Chile. To at least 1,500 m elevation.
Status. Uncommon in rainforest. *L. blossevillii* is common in dry forests south of Amazonia. All rainforest species widespread except probably *L. egregius*. *L. ebenus* is known only from the type, from the Atlantic forest of Brazil.
References. Shump, K. A., Jr., and A. U. Shump. 1982. *Lasiurus borealis*. Mammalian Species, no. 183.
    Fazzolari-Corréa, S. 1994. *Lasiurus ebenus,* a new vespertilionid bat from southeastern Brazil. *Mammalia* 58:119–23.

Map 84

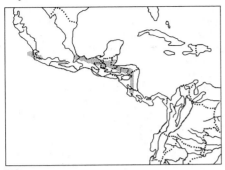

Central American long-eared bat, *Bauerus dubiaquercus*

Figure 5. Central American long-eared bat, *Bauerus dubiaquerqus.*

Central American Long-eared Bat
Bauerus dubiaquercus
Figure 5, map 84
Identification. Measurements: HB = 58–76; T = 49–57; HF = 12–14; E = 21–27; FA = 50–57; WT = 14–17 g. Females larger than males.
Upperparts warm pale brown, near chestnut, hair bicolored dark on basal half, pale on distal half, **fur long and dense;** wings and tail membrane brown. **Muzzle broad, swollen between nose and eye,** with punctuations from vibrissae, fur on this area slanting forward; chin almost naked, with two smooth pads on point; **ears very long, inner rim with backward curve, outer rim straight. Canine teeth large and robust, lowers occlude in front of uppers.** Thumb and foot long and robust. **Tail membrane much longer than legs, ending in a point at tail tip, calcar nearly twice as long as** foot. Underparts tawny brown, slightly paler than back.
Similar species. Similar only to the closely allied desert bat *(Antrozous pallidus),* which is very pale, whitish or yellowish. No other bat of this size in Central America has very large ears and no noseleaf, and tail entirely within tail membrane.
Natural History. Presumably feeds on insects. Little is known of the ecology of this bat; the diet and roosts are undescribed. Found in lowland and montane tropical rainforests, tropical deciduous forests, and moist pine-oak forests.
Geographic Range. Central America: from Nayarit, Mexico, south to Costa Rica.
Status. Known from only a few, widely scattered localities.
References. Engstrom, M. D., T. E. Lee, and D. E. Wilson. 1987. *Bauerus dubiaquercus.* Mammalian Species, no 282.

Free-tailed or Mastiff Bats (Molossidae)

This is one of the most distinctive families of bats. They are medium-sized to large bats with long, thick, naked tails that extend free for about half of their length beyond the edge of the tail membrane. They have short, velvety fur; leathery wings and tail membranes; short, broad, hairy feet; rectangular, flattened bodies; very narrow wings; doglike faces with no noseleaf; wide mouths; and medium-sized black eyes. Their ears are complexly folded and distinctive: they are divided by a longitudinal fold into two open compartments, one facing outward and the other inward, over the eye; the forward or upper edge of the ear extends as a thick fold across the brow; the lower edge of the ear forms a fold that curves around forward under the ear, and this fold has a separate lobe, called the antitragus, that stands up in front of the ear opening and hides the true tragus, which lies behind it. The precise shapes of both the antitragus and the upper edges of the ear are important characters for identification of genera and species. Bats of this family chiefly feed on large insects such as moths and beetles. They can be recognized by their strong, swift, erratic flight and narrow wings. All species seem to fly very high, and they are rarely captured except from roosts. For this reason their distributions are poorly known, and they are often missing from inventories of areas where they should occur. Some species have echolocation calls in the frequency range audible to humans, and they can be clearly heard uttering musical single chirps as they cruise and pulses of sound as they dive to capture prey. The flattened bodies of these bats allow them to roost in narrow crevices in caves, between rocks, under bark, or in tree holes. Worldwide, they have become commensal with man, roosting by the hundreds tightly packed into tiny spaces in the roofs of buildings. They are muscular and agile at crawling or scuttling quadrupedally. During the day in the roost they become torpid, with body temperature near ambient levels. They are able to open their mouths extremely wide, and when threatened they draw back their lips and gape in a ferocious-looking threat. These are the common large bats that fly swiftly over the rooftops of tropical towns at dusk. Large populations in caves feed on enormous numbers of insects and produce guano, and thus are particularly beneficial to man. There are about 14 genera and 90 species worldwide, with 10 genera in the New World.

Dog-faced Bats
Molossops spp.
Plate G, map 85

Identification. Measurements: HB = 43–66; T = 21–29; HF = 5–11; E = 12–15; FA = 30–38; WT = 5–16 g.
Upperparts dark brown, fur pale at base. Fur on forehead diminishes gradually down face between eyes. **Muzzle pointed; lips without folds,** upper lip angles back steeply from nose to undershot mouth; nostrils set in slightly raised pads with hard upper rim; center of muzzle flat, without ridge. **Ears simple, triangular, pointed, upper edges widely separated on crown, reach forward of eye when flattened, division into two longitudinal compartments rudimentary, accordion-like slight flexible fold where upper edge of ear meets head, lower edge of ear not folded back. Tail about 40–50% of length of head and body, free from membrane for less than half its length.** Round gland on throat present in males. Underparts like back or frosted pale, or gray-brown or cinnamon; throat sometimes white.
Species. The two species are found in lowland rainforest: *M. neglectus* is larger (FA = 36–38, WT = 15–16 g) and has dark underparts; *M. temminckii* is tiny (FA = 30–32, WT = 4–9 g) and has underparts frosted or paler than back.
Similar species. Flat-headed bats *(Neoplatymops matogrossensis)* have bumps on the forearm and highly flattened body and head; doglike bats *(Cynomops* spp.) have no extra fold at upper edge of ear that folds down strongly, forming separate compart-

Map 85

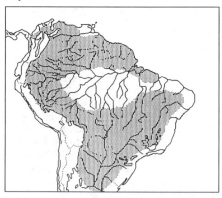

▦ Dog-faced bats, *Molossops* spp.

ment over eye, and have hair of crown that ends abruptly on a line; mastiff bats *(Molossus* spp.) have the top of the muzzle raised in a ridge.
Natural history. *M. temminckii* roosts in groups of two or three in hollow trees, rock crevices, buildings, and fence stakes. They choose holes 1–3 m from the ground with tiny entrances, often slits less than 2 cm wide or holes barely large enough for the bats to squeeze through. They are silent in the roost, so their presence is difficult to detect. They fly out at dusk for short foraging bouts, after which they return and rest in the roost. They frequent mature and highly disturbed rainforest, dry forest, and areas around human habitations and agricultural areas.
Geographic range. South America: Guyana, Venezuela, Colombia, Peru, Bolivia, Argen-

tina, Paraguay, and Brazil south of the Amazon Basin *(M. temminckii),* or in the Amazon basin *(M. neglectus).*
Status. Locally common to rare; *M. neglectus* is known from only 10 individuals; both species widespread.
References. Vizotto, L. D., and V. A. Taddei. 1976. Notas sobre *Molossops temminckii temminckii* e *Molossops planirostris* (Chiroptera-Molossidae). *Naturalia* 2:47–59.
 Myers, P., and R. M. Wetzel. 1983. Systematics and zoogeography of the bats of the Chaco Boreal. Misc. Pub. Mus. Zool. Univ. Mich., no. 165.

Doglike Bats
Cynomops spp.
Plate G, map 86
Identification. Measurements: HB = 44–89; T = 22–40; HF = 8–13; E = 14–20; FA = 30–48; WT = 10–42 g.
Upperparts blackish brown, chestnut, or orange-brown, fur short, dark or pale at base; **hair of crown ends abruptly, usually on line between ears. Muzzle broad, chin broad, rounded in profile; upper lip without vertical wrinkles, slightly puckered in to meet lower lip; nostrils like holes punched in surface,** not on raised pad, no sculptured surface around or between nostrils, **top of muzzle between eyes and nostrils flat, without raised central ridge. Ear rounded, folded longitudinally to form compartment over eye, when flattened forward reaches midway from eye to nose; upper edges of ears do not meet in center of forehead; rear edge of ear slightly folded backward,** antitragus almost rectangular, tragus a narrow spike. Throat gland present. Tail usually 40–50% of head and body length. Underparts dark, slightly grayer and paler than back or throat, and belly whitish.
Species. There are about five species; three occur in lowland rainforest: *C. abrasus,* medium-sized, often chestnut; *C. greenhalli,* small and blackish; and *C. planirostris,* medium-sized, with pale fur base and whitish underparts. These species are sometimes placed in the genus *Molossops.*
Similar species. Dog-faced bats *(Molossops* spp.) have pointed ears and fur decreasing gradually down face between eyes; mastiff bats *(Molossus* spp.) have a raised ridge down center of face.

Map 86

Doglike bats, *Cynomops* spp.

Natural history. Feed on insects. Doglike bats roost in small groups of up to eight in small holes in rotting trees in swamps, hollow branches, fence posts, and buildings. *M. planirostris* roosts are 1.5–5 m above the ground; the bats are noisy in the roost, twittering and scrambling around, especially just before they emerge at night. They often seem to forage around or over ponds or swamps or in clearings. Found in rainforest, gallery forest, llanos, and savannas.
Geographic range. Central and South America: W coast of Mexico south through all Central and South American countries to Paraguay.
Status. Common to rare, widespread.
References. Vizotto, L. D., and V. A. Taddei. 1976. Notas sobre *Molossops temminckii temminckii* e *Molossops planirostris* (Chiroptera-Molossidae). *Naturalia* 2:47–59.

Flat-headed Bat
Neoplatymops matogrossensis
Plate G, map 87
Identification. Measurements: HB = 46–54; T = 22–29; HF = 7–8; E = 13–15; FA = 27–31; WT = 7–9 g.
Upperparts dark brown variegated over pale yellow, hairs yellow with brown tips; fur short, sparse; **forearm skin above sprinkled with small bumps. Head flat, muzzle long, pointed, upper lip under nose angled steeply back to undershot mouth; ears sharply directed forward, upper edges do not meet in center of crown. Body flattened. Throat and chest pale orange to dirty yellow, forming a V on belly sharply**

Map 87

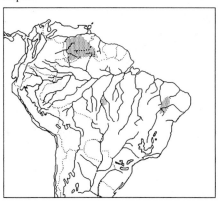

Flat-headed bat, *Neoplatymops mattogrossensis*

Map 88

Brazilian free-tailed bat, *Tadarida brasiliensis*

demarcated from brown sides under wings.
Base of throat with round gland in both sexes.
Tiny, flat bats with forward-pointing faces.
Variation. This species is sometimes placed
in the genus *Molossops*.
Similar species. These are the only tiny free-
tailed bats with bumps on the forearm.
Natural history. Roosts under slabs of rock
and in rock crevices. The geographic distribution of flat-headed bats may be in part
determined by the presence of appropriate
roost sites. Found in arid and caatinga habitats, but also on some rare rock mountains in
rainforest.
Geographic range. South America: Venezuela, Guyana, and Brazil in Rio Branco
and from Mato Grosso to Bahia; distribution
probably poorly known.
Status. Common to rare.
References. Willig, M. R., and J. K. Jones,
Jr. 1985. *Neoplatymops matogrossensis.*
Mammalian Species, no. 244.

Brazilian Free-tailed Bat

Tadarida brasiliensis
Plates G, 7; map 88
Identification. Measurements: HB = 49–
66; T = 29–43; HF = 8–11; E = 16–20;
FA = 36–46; WT = 11–13 g.
Upperparts dark brown to gray-brown,
fur usually uniform to base, sometimes
base paler. Muzzle upturned, pointed; upper
lip with deep vertical folds, hangs over
lower lip; chin and jaw flat, face and chin
with short, stiff, blunt bristles; nose backed
by hard flange. Six lower incisor teeth. Ears
meet, or almost meet, on center of crown
but are not joined; rear edge of ear not
folded back, ear when flattened reaches
tip of nose. Tail 50–70% as long as head
and body (usually about 60%). Underparts
lightly frosted slightly paler than back,
throat darker than abdomen.
Variation. There is some geographic variation in size.
Similar species. See broad-eared free-tailed
bats (*Nyctinomops* spp.).
Natural history. Feeds mainly on moths,
beetles, flies, and hymenopterans. These bats
form renowned colonies of millions in caves
in the southwestern United States, but in the
tropics they roost in more modest numbers in
tree holes, caves, rock crevices, and the roofs
of houses. When foraging, they fly high and
swiftly in open spaces. This is not a rainforest species, but it is found in dry or montane
habitats on the fringes of the entire rainforest region, and some records appear to be
from within it, perhaps from cleared areas
or savannas.
Geographic range. North, Central, and
South America: the southern half of the
United States to N Argentina.
Status. Locally common and widespread.
Some populations are threatened and declining due to overuse of toxic pesticides and disturbance of roosting caves by guano miners.
References. Constantine, D. G. 1967. *Activity patterns of the Mexican free-tailed bat.*
Univ. N. Mex. Pub. Biol., no. 7.

Broad-eared Free-tailed Bats

Nyctinomops spp.
Plate G, map 89

Map 89

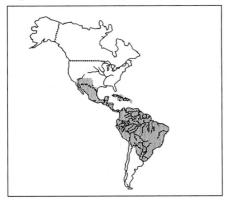

Broad-eared free-tailed bats, *Nyctinomops* spp.

Identification. Measurements: HB = 52–
87; T = 34–56; HF = 8–13; E = 13–30;
FA = 40–64; WT = 10–30 g.
Upperparts dark brown to reddish, fur
either pale at base or uniform; **wings and
tail membrane semitranslucent.** Muzzle
pointed, upturned, **upper lip deeply wrinkled with vertical folds, overhangs lower
lip; chin and jaw flat,** face and chin with a
few long, slender bristles; nose upturned,
nostrils raised on little tubes, backed by
hard ridge. **Four lower incisor teeth. Ears
large, upper edges join on forehead on
fleshy stalk,** rear edge of ear not folded
back, **ears when flattened forward reach
tip of nose.** Tail about 50–80% of head and
body length. **Throat usually brown, darker
than slightly frosted abdomen.**
Species. Three of the four species occur in
lowland rainforest: *N. aurispinosus* has FA
> 47; *N. laticaudatus* has FA = 40–46; *N.
macrotis* has long, lax fur and huge ears and
is the largest (FA = 58–64). These species
are placed by some in the genus *Tadarida*.
Similar species. Brazilian free-tailed bats
(Tadarida brasiliensis) have ears that meet
but do not join on forehead; other free-tailed
bats do not have wrinkled lips.
Sounds. Foraging *N. macrotis* often emit a
loud, piercing chatter. Both echolocation
pulses and communication calls are audible.
Natural history. Feed on insects, mainly
moths. Broad-eared free-tailed bats roost in
colonies of up to scores in caves, rock crevices, hollow trees, crevices between palm
leaves, and buildings. They occupy many
habitats from rainforest to arid scrub.

Geographic range. North, Central, and
South America: SW United States to N Ar-
gentina and Uruguay.
Status. Apparently uncommon, but wide-
spread.
References. Barbour, R. W., and W. H.
Davis. 1969. *Bats of America.* Lexington:
University Press of Kentucky.

Bonneted Bats
Eumops spp.
Plate G, map 90
Identification. Measurements: HB = 60–
125; T = 27–68; HF = 9–20; E = 18–44;
FA = 37–83; WT = 10–60 g.
Upperparts blackish, pale brown, cinnamon,
gray-brown, or chestnut; fur usually pale at
base. **Muzzle pointed, upper lip without
vertical folds, slanting sharply back from
nose; nostrils bounded behind by horny,
raised, semicircular ridges that join to
form a central keel between nostrils; no
raised ridge down center of face; ears
large, flattened forward like a hat pulled
down over eyes, upper edges joining in
midbrow; highly developed deep longitudinal fold or keel separating the two compartments of ear; ear when flattened** forward reaches more than midway from eye
to nose, sometimes to tip of nose; antitragus
large, semiovate, broadest at base. Underparts like back or frosted beige; throat with
large round gland in males. Small to very
large bats.
Species. Seven of the nine species have been
found in lowland rainforest. Two species
are noteworthy for their very large size: *E.
perotis* (FA = 75–83) and *E. underwoodi*
(FA ≈ 71).
Similar species. Broad-eared free-tailed bats
(Nyctinomops spp.) have wrinkled lips; no
other free-tailed bats have ears broadly connected on midbrow and reaching forward of
eye when flattened forward.
Sounds. All species emit audible echolocation chirps.
Natural history. Feed on large insects, for
which the audible echolocation is an adaptation; *E. perotis* specializes on butterflies and
moths. Bonneted bats roost in small groups
in tree holes, cliffs, and roofs of buildings.
Because of their large size and narrow wings,
the largest species must fly swiftly to remain
airborne, and they apparently are unable to
take flight from a level surface. They roost

Map 90

Map 91

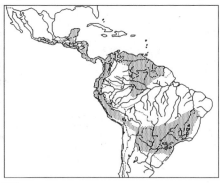

Crested mastiff bats, *Promops* spp.

Bonneted bats, *Eumops* spp.

high enough to allow them to gain speed as they drop from the roost. These bats forage at great heights. Found in many habitats from rainforest to arid scrub and in towns.

Geographic range. North, Central, and South America: SW United States to central Argentina.

Status. Most species are rarely encountered; they may be rare or simply difficult to capture because of their high flying and roosting behaviors.

References. Eger, J. L. 1977. Systematics of the genus *Eumops* (Chiroptera: Molossidae). Roy. Ontario Mus. Life Sci. Contrib., no. 110.

Vaughn, T. A. 1959. Functional morphology of three bats; *Eumops, Myotis, Macrotis*. *Univ. Kans. Pub. Mus. Nat. Hist.* 12:1–153.

Crested Mastiff Bats
Promops spp.
Plate G, map 91
Identification. Measurements: HB = 68–89; T = 50–55; HF = 12–13; E = 13–18; FA = 46–57; WT = 14–25 g.
Upperparts dark brown to chestnut, head and/or neck usually darker than back, blackish; fur velvety. **Head short, broad, crown with slightly longer tuft of hair between ears; ears rounded, reach midway from eye to ear when laid forward,** inner edges meet with fold near horizontal; **lower edge of ear slightly folded back; lips with no vertical wrinkles;** nostrils backed by slight horny ridge that does not descend between nostrils; **top of muzzle from between ears to nose with raised central ridge;** chin

rounded; **palate between the upper teeth deeply concave.** Throat with round gland in both sexes, more highly developed in males. **Four lower incisors.** Tail 60–75% of length of head and body; tail membrane covers more than half of tail. **Underparts paler than back, uniform brown, not frosted,** throat and chest darker than abdomen.

Species. The two species are both found in rainforest: *P. centralis* is larger (FA = 51–57), and *P. nasutus* is smaller (FA = 46–50).

Similar species. Difficult to distinguish from mastiff bats (*Molossus* spp.), which have throat paler than abdomen, two lower incisors, and do not have a deeply concave palate.

Natural history. Crested mastiff bats roost in small groups of about six in holes in rotting trees, in houses *(P. nasutus),* or under palm leaves *(P. centralis).* Found in openings in rainforest or in dry forest.

Geographic range. Central and South America: S Mexico to N Argentina. To at least 1,000 m elevation.

Status. Both species apparently rare but widespread.

References. Sazima, I., and W. Uieda. 1977. O morcego *Promops nasutus* no sudeste Brasileiro (Chiroptera, Molossidae). *Ciencia e Cultura* 29:312–14.

Mastiff Bats
Molossus spp.
Plate G, map 92
Identification. Measurements: HB = 59–89; T = 29–51; HF = 8–15; E = 12–19; FA = 36–53; WT = 9–38 g.

Map 92

Mastiff bats, *Molossus* spp.

Upperparts black, dark brown, or chestnut; fur often paler at base than at tip; fur
short and velvety; wings and tail membrane
black. **Muzzle broad, with large rounded
chin; lips without deep vertical wrinkles,
slightly puckered so that upper does not
overhang lower, upper lip wide below
nose, not steeply angled backward;** nostrils
simple, without upstanding hard ridge behind; **center of muzzle between ears and
nose raised in a ridge. Ears short, rounded,
upper edges almost, but do not quite, meet
in center of crown at horizontal angle;** rear
edge of ear folded back; ears when flattened
reach midway from eye to nose. **Two lower
incisors. Palate between upper tooth rows
almost flat.** Tail usually about 50–60% of
head and body length. Throat with round
gland at base in both sexes; underparts like
back, not frosted, throat often slightly paler
than belly, with hairs paler at base.
Species. All of the about seven species occur in lowland rainforest. The most widespread and common are *M. rufus* (formerly
called *M. ater*), the largest (FA = 47–53,
WT = 25–38 g), black or red, with a naked
tail membrane; and *M. molossus* (FA = 36–
41, WT = 12–15 g), with pale base of hairs
visible through fur. There has been some

confusion about the names and status of several species.
Variation. Blackish and bright rusty individuals of *M. rufus* occur in the same
populations.
Similar species. Difficult to distinguish in
the hand from crested mastiff bats (*Promops*
spp.), which have a deeply concave palate
and four lower incisors; no other free-tailed
bats have the midline of the muzzle raised in
a ridge.
Sounds. Colonies in roofs become active
with much twittering, bumping, and scrambling about at nightfall (but most bats twitter
and squawk in the roost).
Natural history. Feed on insects, chiefly
moths, beetles, and flying ants. Mastiff bats
roost in tree holes, rotting trees, rockpiles,
and buildings. They are often found roosting in large colonies of hundreds in narrow,
closed spaces under the roofs of houses, often with both *M. rufus* and *M. molossus* in
the same roost. These are the common nar-
row-winged bats that fly rapidly like swifts
over the rooftops of tropical cities and towns
at dusk. *M. rufus* leave the roost at nightfall
and forage for about an hour before returning; only a few leave the roost again to forage for a second time before dawn. Found in
rainforest and many types of drier habitats,
and in towns and cities. These bats rarely
carry rabies, and colonies in roofs are more
of a nuisance than a threat to health. Their
consumption of large quantities of insects is
beneficial.
Geographic range. Central to South America: N Mexico to N Argentina and Uruguay.
Status. Common to uncommon, widespread.
References. Marques, S. A. 1986. Activity
cycle, feeding and reproduction of *Molossus ater* (Chiroptera: Molossidae) in Brazil.
Bol. Mus. Paraense Emilio Goeldi, Zool.
2:159–79.

Freeman, P. W. 1981. A multivariate study
of the family Molossidae (Mammalia, Chiroptera): Morphology, ecology, evolution.
Fieldiana Zool., n.s., no. 7.

Monkeys (Primates)

The three families of New World monkeys form a distinct group (Platyrrhini) that differs in many features from the Old World monkeys and apes (Catarrhini). The most obvious external difference is in the structure of the nose, which has close-set, downward-pointing nostrils in Old World primates (including ourselves) and wide-apart, sideward-pointing nostrils in New World monkeys. New World monkeys have short muzzles and flat, naked faces; large, forward-facing eyes; short ear pinnae; short necks and torsos; long hindlimbs; long prehensile digits; a plantigrade stance; and long tails in all but two species. All are primarily arboreal, and they descend to the ground only to cross an open space. A few species will forage on the ground when food is scarce in the trees. Neotropical monkeys show great geographic variation in color patterns. This variation poses problems when we try to sort them into species (see Appendix C), and there is still much uncertainty about the classification of species in some genera. This edition continues to present the conservative view of a number of genera for which, despite recent revisions, the taxonomy still appears unconvincing. There has been a recent trend toward describing new species within variable genera on the basis of color characteristics alone, without analysis of the species group as a whole or demonstration that the new forms are discrete. Diagnoses for the species do not seem to work throughout the putative ranges. For those who wish to use "split" taxonomies, we map the ranges of the forms considered to be species by other authorities. These ranges are often incomplete, and for accurate identification with use of the most speciose taxonomies, specimens should be collected, as many of the forms cannot be readily diagnosed with field observations. We expect that many species-level changes will be made in future classifications of the monkeys. All New World monkeys not listed under CITES Appendix I are listed under Appendix II.

Marmosets and Tamarins (Callitrichidae)
Goeldi's Monkey (Callimiconidae)

Dental formula: I2/2, C1/1, P3/3, M2/2 = 32 (Callithrichidae); I2/2, C1/1, P3/3, M3/3 = 34 (Callimiconidae). The marmosets (*Callithrix* spp., *Cebuella*) and tamarins (*Saguinus* spp., *Leontopithecus* spp.) are distinguished from other New World monkeys by tiny size (about 100–600 g; the largest weighs as much as a large squirrel); long, nonprehensile tails that can be tightly coiled; heads decorated with a variety of ear tufts, tassels, ruffs, manes, mustaches, or mantles of long hair; hands and feet with claws instead of nails; frequent twins instead of single young; and a variety of internal features. There are two major groups of callitrichids: the marmosets and the tamarins. The main difference between them is that the former have specialized lower jaws and teeth (small "short-tusked" canines and procumbent incisors lacking lingual enamel) that they use to dig holes in the bark of trees and vines. Marmosets feed extensively on the plant exudates (sap, gum, resin) that collect in these holes. Tamarins have larger, "long-tusked" lower canines (and more vertical incisors with lingual enamel) and do not gouge holes in bark, although they will feed on exudates they find. Members of both groups also feed on fruits and extensively on insects. All callitrichids have quite similar behavior when seen in the field: they bound or gallop lightly along branches, rapidly springing between trunks or stems with "vertical cling-and-leap" jumps—taking off and landing with the head and body held vertical. While active, they constantly bob and turn their heads, peering in every direction. They call with similar birdlike whistles and soft chirping sounds. These monkeys can be wary and stealthy, hiding quickly behind trunks, but they are often heard and can easily be found if their calls are recognized. All emit loud "long" calls, which are involved in territorial behavior. Those species that have been studied live in small families consisting of a single breeding female, one or more males (they can be polyandrous or monogamous), two to four subadults, and carried juveniles. Young are usually carried on the back of a male. Females may have young twice a year. Most species favor dense vegetation with festoons of vines, where their insect prey is numerous; they thus adapt well to secondary and disturbed forest. Because they are too small to be prey for human hunters and thrive in secondary vegetation, callitrichids are usually the most common monkeys near villages or towns, and some are even found in city parks. There are four genera and about seventeen species of Callitrichidae, all in the Neotropics.

Goeldi's monkeys (Callimiconidae) are much like tamarins in size and external appearance, but differ from them in many internal anatomical features. They have only one young at a time. The family contains only one species.

The major reference for all members of these two families is P. Hershkovitz, *Living New World Monkeys (Platyrrhini),* vol. 1 (Chicago: University of Chicago Press, 1977). We mainly follow the taxonomy of A. Rylands, A. Coimbra-Filho, and R. A. Mittermeier, "Systematics, geographic distribution, and some notes on the conservation status of the Callithrichidae" (in A. B. Rylands, ed., *Marmosets and tamarins: Systematics, behaviour, and ecology,* 11–77. Oxford: Oxford University Press, 1993).

Pygmy Marmoset
Cebuella pygmaea
Plate 9, map 93

Map 93

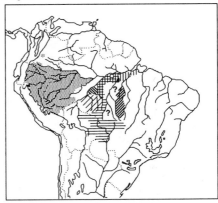

Identification. Measurements: HB = 117–152; T = 172–229; HF = 36–46; E = 15–22; WT = 85–141 g. Females slightly larger than males.

Head and forequarters tawny gold-gray, with fine striations visible at close range; back and **hindquarters coarsely variegated with tawny buff and gray. Back of head, neck, and shoulders with a mane of longer hair.** Face thinly haired, with a pale stripe from between eyes to nose, muzzle sometimes grizzled whitish; eyes large. Hands and feet tawny. **Tail slender, tapered at the tip, tawny, faintly banded with black above.** Throat and chest tawny-buff, belly yellowish white or buff; genital area of adults black. Young grayer than adults and without pale face stripe or black genital tuft. **A tiny monkey with a large head, from a distance looks pale buffy gray.** The world's smallest true monkey.

Similar species. Tamarins (*Saguinus* spp.) and Goeldi's monkeys *(Callimico goeldii)* in the same region are either mostly red and/ or black or have striking white mustaches or muzzles; small squirrels are brown-olivaceous, with short legs and bushy tails.

Sounds. Alarm a soft raspy chirp like the sound of a katydid. Several monkeys usually call together. Long call of two to seven high-pitched, slightly rising whistles, about half a second each.

Natural history. Diurnal; arboreal; small groups of 2–9. Feed on tree exudates and insects, with occasional fruits, buds, and nectar. Pygmy marmosets bite many small,

▨ Pygmy marmoset, *Cebuella pygmaea*
▤ Silvery marmoset, *Callithrix argentata melanura*
▥ *C. a. argentata*
▧ *C. a. nigriceps*
▪ *C. a. leucippe*
▨ *C. a.* ssp.
▨ Snethlage's marmoset, *C. emiliae*
▦ Tassel-ear marmoset, *C. humeralifera*
▢ Rio Maués marmoset, *C. mauesi*

shallow, round to oblong, 10–20 cm holes in the bark of a tree or vine and return day after day to feed on the exudate that seeps into the wounds. Each group has a few such trees, some with hundreds of old and currently used feeding holes. Only certain species of trees are used; these differ in different areas, but *Spondias, Inga, Parkia, Vochysia,* and *Qualea* often dominate the diet. Pygmy marmosets have small home ranges and short daily movements, and sleep at night hidden at 7–10 m height in the central branches of medium-sized trees. They stay low in the forest, usually below 20 m. They move quietly and are shy and difficult to see. The surest way to find them is to wait discreetly in early morning or late afternoon near a feeding tree with freshly made holes. Tamarins may also come to "steal" gum, especially in the dry season. Throughout most of their geographic range these marmosets seem restricted to seasonally flooded forest, river margins, floodplains, or streamsides; rarely they are seen in terra firme forest far from water. Found with a patchy distribution in mature and secondary lowland rainforest.

Geographic range. South America: east of the Andes in Colombia, Ecuador, Peru, and Brazil (Acre), from the base of the Andes east to the Rio Madeira. Limits of range not precisely known.

Status. CITES Appendix II. Locally common to rare; patchy. Not hunted, but potentially at local risk from selective deforestation of floodplain forests.

Local names. Sagüi, leãozinho (Br); mono de bolsillo, leoncito, leoncillo (Pe, Ec, Co); leoncillo, chambira chichico (Ec); chichíco (Co).

References. Soini, P. 1988. The pygmy marmoset, genus *Cebuella*. In R. A. Mittermeier, A. B. Rylands, A. F. Coimbra-Filho, and G. A. B. da Fonseca, eds., *Ecology and behavior of Neotropical primates,* vol. 2, 79–129. Washington, D.C.: World Wildlife Fund.

Silvery Marmoset
Callithrix argentata
Plate 8, map 93
Identification. Measurements: HB = 180–280; T = 265–380; HF = 52–72; E = 25–31; WT = 420 g.

Upperparts silvery white, often with brownish or sooty gray tinge to lower back; face reddish or white, eye black; **or pale brown with** whitish or tan neck and shoulders, face brown, hindfeet brown, forefeet tan, **front of thigh with large pale yellowish or whitish patch from hip to knee. Ears naked, without tufts, exposed. Tail black or brown.** Underparts white to yellowish, tan in brown races. External genitalia naked, white.

Variation. Five color forms of this marmoset are recognized by some as species: a pale, silvery race *(C. a. leucippe)* is found in the northern part of the species range in Pará, Brazil. Beige to gray-white animals from be-

tween the Rios Tapajós and Cupari may be completely white, with black, white, yellow, or brownish tails *(C. a. argentata).* Animals from Rondônia, Mato Grosso, and Bolivia *(C. a. melanura)* are brown with a striking pale thigh stripe. A race from Rondônia *(C. a. nigriceps)* has forequarters pale gray-yellow, crown blackish with white patch between and above eyes, hands blackish, feet blackish or rusty, rump darker brownish gray, thighs rusty, and tail pitch black. Another form, *C. a. intermedia,* seems intermediate between *C. argentata* and *C. humeralifera* and was previously classified with the latter. Molecular genetics suggests that *C. argentata* and *C. humeralifera* may be conspecific. *C. a. intermedia* has slightly tufted ears; forequarters silvery gray; hindquarters variegated with blackish; hindlimbs and hip patches orange; forelimbs paler orange; throat and chest whitish or pale buff; belly and inguinal region orange; tail reddish brown at base becoming gray distally, with dark bands or spots.

Similar species. Tassel-ear marmosets *(C. humeralifera, C. mauesi)* have large ear tufts, a banded tail, and mottled or grizzled hindquarters; saddleback tamarins *(S. fuscicollis)* have blackish forequarters and red hindquarters in the area of overlap; and golden-handed tamarins *(Saguinus midas)* are completely blackish where they occur with silvery marmosets.

Sounds. Soft trills and birdlike chirps; long, thin, unmodulated high-pitched whistles slightly longer than half a second each.

Natural history. Diurnal; arboreal; groups of 5–14. Feeds on fruit, insects, and plant exudates. Silvery marmosets gouge oblong horizontal holes on tree trunks and feed from them on exudate. The pale color of these marmosets makes them hard to see against the pale bark of deciduous trees, which are common in parts of their range. To move from one clump of trees to the next in savanna, these marmosets descend to the ground and scamper across open grassland. They seem to favor dense vegetation of forest and river edges, and occupy primary and secondary forest from evergreen rainforest to deciduous gallery forest, dry forest, and clumps of trees in cerrado.

Geographic range. South America: Brazil, Bolivia, and Paraguay, south of the Amazon from the Rio Madeira east to the Toncantins

and south to N Paraguay. SE limits poorly known.
Status. CITES Appendix II. IUCN Red List vulnerable, *C. a. leucippe* and *C. a. nigriceps* only. Locally common and geographic range large.
Local names. Leoncito, tití blanco (Bo); sauim, sagüi-branco (Br).
References. Rylands, A. B., and D. S. de Faria. 1993. Habitats, feeding ecology, and home range size in the genus *Callithrix.* In A. B. Rylands, ed., *Marmosets and tamarins: Systematics, behaviour, and ecology,* 262–72. Oxford: Oxford University Press.
　　Rylands, A. B. 1981. Preliminary field observations on the marmoset *Callithrix humeralifera intermedius* (Hershkovitz, 1977) at Dardanelos, Rio Aripuaná, Mato Grosso. *Primates* 22:46–59.

Snethlage's Marmoset
Callithrix emiliae
Map 93
Identification. Measurements: HB = 198–230; T = 350–360; HF = 62–67; E = 27–30; WT = 340–416 g.
Upperparts pale gray-brown; forequarters silvery gray, hindquarters darker brown-gray, without pale stripe on hip. Crown blackish, muzzle white, face pink; eyes yellowish brown; ears without tufts. Limbs gray-brown externally, whitish to ochraceous on inner sides; hands and feet blackish. **Tail black.**
Variation. Much like dark forms of silvery marmoset *(C. argentata),* of which they are possibly a subspecies, but there is a report of the two forms occurring at the same site. The name *C. emiliae* has also been used for populations of brownish marmosets from Rondônia—here considered to be forms of *C. a. melanura.*
Similar species. Silvery marmosets *(C. a. argentata)* of the same region are entirely silvery gray, with pale heads; brown forms from western Amazonia *(C. a. melanura)* have a striking pale stripe on the top of the thigh.
Natural history. Presumably similar to other marmosets. A group included 8–10 individuals. The geographic range suggests a largely dry forest or cerrado habitat, but there is little information.
Geographic range. South America: Brazil, Pará between the Rios Iriri and Xingu, S to

the upper Rio Teles Pires-Peixoto do Azevedo, Mato Grosso. Possibly also W of the lower Rio Iriri. Known from only two collection localities.
Status. CITES Appendix II. Poorly known, but geographic range seems small.
Local names. Sagüi, sauim (Br).
References. Ávila-Pires, F. D. 1986. On the validity and geographical distribution of *Callithrix argentata emiliae* Thomas, 1920 (Primates, Callithrichidae). In M. T. Mello, ed., *A primatologia no Brasil—2.* Brasília: Sociedade Brasileira de Primatologia.

Tassel-ear Marmoset
Callithrix humeralifera
Plate 8, map 93
Identification. Measurements: HB = 198–300; T = 298–398; HF = 53–75; E = 28–30; WT = 300–400 g.
The two geographic races are described individually below; they are sometimes considered separate species. General features are: **ear hidden by long, yellow-white tufts of hair growing from both in front of and behind pinna; tail banded with faint or distinct bands.**
　　C. h. humeralifera. **Head blackish, face gray; forequarters silvery gray** frosted with black; **hindquarters variegated black and yellow-white;** hands and feet blackish; **tail gray** with more or less distinct black bands; throat yellowish; underparts from chest to thigh orange; hip and upper thigh often with a pale spot of silvery buff.
　　C. h. chrysoleuca. **Head and body pure yellow-white;** limbs, feet, and tail golden above and below, tail faintly banded with darker gold. Fur silken and shiny.
Similar species. Silvery marmosets *(C. argentata)* have no ear tassels and no banding on tail; tassel-ear and silvery marmosets may have no geographic range overlap, but this is still uncertain. Rio Maués marmosets *(C. mauesi),* which may be a subspecies, have dark brown-and-white marbled back and blackish ear tufts.
Sounds. Birdlike whistles and twitters; alarm call a repeated "tsik."
Natural history. Diurnal; arboreal; groups of 5–15. Feeds on fruits, insects, small vertebrates, and tree exudates. Tassel-ear marmosets prefer or require areas of dense viny vegetation such as those created by second growth into natural or human-caused

disturbance and river edges. They make holes
in trees to produce exudate flow, but seem
to use exudates to a far lesser extent than
tufted-ear or pygmy marmosets. Sometimes
they follow army ant swarms and forage for
flushed insects. They have small home ranges
of about 13 ha. At night they sleep in vine-
covered trees or, rarely, in tree holes. Found
in mature and secondary rainforest.
Geographic range. Brazil: south of the
Amazon in a narrow strip between the Rio
Tapajós and lower Rio Madeira and Rio
Roosevelt. *C. h. humeralifera,* between the
Rios Tapajós and Canumã, southern limit
unknown; *C. h. chrysoleuca,* between the
Rios Madeira, Aripuanã, and Canumã.
Status. CITES Appendix II. IUCN Red List
vulnerable, *C. h. chrysoleuca* only.
Local names. Sauim, sagüi (Br).
References. Rylands, A. B. 1986. Ranging
behaviour and habitat preference of a wild
marmoset group, *Callithrix humeralifer*
(Callitrichidae, Primates). *J. Zool.* (Lond.)
210:489–514.

Rio Maués Marmoset
Callithrix mauesi
Map 93
Identification. Measurements: HB =
198–226; T = 339–376; HF = 59–66;
E = 30–37; WT = 315–405 g.
**Upperparts dark brown variegated (mar-
bled) with streaks of white, no pale shoulder mantle.** Crown dark brown centrally,
bordered by pale patches of silvery brown;
face pink; **ears below prominent, erect, silvery brown tufts. Tail black, faintly banded
with gray.** Hindlimbs frosted silver-gray,
front of thigh with pale patch from hip to
knee; forelegs dark brown frosted with silver. Underparts buff with orange tints; scrotum pinkish white.
Variation. This newly described species is
known from only a few individuals, most in
captivity. It may be a subspecies of tassel-ear
marmoset *(C. humeralifera).*
Natural history. Found in dense mature
rainforest.
Geographic range. South America: known
only from near the type locality on the W
bank of the Rio Maués-Açú, Amazonas,
Brazil.
Status. CITES Appendix II. Unknown.
Local names. Sauim.

Map 94

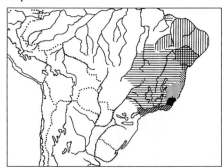

Tufted-ear marmoset, *Callithrix jacchus geoffroyi*
White-tufted-ear marmoset, *C. j. jacchus*
Black-tufted-ear marmoset, *C. j. penicillata*
Buffy-tufted-ear marmoset, *C. j. aurita*
Buffy-headed marmoset, *C. j. flaviceps*
Wied's black-tufted-ear marmoset, *C. j. kuhli*
Callithrix j. hybrids

References. Mittermeier, R. A., M. Schwarz,
and J. M. Ayres. 1992. A new species of
marmoset, genus *Callithrix* Erxleben, 1777
(Callitrichidae, Primates) from the Rio
Maués region, state of Amazonas, Central
Brazilian Amazonia. *Goeldiana Zoologia,*
no. 14.

Tufted-ear Marmoset
Callithrix jacchus group
Plate 8, figure 6, map 94
Identification. Measurements: HB =
190–248; T = 270–350; HF = 52–68;
E = 21–31; WT = 261–323 g; females
usually below 300 g, males heavier.
The distinct geographic forms of this group
of marmosets are individually described below. General features are: **ears with tufts or
tassels; posterior half of back finely banded
or streaked with gray or dull orange and
black, over orange fur base on hindquarters; tail banded with black, often sharply.**

White-tufted-ear marmoset, *C. j. jacchus.* **Head dark brown dominated by enormous whitish tufts** surrounding ear; forehead with white spot; shoulders grizzled
pale gray or gray and buff; hindquarters and
tail banded gray and white; underparts and
feet gray.

Black-tufted-ear marmoset, *C. j. penicillata.* As above, but **ear tufts large, black;
muzzle and mask around eyes blackish;
forehead blackish, with pure white central**

Figure 6. Head patterns of subspecies of **tufted-ear marmosets,** *Callithrix jacchus: (a) C. j. jacchus;* *(b) C. j. penicillata; (c) C. j. geoffroyi; (d) C. j. flaviceps.*

spot; **black cape and collar around neck and chest;** hindlegs buff-gray; feet gray to black.

Wied's black-tufted-ear marmoset, *C. j. kuhli.* Like above form, but **crown black; ear tufts black; brow and cheeks whitish or grizzled gray, cape grizzled, not pitch black.**

Tufted-ear marmoset, *C. j. geoffroyi.* As above, but **face and crown pure white, sharply demarcated from pitch black hood;** ear tufts black, growing from in front of ear; back banded gray and black variegated bright rust on lower back; underparts dark brown or black except throat white.

Buffy-headed marmoset, *C. j. flaviceps.* **Face with dark central mask surrounding nose and eyes; rest of head ochraceous,** including long tuft growing from inside of ear pinna; back banded orange and black; hands and feet yellowish orange; underparts buffy with black stripe down chest, belly, and inguinal region. Considered a subspecies of the following form by those who recognize the other forms as species.

Buffy-tufted-ear marmoset, *C. j. aurita.* **Face, forehead, and long tuft growing from inside of ear pinna whitish** to pale buff; **sides of head black;** crown buff; back blackish brown; tail grizzled yellowish gray-buff with sharp black bands; underparts blackish buff with black midline on chest and belly.

Variation. The forms in this group are now often treated as species, but natural hybrids occur between populations of *C. j. flaviceps* and both *C. j. geoffroyi* and *C. j. aurita; C. j. kuhli* and *C. j. penicillata.* The vocalizations of *C. j. jacchus, C. j. geoffroyi,* and *C. j penicillata* are said to be identical. Molecular genetics shows little divergence between the forms. The taxonomy requires further clarification. There can be much variation within a given form: for example, some *C. j. penicillata* lack a white brow spot and black collar.

Similar species. Lion tamarins (*Leontopithecus* spp.) are larger and red-gold and/or black, with no grizzling or bands on body.

Sounds. High-pitched whistles, chirps, and twitters; mobbing alarm a repeated "tsik."

Natural history. Diurnal; arboreal; small groups of 2–13. Feeds on fruits, insects, and the exudates of trees. Tufted-ear marmosets bite holes in the bark of trees, then lick the exudate. Frequently used trees are riddled with hundreds of horizontal slitlike to round holes, 1–1.5 cm wide by 2–20 cm long. About a third of the total day's activity is spent feeding on the gum that seeps into these holes. Groups occupy small territories of 0.5–5 ha. At night members of a group sleep together in a dense tangle of vegetation or on a branch. Found in evergreen and deciduous forests, gallery forests, and upland scrub; in mature and secondary forests and plantations, even city parks.

Geographic range. South America: Brazil, E forests from Ceará to São Paulo. *C. j. jacchus,* in Ceará, Piauí, Pernambuco; *C. j. penicillata,* in dry and upland forest and scrub from Bahia and Goiás to São Paulo; *C. j. geoffroyi,* in coastal and lowland Minas Gerais and Espírito Santo; *C. j. flaviceps,* at higher elevations in SE Minas Gerais; *C. j. aurita,* in coastal and lowland Minas Gerais and São Paulo.

Status. CITES Appendix I, US-ESA endangered, IUCN Red List endangered, *C. j. aurita* and *C. j. flaviceps* only; others CITES Appendix II; IUCN Red List vulnerable *C. j. geoffroyi* only. Many populations are fragmented and highly disturbed.

Local names. Sauim, sagüi, mico-estrela.

References. Stevenson, M., and A. B. Rylands. 1988. The marmosets, genus *Callithrix.* In R. A. Mittermeier, A. B. Rylands, A. F. Coimbra-Filho, and G. A. B. da Fonseca, eds., *Ecology and behavior of Neotropical primates,* vol. 2, 131–222. Washington, D.C.: World Wildlife Fund.

Faria, D. S. de. 1985. Tamanho, composição de um grupo social e área de vivência (home range) do sagüi *Callithrix jacchus penicillata* na mata ciliar do córrego capetinga, Brasília, DF. *A Primatologia no Brasil* 2:87–105.

Saddleback Tamarin

Saguinus fuscicollis

Plate 9, map 95

Identification. Measurements: HB = 175–270; T = 250–383; HF = 77; E = 21–35; WT = 338–436 g.

Upperparts divided into three color zones: shoulders and forelegs pure black, dark brown, or mahogany red-brown; back from behind shoulders to hip a "saddle" of variegated mixture of blackish and yellowish or

Map 95

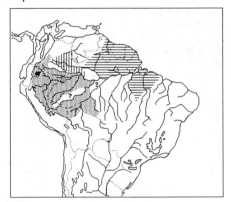

▨	Saddleback tamarin, *Saguinus fuscicollis*
☰	Golden-handed tamarin, *S. midas*
⦀	Mottled-face tamarin, *S. inustus*
◼	Golden-mantle tamarin, *S. tripartitus*

orange striations; hindlegs dark brown to red-brown or deep red, sometimes the same color as forelegs, sometimes much redder. Head black, muzzle white, with short hairs, some races with a white eyebrow band across forehead; eyes brown. Tail dark brown or blackish. Underparts dark reddish.

Variation. There are twelve subspecies. The most distinctive is *S. f. melanoleucus,* which is **creamy white,** sometimes washed with yellowish or buff, except ears black and face with dark skin. These tamarins are highly variable; different color morphs can occur within the same troop, and they intergrade with neighboring subspecies where they come into contact. Pale forms occur in the headwaters from the east bank of the upper Rio Juruá to the upper Rio Purus and Rio Madeira drainages in Acre, Brazil, and Pando, Bolivia.

Similar species. Black-mantle tamarins *(S. nigricollis)* have no distinct saddle and black forequarters (saddlebacks have red forequarters where the species overlap); mustached tamarins *(S. mystax)* are black with long white mustaches; emperor tamarins *(S. imperator)* have long, drooping white mustaches; red-chested tamarins *(S. labiatus)* have sharply contrasting orange or red underparts; golden-mantle tamarins *(S. tripartitus)* have a golden ruff on neck; Goeldi's monkeys *(Callimico goeldii)* are entirely black, with a crown of long hairs.

Sounds. Birdlike twitters, trills, and whistles; long call of loud whistles repeated 7–10 times, each with a modulated up and down ending, and calls gradually increasing in length within the series.

Natural history. Diurnal; arboreal; groups of 2–12. Feeds on small fruits, nectar, and insects. Saddlebacks most often use the middle and lower levels of the forest, below the canopy, and are usually found in dense, viny habitats. They may be wary and difficult to see, but often a soft trill betrays their presence nearby, or at a distance, loud whistles. They run actively through the vines and lower branches of large trees, making many quick jumps across gaps, or they sit in a group, with their tails coiled forward or hanging straight down below. Saddlebacks spend most of their insect-foraging time investigating knotholes and crevices on the trunks of large trees, often near the ground. Groups consist of one breeding female, one or more adult males, and their young. They have large home ranges of 16–100 ha. At night they sleep in a dense festoon of vines surrounding a large tree, or in a tree hole. A group of this species sometimes travels and feeds with a group of another tamarin species that occupies the same territory (emperor, mustached, or red-chested, depending on the region). This is the most commonly seen tamarin in Peru, and it is the most widespread species; it overlaps the geographic ranges of five others. Occupies mature, disturbed, and secondary lowland rainforests.

Geographic range. South America: east of the Andes in Colombia, Ecuador, Peru, Bolivia, and the upper Amazon Basin of Brazil, west of the Rio Jiparaná and south of the Caquetá/Japurá.

Status. CITES Appendix II. Widespread and common.

Local names. Chichilo, tití (Bo); sauim, sagüi (Br); bebeleche (Co); chichíco, leoncito (Ec, Bo); pichico (Pe).

References. Terborgh, J. 1983. *Five New World primates.* Princeton, N.J.: Princeton University Press.

Terborgh, J., and A. W. Goldizen. 1985. On the mating system of the cooperatively breeding saddlebacked tamarin *(Saguinus fuscicollis). Behav. Ecol. Sociobiol.* 16: 293–99.

Golden-mantle Tamarin
Saguinus tripartitus
Plate 9, map 95
Identification. Measurements: HB =
218–240; T = 316–341; E = 31–32.
Back variegated grayish and white or
orange; limbs orange. Head black, with
black collar of hair continuous under
throat; muzzle and sometimes face pure
white, neck with a ruff of bright golden
to creamy fur dorsally, sharply contrasting with black crown. Tail black above, orange on proximal part of undersurface. Feet
and hands blackish orange. Underparts
orange.
Variation. This tamarin is considered by
some to be a subspecies of saddleback tamarin. However, it overlaps saddleback tamarins
in geographic range and is highly distinct
from all neighboring races of saddlebacks.
Similar species. See saddleback tamarin,
S. fuscicollis.
Natural history. Diurnal; arboreal; groups
of 4–10. Eats fruit and probably insects.
Found in lowland evergreen rainforest that is
a mixture of high points surrounded by seasonally flooded flats. May occur in different
habitat types than saddleback tamarins.
Geographic range. South America: east
of the Andes in Ecuador and NE Peru; from
Napo/Pastaza, Ecuador, and Loreto, Peru,
apparently in the wedge between the Ríos
Napo, Curaray, and perhaps Nanay. Range
poorly known.
Status. CITES Appendix II. Can be the
most common primate where it occurs, but
known from only a few records in a small
region.
Local names. Leoncito, chichico (Ec);
pichico (Pe).
References. Thorington, R. W., Jr. 1988.
Taxonomic status of *Saguinus tripartitus*
(Milne-Edwards, 1878). *Am. J. Primatol.*
15:367–71.
    Albuja, L. 1994. Nuevos registros de
Saguinus tripartitus en la Amazonia Ecuato-
riana. *Neotropical Primates* 2(2):8–10.

Golden-handed or Midas Tamarin
Saguinus midas
Plate 9, map 95
Identification. Measurements: HB =
206–282; T = 316–440; HF = 57–80;
E = 28–40; WT = 415–665 g.

Face, head, and forequarters entirely
black; back black variegated with yellowish or buff. Hands and feet bright golden
yellow or orange, or black. Tail and underparts black. Young may have pale areas
around eyes and mouth.
Variation. Animals from north of the Rio
Amazonas have golden paws and yellowish
variegation on back, and usually weigh
500–600 g; those from south of the Amazonas have black paws and backs frosted
with buff *(S. m. niger),* larger ears, and may
be smaller (usually < 500 g).
Similar species. Yellow-handed titis *(Callicebus torquatus)* are larger, with a pale
collar on throat; bare-face tamarins *(Saguinus bicolor)* are whitish or yellowish with
orange on the tail; silvery marmosets *(Callithrix argentata)* have whitish forequarters.
This is the only callitrichid in most of its geographic range. There is slight overlap with
C. argentata east of the Xingu.
Sounds. Birdlike whistles; mild alarm a
chorus of soft cricketlike, froglike, or birdlike chirps.
Natural history. Diurnal; arboreal; groups
of 2–6. Feeds on fruit and insects. These
tamarins are generally seen in smaller groups
than other callitrichids. Unlike other species,
they tend to frequent open, high forest formations, where they forage at heights of
5–25 m, and they often travel in the canopy.
They also favor dense, viny habitats such as
swamps and streamsides. Found in mature
and secondary terra firme rainforest.
Geographic range. South America: Brazil
and the Guianas, the eastern Amazon Basin
east of the Rio Negro and the lower Rio
Xingu.
Status. CITES Appendix II. Widespread
and common.
Local names. Sauim, sagüi (Br); sapajou
noire (FG); sagoewijntje, sagoewenki (Su).
References. Thorington, R. W. 1968. Ob-
servations of the tamarin *Saguinas midas.*
Folia Primatol. 9:95–98.
    Mittermeier, R. A., R. C. Bailey, and
A. F. Coimbra-Filho. 1977. Conservation
status of the Callitrichidae in Brazilian
Amazonia, Surinam, and French Guiana.
In D. Kleiman ed., *The biology and con-
servation of the Callitrichidae,* 137–46.
Washington, D.C.: Smithsonian Institution
Press.

Map 96

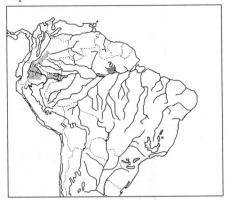

Black-mantle tamarin, *Saguinus nigricollis*
Brazilian bare-face tamarin, *S. bicolor*

Black-mantle Tamarin
Saguinus nigricollis
Plate 9, map 96
Identification. Measurements: HB =
210–251; T = 308–361; HF = 63–72;
E = 26–97; WT = 419–505 g.
**Forequarters black, lower back and hind-
limbs dark red; or forequarters blackish
grizzled olivaceous and hindquarters
blackish yellow-olivaceous (agouti). Crown
black, sides of head dark to pale brown;
face thinly haired, black; muzzle grizzled
white,** with short hairs; ears black, inconspicuous. Hands, feet, and tail black. Underparts
dark brown or black, including inguinal region. From a distance looks entirely grayish
or black.
Variation. Animals from west of Iquitos are
grizzled olivaceous; those from east of Iquitos (north of the Amazonas/Solimões) are
black with dark red hindquarters.
Similar species. Where they overlap, saddleback tamarins *(S. fuscicollis)* have dark
red or orange forequarters and hindquarters
and a grizzled gray saddle on the middle of
the back; they are difficult to distinguish in
poor light in the field. Pygmy marmosets
(Cebuella pygmaea) are much smaller and
pale colored; Goeldi's monkeys *(Callimico
goeldii)* are entirely black, with black muzzle
and a ruff of long hair.
Sounds. Whines and birdlike chirps; long
call of 3–10 (usually 4–6) long whistles
descending on final pulse.

Natural history. Diurnal; arboreal; groups
of 4–8, which may join together to form
large, temporary troops of up to 40. Feeds
largely on insects, especially large orthopterans, and on fruit, nectar, and plant gum.
Black-mantle tamarins favor dense viny
vegetation, where they forage at all levels.
They are the only callitrichids known to
sometimes travel in large groups. The larger
groups are noisy and resemble squirrel
monkey troops as they jump and rummage
through the forest searching for insects in
a broad band from near the ground to the
canopy. Found in mature, disturbed, and
dense secondary rainforest.
Geographic range. South America: east
of the Andes in Ecuador, Colombia, and
N Peru; between the Ríos Putumayo, Napo,
and the headwaters of the Putumayo. Southeastern limits of range unknown. Apparently
absent from the wedge between the Ríos
Napo and Curaray. To 1,000 m elevation.
Status. CITES Appendix II. IUCN Red
List vulnerable, *C. n. hernandezi* only. Little
hunted and locally common.
Local names. Sauim (Br); bebeleche (Co);
leoncito, chichico (Ec); pichico (Pe).
References. Izawa, K. 1978. A field study
of the ecology and behavior of the black-
mantle tamarin *(Saguinus nigricollis)*. *Primates* 19:241–74.

Red-chested Mustached Tamarin
Saguinus labiatus
Plate 9, map 97
Identification. Measurements: HB =
234–300; T = 345–410; HF = 69–78;
WT = 500–650 g. **Back and hindlegs
black, heavily frosted or variegated with
silvery gray. Head and face black with
white nose and lips in a prominent upside-
down V; ear with pale stripe of white, bare
skin in front;** crown sometimes with a narrow white or copper stripe down center; **nape
with white triangular patch.** Tail, hands,
and feet black. Chin and sometimes throat
black, rest of **underparts a sharply contrasting, bright, pure rust-red.**
Variation. Populations just south of the Rio
Solimões have a rusty red or orange patch
on crown of head; animals from north of the
river have the upper part of the chest black.
Similar species. See black-chested mustached tamarin.

Map 97

▨ Black-chested mustached tamarin, *Saguinus mystax*
▥ Red-chested mustached tamarin, *S. labiatus*
▤ Emperor tamarin, *S. imperator*

Sounds. Birdlike whistles and chirps. Long call of short, up-and-down-modulated whistles gradually changing to longer, rising whistles.
Natural history. Diurnal; arboreal; groups of 2–13. Feeds on fruits and insects. Red-chested tamarins chiefly use the middle layers of the forest, where they forage for small insects on twigs and leaves. Groups include one breeding female, one to five adult males, one to three nonbreeding females, and young. They often travel in mixed groups with saddleback tamarins. Found in mature and secondary evergreen and deciduous forest.
Geographic range. South America: east of the Andes in W Brazil, NE Bolivia, lowlands between Rios Purus and Madeira, and possibly north to the Rio Solimões. Range limits poorly known; the distribution of the three mustached tamarins in the river headwater regions is complex, but they probably are nowhere sympatric.
Status. CITES Appendix II. Locally common.
Local names. Leoncito, chichilo (Bo); sauim (Br); pichico (Pe).
References. Yoneda, M. 1984. Comparative studies on vertical separation, foraging behavior and traveling mode of saddlebacked tamarins *(Saguinus fuscicollis)* and red-chested moustached tamarins *(Saguinus labiatus)* in northern Bolivia. *Primates* 25:414–22.

Pook, A. G., and G. Pook. 1982. Poly-specific association between *Saguinus fuscicollis, Saguinus labiatus, Callimico goeldii* and other primates in north-western Bolivia. *Folia Primatol.* 38:196–216.

Black-chested Mustached Tamarin
Saguinus mystax
Plate 9, map 97
Identification. Measurements: HB = 235–280; T = 365–435; HF = 66–79; E = 26–31; WT = 536–700 g.
Forequarters black or dark blackish brown; back and hindquarters blackish brown frosted or grizzled with gray or dark buff. **Head black with nose and region around mouth white, short** tufts of longer **white** hair forming **mustaches at sides of mouth** that part away from chin; spot of white, bare skin in front of ear; **crown black or bright, dark rust-red.** Hands, feet, and tail black. **Underparts dark brown,** in some populations with white genital area.
Variation. *S. m. pileatus,* with red crown and gray-frosted back, is from between the lower Rios Juruá and Purus.
Similar species. Red-chested mustached tamarins *(S. labiatus)* have red underparts; emperor tamarins *(S. imperator)* have long, down-curling white mustaches and orange tail; saddleback tamarins *(S. fuscicollis)* are smaller and have reddish hindquarters, a saddle of mottled hair, less prominent whitish muzzle; Goeldi's monkeys *(Callimico goeldii)* are entirely black.
Sounds. Birdlike whistles and chirps. Long call of short, warbling whistles gradually increasing to long, descending whistles with a slight initial rise.
Natural history. Diurnal; arboreal; groups of 2–6. Feeds on fruit and insects. These tamarins sometimes travel in mixed groups with saddleback tamarins; they use a higher level of the forest than the saddlebacks. They tend to run along branches, rarely using the vertical cling-and-leap mode of locomotion that is typical of saddlebacks. Found chiefly in mature rainforest, occasionally in old secondary forest.
Geographic range. South America: east of the Andes in Peru and Brazil, south of the Solimões/Amazonas, west of the Rios Purus and upper Madeira, and south of the Japurá. Range limits poorly known; a report from

Pando, Bolivia, requires verification with a specimen.

Status. CITES Appendix II. Locally common; seems less adaptable to disturbed habitat than other tamarins and, because of its larger size, is sometimes hunted for meat.

Local names. Sauim (Br); pichico (Pe).

References. Castro, R., and P. Soini. 1977. Field studies on *Saguinus mystax* and other callitrichids in Amazonian Peru. In D. Kleiman, ed., *The biology and conservation of the Callitrichidae,* 73–78. Washington, D.C.: Smithsonian Institution Press.

Ramirez, M. 1984. Ecology and conservation of the moustached tamarin in Peru. *Primate Conservation* 4:22–23.

Emperor Tamarin
Saguinus imperator
Plate 10, map 97

Identification. Measurements: HB = 230–258; T = 350–415; HF = 68–72; E = 28; WT ≈ 400 g.

Forequarters, back, and hindquarters entirely grizzled pale gray washed with buff on sides. **Head and ears blackish, face dominated by long, down-curling white mustaches extending far below chin;** brow with a whitish band above eyes, eyes brown. **Tail bright rusty orange,** rust extending onto rump around tail base, tip of tail blackish; **or tail black above, orange below, with tip entirely black.** Hands and feet blackish. Throat and chest grizzled gray or brown, belly and insides of limbs bright orange or brown. Young have short mustaches.

Variation. Animals from the upper Rio Juruá have gray underparts; those from Peru and the upper Rio Juruá have a thin beard of long white hairs as well as mustaches.

Similar species. No other Neotropical mammals have long white mustaches. See saddleback tamarin *(S. fuscicollis).*

Sounds. Birdlike chips, quavering whistles, and whistles in a long, descending series.

Natural history. Diurnal; arboreal; groups of 2–10. Feeds on fruit, insects, and flower nectar in the dry season. Emperor tamarins mainly forage for insects on leafy foliage and twigs, where they catch active prey with catlike stealth. They often form a mixed troop with a group of saddleback tamarins, with which they may defend a common territory. Emperor tamarins are shy and secretive and difficult to observe; they travel quietly. They prefer dense viny vegetation, where they spend several hours a day resting inconspicuously. A quavering whistle or chirp sometimes reveals their location in a thicket. At night they sleep in large, vine-draped trees, especially ones isolated from the surrounding forest. They are rarer in their habitat than are saddleback tamarins. Found in mature and disturbed lowland rainforest, especially riversides and viny habitats.

Geographic range. South America: east of the Andes in Brazil, Peru, and perhaps Bolivia; SW Acre, Brazil, Pando, Bolivia, and north of the Río Madre de Dios, Peru. Range poorly known.

Status. CITES Appendix II. IUCN Red List vulnerable, *S. i. imperator* only. Generally uncommon and apparently patchy.

Local names. Pichico, chico chico (Pe); mono Nicolás Suárez, chichilo (Bo); bigodeiro, sagüi-imperador (Br).

References. Terborgh, J. 1983. *Five New World primates.* Princeton, N.J.: Princeton University Press.

Brazilian Bare-face Tamarin
Saguinus bicolor
Plate 10, map 96

Identification. Measurements: HB = 208–283; T = 335–420; HF = 63–83; E = 25–31. There are three distinctive subspecies. General features are: **face and head to level of ears naked, tail bicolored blackish or brown above, orange below; large, naked ears.**

Pied bare-face tamarin, *S. b. bicolor.* **Forequarters white; back and hindquarters pale brown;** naked skin of face and head entirely black, or mottled with white patches; tail black above, rust-gold below; throat and chest white; belly brown mixed with buff; inner thighs and lower abdomen bright rust-red.

Martin's bare-face tamarin, *S. b. martinsi.* **Sides and hindlegs grizzled cinnamon; forearms orange;** naked skin of face black; crown to base of tail brown, in a broad band; tail black above, orange below; underparts orange.

Ochraceous bare-face tamarin, *S. b. ochraceus.* **Upperparts entirely pale brown** with buff tinge, paler on forequarters;

legs like back or more orange; naked skin of face and head blackish; tail dark brown above, gold below; underparts entirely orange to golden brown.

Similar species. Golden-handed tamarins *(S. midas)* are blackish with gold paws.

Sounds. Birdlike chirps and whistles. Long call of long whistles of constant frequency and length, each pulse about one second long.

Natural history. Diurnal; arboreal; small family groups of 2–9. Feeds on fruit, insects, and exudates. Pied bare-face tamarins favor dense viny vegetation of secondary forests and forest edges, where they feed on small, juicy fruits of small-crowned undercanopy trees. In the dry season they eat gum from trees of Vochysiaceae and other sources, but they do not gouge holes in bark. They forage for arthropods mostly on the trunks of trees. Small populations live in isolated secondary forest plots in the city of Manaus, including the grounds of the Instituto Nacional de Pesquisas da Amazônia (INPA). In these places it is the only tamarin present, but it is said to occur together with the golden-handed tamarin in a forest outside the city. Found in disturbed and secondary lowland rainforest.

Geographic range. South America: Brazil, north of the Rio Amazonas in a narrow, discontinuous strip near the river from the Rio Negro east to the lower Cuminá. Pied bare-faced tamarins are restricted to a 30 km radius of Manaus; the other two subspecies probably range between the lower Rios Uatumã and Cuminá (Erepecurú). Range limits not precisely known.

Status. CITES Appendix I, US-ESA endangered, IUCN Red List endangered, *S. b. bicolor* only. Poorly known, perhaps at risk. This species has a tiny geographic range and lives in a densely inhabited region. Although it thrives in secondary habitats, populations may already be highly fragmented, and several groups in Manaus have recently been extirpated by development.

Local names. Sauim, sagüi-de-cara-nua, sagï-de-duascores.

References. Egler, S. G. 1992. Feeding ecology of *Saguinus bicolor bicolor* (Callitrichidae, Primates) in a relict forest in Manaus, Brazilian Amazon. *Folia Primatol.* 59:61–76.

Silvery-brown Bare-face Tamarin
Saguinus leucopus
Plate 10, map 98

Identification. Measurements: HB = 223–263; T = 347–417; HF = 68–80; E = 24–30.

Upperparts pale silvery brown, or yellowish white variegated with brown. Face almost naked, thinly haired with white; **brow and crown to ears whitish; neck and top of head from behind ears with ruff of sharply contrasting darker brown;** ears thinly haired, dark. **Tail brown,** distinctly darker than body, frosted silver toward tip, **extreme tip usually white.** Legs and feet whitish or silvery; ankle and sometimes foot with red-brown spot(s). **Underparts dark to bright rusty orange.**

Similar species. Cotton-top tamarins *(S. oedipus)* have long plumes of white hair on head, sharply bicolored body with blackish back and white limbs, and pale underparts.

Sounds. Birdlike warbles, chirps, and trill; a shrill "tee-tee."

Natural history. Diurnal; arboreal; groups of up to 15. Probably feeds on fruits, particularly undergrowth berries, and insects. This species uses all heights of the forest and favors edge habitats such as streamsides. It thrives in second-growth vegetation, where it is the only remaining primate in much of its geographic range. Habits reported to be similar to those of cotton-top tamarins. Found in disturbed lowland and lower montane rainforest.

Geographic range. South America: Colombia east of the Andes, between the eastern bank of the lower Río Cauca and the western bank of the middle Río Magdalena, and foothills of the central Andes. One of the smallest geographic ranges among tamarins.

Status. CITES Appendix I, US-ESA threatened, IUCN Red List vulnerable. Most of the geographic range has been deforested.

Local names. Titi, titi gris.

References. Hernández-Camacho, J., and R. W. Cooper. 1976. The non-human primates of Colombia. In R. W. Thorington and P. G. Heltne, eds., *Neotropical primates,* 35–69. Washington, D.C.: National Academy of Sciences.

Green, K. M. 1978. Neotropical primate censusing in northern Colombia. *Primates* 19:537–50.

Map 98

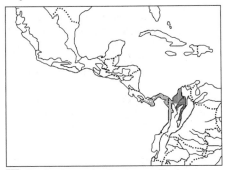

▨ Geoffroy's tamarin, *Saguinus geoffroyi*
▨ Cotton-top tamarin, *S. oedipus*
▨ Silvery-brown bare-face tamarin, *S. leucopus*

Mottled-face Tamarin
Saguinus inustus
Plate 10, map 95
Identification. Measurements: HB =
208–259; T = 330–410; HF = 70–75.
**Entirely blackish, except hindquarters
dark red. Face almost naked, with pure
white patch of skin below nose; cheeks
and area around eyes naked, white, mottled with variable amounts of black;** ears
naked, black, sometimes with white patches.
External genitalia naked, white.
Similar species. Golden-handed tamarins
(S. midas) have contrasting yellow hands
and feet; saddleback tamarins *(S. fuscicollis)*
have a saddle of tawny variegated fur in midback; Goeldi's monkeys *(Callimico goeldii)*
have a completely black face.
Natural history. Undescribed.
Geographic range. South America: SE
Colombia and NW Brazil, from a small,
remote region between the Ríos Guaviare,
Japurá, and upper Rio Negro, range proba-
bly incompletely known.
Status. CITES Appendix II. Unknown.
Local names. Sauim, sagüi (Br); mico dia-
blo, diablito, tití diablito (Co).

Cotton-top Tamarin
Saguinus oedipus
Plate 10, map 98
Identification. Measurements: HB =
205–255; T = 307–411; HF = 66–72;
E = 20–24; WT = 350–510 g; males
slightly larger than females.

Back brown, sometimes frosted silvery or
yellowish; hair of back overhangs sides in
a long fringe. **Head dominated by crest
of long white hairs** growing in a narrow
peak between areas of naked skin on crown,
widening backward to cover entire top
and sides of neck; **face,** chin, and sides of
crown to behind ear almost naked, black;
**with thin white hairs especially in a
fringe over eyebrows and from corner
of eye to jowl. Tail dark chestnut-red for
first third,** chestnut extending onto area
around tail base and hind thighs, distal
two-thirds black. **Limbs and feet cream
to yellowish white. Underparts cream** to
yellowish white. Young have short white
hairs on "crest."
Similar species. Silvery-brown tamarins *(S.
leucopus)* are uniform colored, with a short
brown crest; Geoffroy's tamarins *(S. geoffroyi)* have a dark chestnut neck and white
triangle of short hairs on crown (no long
white plumes).
Sounds. Birdlike chirps; long call of slightly
ascending, then descending, whistles of over
a second in length.
Natural history. Diurnal; arboreal; groups
of 2–13. Feeds on small fruits, insects, and
occasional buds and leaves. Cotton-top tamarins defend small home ranges or territories
of as little as 8 ha. They arise an hour after
dawn and, like other tamarins, retire well
before dark. They travel 1–2 km a day, foraging for insects, feeding on fruit, and seeking favorite sleeping spots, which are high in
the forks of large trees, dense vine tangles,
or tufts of leafy vegetation. Found in evergreen and deciduous mature and secondary
forests.
Geographic range. South America: N
Colombia east of the Río Atrato and west
of the Ríos Cauca and lower Magdalena.
To 400 m elevation.
Status. CITES Appendix I, US-ESA endangered, IUCN Red List endangered. More
than three-fourths of the original habitat of
the cotton-top has been deforested, much of
it for cattle pasture. The species is now restricted to a few isolated forest fragments.
Local names. Titi, titi blanco, titi leoncito,
titi pielroja.
References. Neyman, P. F. 1977. Aspects
of the ecology and social organization of
free-ranging cotton-top tamarins *(Saguinus*

oedipus) and the conservation status of the species. In D. Kleiman, ed. *Biology of the Callitrichidae,* 39–71. Washington, D.C.: Smithsonian Institution Press.

Geoffroy's Tamarin
Saguinus geoffroyi
Plate 10, map 98
Identification. Measurements: HB = 200–287; T = 315–423; HF = 64–82; E = 20–31; WT = 453–520 g.
Back and upper thighs black variegated (almost spotted) with short stripes of pale dirty yellow; forelegs, chest, and feet yellow-white; lower hindlegs and feet yellowish gray; hair of back overhangs belly in long fringe. **Face** and sides of head to ear **naked, black, thinly haired with white,** especially prominent on eyebrow; stripe from corner of eye to below ear; **crown topped with narrow triangle of short, white, flat-lying hair; neck** to shoulder **dark, chestnut-red. Tail dark chestnut-red at base, distal half black. Underparts pale yellow** with buff tinge.
Variation. Considered by some to be a subspecies of cotton-top tamarin.
Similar species. Cotton-top tamarins *(S. oedipus)* have long white plumes on head; Central American squirrel monkeys *(Saimiri oerstedii)* are yellowish with black crown and muzzle, pale mask around eyes.
Sounds. Birdlike chirps, twitters; long call of ascending then descending whistles repeated two to four times, each half a second to over a second in length.
Natural history. Diurnal; arboreal; groups of up to 14. Feeds on fruit, insects (chiefly orthopterans), plant exudates, and small amounts of leaves and shoots. Geoffroy's tamarins are most common in areas of dense vegetation, such as river edges and disturbed or secondary forest, where they forage for insects on thin vines and branches in the low shrub layer of the understory. They often hang by their hindfeet to reach a prey item. At night they sleep in large emergent trees. Found in a wide range of habitats from pluvial rainforest to seasonal dry forest.
Geographic range. Central and South America: S Costa Rica, Panama, and Colombia west of the Andes to the Río San Juan.
Status. CITES Appendix I. Threatened by deforestation.

Local names. Titi (Co, Pn); bichichi (Co); marmoseta, tamarín (CR).
References. Garber, P. A. 1984. Proposed nutritional importance of plant exudates in the diet of the Panamanian tamarin, *Saguinus oedipus geoffroyi. Int. J. Primatol.* 5:1–15.
———. 1984. Use of habitat and positional behavior in a Neotropical primate, *Saguinus oedipus.* In J. G. H. Cant and P. S. Rodman, eds., *Adaptations for foraging in non-human primates.* New York: Columbia University Press.

Golden Lion Tamarin
Leontopithecus rosalia
Plate 8, map 99
Identification. Measurements: HB = 225–280; T = 300–400; HF = 67–90; E = 24–26; WT = 410–650 g. Males larger than females.
Upperparts entirely gold, hair long, shiny and silky; head with a mane of long hair parted in the middle on the top and extending completely around the head under the chin; face naked, pinkish gray; tail sometimes faintly barred with darker gold, or white-tipped. Hands and fingers exceptionally long and narrow. Feet and tail with occasional black patches.
Similar species. Most tufted-ear marmosets *(Callithrix jacchus)* are gray, with banded tails, striations on back, and usually white markings on head.
Sounds. A large variety of birdlike calls: trills, whines, whistles, and clucks. Long call a long series of ascending then descending modulated whistlelike pulses, gradually increasing in length from about nine to two per second.
Natural history. Diurnal; arboreal; groups of 3–11. Feeds on fruit and insects. The long, narrow hands of lion tamarins are used for reaching into tree hollows and crevices to probe for hidden insects. Lion tamarins are usually active at heights of 5–10 m; they favor areas of dense primary vegetation, especially those with many bromeliads. At night the whole group sleeps together in a tree hollow, generally one with a small entrance. Found in humid to seasonal Atlantic forests, where they seem best adapted to mature or primary vegetation. Lion tamarins do not show the adaptability of tufted-ear marmosets to disturbed or man-made habitats and may always have been uncommon or rare.

Map 99

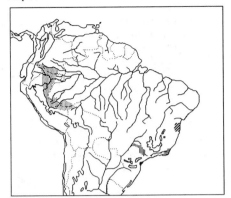

▨ Goeldi's monkey, *Callimico goeldii*
▤ Golden lion tamarin, *Leontopithecus rosalia*
▥ Golden-rumped lion tamarin, *L. chrysopygus*
▨ Golden-headed lion tamarin, *L. chrysomelas*
▣ Black-faced lion tamarin, *L. chrysopygus caissara*

Geographic range. South America: SE Brazil, remnant forest fragments in the state of Rio de Janeiro.
Status. CITES Appendix I, US-ESA endangered, IUCN Red List critically endangered. Over 90–95% of the original habitat of this species has been deforested. Fewer than 1,000 individuals remain in the wild, with remaining populations mostly in secondary or degraded habitats. The species adapts poorly to disturbance. Efforts are being made to preserve remaining populations, introduce captive-bred groups to the wild, and restore forest habitat.
Local names. Sauim vermelho, mico-leaõ-dourado.
References. Coimbra-Filho, A. 1977. Natural shelters of *Leontopithecus rosalia* and some ecological implications (Callitrichidae: Primates). In D. Kleiman, ed., *Biology and conservation of the Callitrichidae,* 79–89. Washington, D.C.: Smithsonian Institution Press.
   Neotropical Primates, 2: Supplement, December 1994 (whole issue on *Leontopithecus* spp.).

Golden-rumped Lion Tamarin
Leontopithecus chrysopygus
Map 99
Identification. Measurements: HB = 247–302; T = 358–406; HF = 74–83; WT = 540–690. Males larger than females.

Similar to golden lion tamarin except in color: **entirely glossy black except rump, upper thighs, and tail base usually golden or reddish,** the size of the golden rump patch is variable, rump **may be entirely black;** outer surfaces of legs reddish or yellowish brown.
Variation. Color pattern highly variable. The newly described form *L. caissara* apparently cannot be distinguished by color pattern from some individuals of *L. chrysopygus,* its nearest geographic neighbor in the genus, and we therefore concur with authorities who consider them subspecies.
Similar species. Marmosets have gray bodies with banded backs and tails.
Sounds. Similar to golden lion tamarin.
Natural history. Diurnal; arboreal; groups of 2–11. Feeds on fruit, insects, nectar, and exudates. Golden-rumped lion tamarins spend most of their time in the middle levels of the forest at 7–15 m. They have large territories that can exceed 100 ha. At night the whole group sleeps together in a tree hole. They occupy semi-deciduous inland forests to humid coastal plain forests and littoral vegetation types.
Geographic range. South America: Brazil, two zones of forest fragments in the state of São Paulo and coastal Paraná. Up to 700 m elevation.
Status. CITES Appendix I, US-ESA endangered, IUCN Red List critically endangered. Probably fewer than 700 individuals of both subspecies remain.
Local names. Sauim preto; sagüi; mico-leão-preto.
References. Rosenberger, A. L., and A. F. Coimbra-Filho. 1984. Morphology, taxonomic status and affinities of the lion tamarins, *Leontopithecus* (Callitrichinae, Cebidae). *Folia Primatol.* 42: 149–79.
   Neotropical Primates, 2: Supplement, December 1994.

Golden-headed Lion Tamarin
Leontopithecus chrysomelas
Plate 8, map 99
Identification. Measurements: HB = 220–260; T = 330–385; HF = 70–82; E = 22–25; WT = 480–700. Males larger than females.
Similar to golden lion tamarin except in color: **crown and fringe around face, forelimbs, and often hindfeet golden, rest of**

body black, except tail may be gold at base, above or below, **tail usually with a narrow red stripe down half its length;** or all black; hindquarters may be reddish.
Variation. This is the largest and most morphologically differentiated of the lion tamarins, and may be the only one that deserves species distinction from the group of other forms.
Similar species. Marmosets have grizzled gray banded backs and tails, and black or white ear tufts.
Sounds. Similar to golden lion tamarin, but pulses of long call truncated, final pulses not drawn out.
Natural history. Diurnal; arboreal; groups of 2–8. Feeds on fruit, insects, nectar, and gum. These tamarins use large home ranges of about 75 ha. They forage for insects in the middle levels of the forest (12–20 m) by probing with their fingers for prey hidden in leaf debris caught in palm crowns, under loose bark, in tree holes and crevices, and in bromeliads. At night they sleep in tree holes. Found in forests that have many bromeliads and bamboo, in both tall, humid evergreen coastal forests and inland semi-deciduous forests. These tamarins seem to require mature forest, although they can use nearby secondary habitats.
Geographic range. South America: Brazil, forest fragments in the state of Bahia, mostly in coastal regions.
Status. CITES Appendix I, US-ESA endangered, IUCN Red List endangered.
Local names. Sauim-una; mico-leão; sagüi.
References. Rylands, A. B. 1993. The ecology of the lion tamarins, *Leontopithecus:* Some intrageneric differences and comparisons with other callitrichids. In A. B. Rylands, ed., *Marmosets and tamarins: Systematics, behaviour, and ecology,* 296–313. Oxford: Oxford University Press.

Goeldi's Monkey
Callimico goeldii
Plate 8, map 99
Identification. Measurements: HB = 210–234; T = 255–324; HF = 68–78; E = 20–26; WT ≈ 480–505 g.
Upperparts, head, legs, and tail entirely glossy black or with slight grizzling of reddish or yellowish on rump, thighs, and tail base; **fur** long and silky, feathery on tail, **looks disheveled.** Face thinly haired, almost

naked, black; **head with long hair in two tiers: a velvety crown on top and sides, with long side-whiskers framing face, and a longer ruff behind crown, from top of head to below ear; ear** naked, **hidden in ruff.** Tail not prehensile, but can be coiled ventrally. **Belly, chest, and insides of limbs thinly haired, almost naked, with pure white skin** visible in extended postures. Young may have grizzled rump and thighs.
Similar species. All tamarins that occur in the same geographic region have white or gray mustaches or muzzles and ears that protrude through the head hair.
Sounds. Alarm call a loud chuck; long call a trill-like fast series of descending pulses of constant frequency, each pulse decreasing in frequency and increasing in length, each trill about 2–3 seconds and including ten pulses per second.
Natural history. Diurnal; arboreal and terrestrial; groups of up to 8. Feeds on invertebrates and fruit. Goeldi's monkeys usually travel close to the ground, below 5 m, with frequent "vertical clinging and leaping" between upright undergrowth trunks. When alarmed, they keep close to the ground while they flee. Goeldi's monkeys sometimes travel in association with tamarins, which usually stay above 5 m. They seem to be habitat specialists, inhabiting scrubby upland (terra firme) forests away from river floodplains. These forests have a broken canopy, dense undergrowth, and much bamboo.
Geographic range. South America: east of the Andes in Colombia, Ecuador, Peru, Bolivia, and Brazil (Acre); in a narrow strip within 400–500 km of the base of the Andes. To 500 m elevation. Range poorly known from few confirmed localities.
Status. CITES Appendix I, US-ESA endangered, IUCN Red List vulnerable. These monkeys seem everywhere rare, with thinly scattered, patchy populations, and they are often unknown to local people. Apparent rarity puts this species at potential risk, but its true status and even its exact geographic range are unknown.
Local names. Pichico (Pe).
References. Heltne, P. G., J. F. Wojcik, and A. G. Pook. 1981. Goeldi's monkey, genus *Callimico.* In A. F. Coimbra-Filho and R. A. Mittermeier, eds., *Ecology and Behavior of Neotropical Primates,* 169–209. Rio de Janeiro: Academia Brasileira de Ciências.

Pook, A. G., and G. Pook. 1981. A field study of the socio-ecology of the Goeldi's monkey *(Callimico goeldii)* in northern Bolivia. *Folia Primatol.* 35:288–312.

Monkeys (Cebidae)

Dental formula: I2/2, C1/1, P3/3, M3/3 = 36. The cebid monkeys range in size from 0.8 to 15 kg. They have five toes on all feet, thumb vestigial in two genera; nails on all digits. The heaviest species all have prehensile tails, the lightest species do not. Monkeys with prehensile tails wrap the tip tightly around branches, hang supported by it, and often carry the tail with the tip in a tight downward coil when they walk along the top of a branch; only prehensile-tailed cebids carry their tail tips coiled (but see marmosets and tamarins). Cebids give birth to one young, which is carried by the mother in large species and small nonmonagamous species, and often by the father in small monogamous species. All species are arboreal (unlike Old World monkeys, which include many terrestrial species). All species eat fruit; some also eat many insects, and others eat more leaves than fruit. There is much geographic variation in color between populations of monkeys in different areas separated by large rivers or mountains, and it is still unclear how many species are represented in several groups. There are 11 genera and a minimum of 31 species. This book generally follows the taxonomy presented in A. F. Coimbra-Filho and R. A. Mittermeier, *Ecology and behavior of Neotropical primates,* vol. 1 (Rio de Janeiro: Academia Brasileira de Ciências, 1988). Members of a genus, such as howlers or sakis, are similar in diet, behavior, and body shape.

Small monkeys. There are three genera of small monkeys that all weigh about 1 kg: night monkeys, titis, and squirrel monkeys. They are not closely related and have completely different ecologies. They are about twice as heavy as the marmosets and tamarins and do not travel by vertical clinging and leaping. They do not have prehensile tails.

Medium-sized monkeys. There are two groups of monkeys that weigh 2–4 kg. The sakis, bearded sakis, and uakaris (Pithecinae) are a closely related group of specialized monkeys. They have bushy, nonprehensile tails that they sometimes wag below a branch, long hair (shaggy in the sakis and uakaris), and large, divergent upper and lower canine teeth that bulge under the lips and pull the mouth into a permanent frown. They use these teeth to split open the husks of unripe fruits, on whose seeds they feed. This group is restricted to the Amazon Basin and the Guianas.

The capuchin monkeys are brown or yellowish with dark caps (one has a black back), and they have prehensile tails that are more weakly developed than those of the larger monkeys: the tip of the tail is hairy below, and the monkeys rarely swing supported only by the tail. They feed on fruit, invertebrates, and small vertebrates.

Large monkeys. The spider, woolly, muriqui, and howler monkeys (Atelinae) all weigh 4 kg or more. They have strongly prehensile tails with a naked gripping surface on the underside of the tip. They feed on fruit and leaves. Spider monkeys and muriquis have long, thin arms and legs and often swing by their arms or tail below branches; woolly monkeys sometimes swing by their arms, but howler monkeys rarely do so. Because of their large size (which gives the highest return per round of ammunition), these are the monkeys most intensively hunted for meat, and they are rare or extirpated in many inhabited areas.

Monkeys are the most important seed dispersers for hundreds of plant species, especially canopy trees and lianas, and they are therefore key elements of rainforest ecosystems. If they are eliminated from the forests by hunting, the plant species composition of the forests will eventually change.

Night or Owl Monkeys
Aotus spp.
Plate 11, map 100
Identification. Measurements: HB = 240–475; T = 220–418; HF = 89–96; E = 27–35; WT = 780–1,249 g.

Upperparts grizzled gray to brown; fur soft, dense, and woolly. Head round, with short fur, **face** slaty brown **rimmed with white, brow patterned with three black stripes: a large central black spot between eyes, white patches above eyes, and two**

Map 100

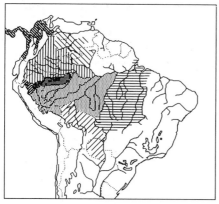

||||| Lemurine night monkey, *Aotus lemurinus*

Night monkey, *A. trivirgatus*

Noisy night monkey, *A. vociferans*

Ma's night monkey, *A. nancymaae*

Black-headed night monkey, *A. nigriceps*

Feline night monkey, *A. infulatus*

Azara's night monkey, *A. azarae*

Andean night monkey, *A. miconax*

narrow black stripes from corner of eyes converging backward and sometimes meeting on the rear of the crown; **eyes large,** round, brown, **set close together; eyeshine bright, reddish orange;** mouth with a slight, enigmatic smile. **Tail nonprehensile,** never coiled, **brown or dark rusty orange frosted with black for first half,** becoming **black at** slightly tufted **tip.** Fingers long and narrow, with expanded pads at tips. Underparts pale yellow to bright orange; these colors either extending onto insides of limbs and prominently onto sides of neck and throat, or not.
Species. Some authorities recognize as many as ten geographically separated species, but a recent work suggests no more than five to seven, as listed below. These fall into two groups, outlined as follows: dull-colored animals with gray sides of neck, inner limbs, and throat and yellowish underparts, from Central America and the Amazon Basin north of the Marañón-Amazonas, include *A. trivirgatus* and *A. vociferans;* more brightly colored animals with orange neck and throat, and often orange inner limbs and tail base, are from south of the Amazon River and include *A. miconax, A. nigriceps, A. nancymaae, A. azarae,* and *A. infulatus.* The forms within each group are often difficult to dis-

tinguish and do not seem to have clear sets of diagnostic characters (for example, some gray-necked populations of *A. nigriceps* occur south of the Amazon); in our opinion, some of the forms are subspecies of about three variable species, but the situation remains unclear. For reliable identification, a museum specimen is needed. *A. nancymaae* has been reported as sympatric with *A. vociferans* at sites near Leticia, Colombia.
Similar species. These are the only monkeys with huge eyes and nocturnal activity. Kinkajous *(Potos flavus)* and olingos *(Bassaricyon* spp.) have no facial markings and have wide-set eyes with yellow eyeshine.
Sounds. Alarm call soft metallic clicks often accompanied by low-pitched, resonating, tonal clunks; on moonlit nights (most often) a three-syllable, low-pitched, owllike hoot series given by a single animal.
Natural history. Nocturnal; arboreal; groups of 2–5. Feed on fruit, insects, and flower nectar. Usually night monkeys are seen in the upper half of the forest. They can be found throughout the forest, but they are most common in denser areas with many vines, such as the edges of rivers and open areas. These little monkeys travel quietly and are relatively sedentary. They may spend several hours in a single fruit tree, moving around from time to time or sitting quietly. The light of flashlights disturbs them, and their eyeshine is often visible for only a few seconds before they turn their heads away or flee. Their most evident feature at night is the long, black-tipped tail hanging straight down from a branch. They are most active and easy to find on moonlit nights, when their hooting betrays their location. By day they sleep in a vine-draped tree, tree hollow, or dense festoon of vegetation. They are monogamous, and the father often carries the young. In the temperate zone of Paraguay where nights are cold, owl monkeys are sometimes active in the daytime. These are one of the commonest monkeys near human settlements. They occupy a wide range of mature, disturbed, and secondary forest habitats from the wettest rainforests to dry scrub in the Chaco.
Geographic range. Central and South America: Panama to the Chaco of Argentina; except absent from the Guianas and Brazil north of the Rio Solimões and east of the Rio Negro. To elevations of 3,200 m.

Status. CITES Appendix II. IUCN Red List endangered, *A. lemurinus grisimembra;* vulnerable, *A. l. lemurinus, A. trivirgatus brumbacki, A. miconax.* Widespread and common. Little hunted except for the laboratory animal trade; adapts well to disturbed forests.

Local names. Mono de noche (Span); mono nocturno, cuatro ojos (Bo); macaco-da-noite (Br); marta, mico dormilón, mico de noche (Co); tutamono (Ec); musmuqui, buri-buri (Pe); jujaná, gná, marteja (Pn); mirikiná (Ar, Gua). Note that some of the same names are used for kinkajous, olingos, and woolly opossums.

References. Wright, P. C. 1978. Home range, activity pattern, and agonistic encounters of a group of night monkeys *(Aotus trivirgatus)* in Peru. *Folia Primatol.* 29:43–55.

Ford, S. M. 1994. Taxonomy and distribution of the owl monkey. In J. F. Baer, R. E. Weller, and I. Kakoma, eds., Aotus: *The owl monkey,* 1–57. San Diego: Academic Press.

Dusky Titi Monkey
Callicebus moloch
Plate 11, map 101
Identification. Measurements: HB = 287–360; T = 331–480; HF = 84–102; E = 26–36; WT = 860–1,400 g.
Upperparts dull brown, red-brown, or finely grizzled **gray;** fur long and dense. Head round, face small, flat, blackish or gray; brow black, white, or gray, often contrasting with crown; sides of head either contrasting bright yellow or orange, or white around ear, or same color as crown; ears inconspicuous. **Tail nonprehensile, thickly furred, gray, brown, or blackish, tail tip often pale,** dirty white or brown for up to half its length, **usually hangs straight downward** or droops over a branch. Underparts pale yellowish to bright or dark red-orange. **Head held low, below level of shoulders; usually sits with feet tucked forward under body close to hands**—a furry ball with a dark face and pendant tail. Thick fur hides the neck and makes the body, limbs, and tail appear chunky, and the almost naked face look small.
Variation. Dusky titis are highly variable in color between populations, and somewhat variable within them. Those from the dryer southern parts of the range in Bolivia, Paraguay, and Mato Grosso are pale gray; those

Map 101

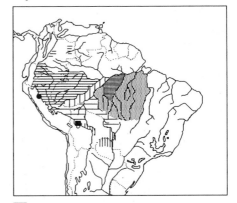

▦ Dusky titi monkey, *Callicebus moloch moloch*
▤ Dusky titi monkey, *C. m. cupreus*
▥ Dusky titi monkey, *C m. donacophilus*
▯ Dusky titi monkey, *C. m. caligatus*
▱ Dusky titi monkey, *C. m. brunneus*
▰ Dusky titi monkey, *C. m. hoffmanni*
◉ Dusky titi monkey, *C. m. modestus* and *olallae*
◼ Dusky titi monkey, *C. m. oenanthe*

from south of the Amazon in Pará and Amazonas have bright orange or yellow cheeks. This taxon has been divided into as many as 10 species, but as in night monkeys, the number of species and their characters and distributions are unclear. Recent molecular work shows little genetic differentiation between the forms, and we treat them as subspecies.
Similar species. Tamarins (*Saguinus* spp.) are small and slender and jump with vertical cling-and-leap locomotion; squirrel monkeys (*Saimiri* spp.) are slender, with short fur and slender tails; sakis (*Pithecia* spp.) are much larger and shaggier and have bushy tails; night monkeys (*Aotus* spp.) have slender tails, large eyes, and three black stripes on head.
Sounds. At dawn and occasionally at other times, pairs give extremely loud duets of prolonged, continuous, rapidly modulated whoops and high screaming. Duets of different subspecies seem indistinguishable. Alarm call a soft grunt or, rarely, tamarin-like chirps and twitters.
Natural history. Diurnal; arboreal; groups of 2–5. Feeds on leaves and fruits: the only largely folivorous small monkey. Dusky titis are most often found in areas of dense vegetation, especially on water edges and in swamp forest, where they use the middle

and lower levels. They like to feed on the leaves of bamboo and are often found in bamboo thickets. Like most arboreal folivores, they are quite inactive and spend many hours each day just sitting, like blobs of fur on a branch. When running along a branch, they have a characteristic bouncing gait, bounding with both forefeet, then both hindfeet moving together. They are monogamous, and the young is usually carried on the father's back. When disturbed, titis will often vanish into a dense tangle of vegetation, where they may remain for hours without betraying their presence by any sound or movement. Where hunted they are wary and stealthy and seldom seen, although they are usually common. Their presence in an area is always shown by their morning duets, often given from the middle of dense thickets. At night they sleep at medium heights (8– 18 m) on branches sheltered by vegetation. Occupies mature and disturbed lowland evergreen rainforest, dry forest, and gallery forest, mostly in dense, viny vegetation types with a low canopy, such as riversides and forest edges. *Geographic range.* South America: east of the Andes in a strip along their base from Colombia to Bolivia and Paraguay, and Brazil south of the Amazon only. To 850 m elevation.
Status. CITES Appendix II. Widespread and common, not usually hunted for meat.
Local names. Mono tití, faca-faca, lucashi, ururó (Bo); zogue-zogue, uapusá (Br); zocay (Co); zocayo, saui, songo songo (Ec); tocón, tití, tocari (Pe); sauhí (TGua).
References. Hershkovitz, P. 1988. Origin, speciation, and distribution of South American titi monkeys, genus *Callicebus* (family Cebidae, Platyrrhini). *Proc. Acad. Nat. Sci. Philadelphia* 140:240–72.

Wright, P. 1986. Ecological correlates of monogamy in *Aotus* and *Callicebus*. In J. G. Else and P. C. Lee, eds., *Primate Ecology and Conservation,* 159–67. Proceedings of the 10th Congress of the International Primate Society, vol. 2.

Yellow-handed Titi Monkey
Callicebus torquatus
Plate 11, map 102
Identification. Measurements: HB = 310–375; T = 420–493; HF = 91–105; E = 23–33; WT = 0.8–1.5 kg.

Map 102

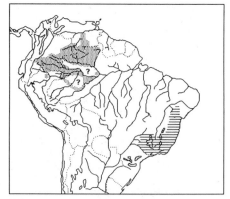

▨ Yellow-handed titi monkey, *Callicebus torquatus*
▤ Masked titi monkey, *C. personatus*

Upperparts red-brown, dark brown, or almost black, thighs blackish to rust-red; fur long and dense. Face thinly haired with whitish; **neck with broad white to buff collar from ear to ear under throat; chin white. Hands bright yellow-white, sharply demarcated from forearms, or black. Tail dark brown or black.** Underparts red-brown to sooty brown. Locomotion and postures like dusky titi.
Variation. Animals from near the Río Orinoco are blackish. The degree of reddish color seems to vary individually within populations near the Río Amazonas. The black-handed race is from Colombia, between the upper Ríos Caquetá and Putumayo.
Similar species. Golden-handed tamarins *(Saguinus midas)* have similar coloration, but no white collar, and are much smaller; no other monkeys are black/brown with yellow hands.
Sounds. Loud whooping duets at dawn; chucks and birdlike chirrups.
Natural history. Diurnal; arboreal; groups of 2–5. Feeds mainly on fruit, with some insects. Yellow-handed titis are generally found in tall forest, where they use the middle and upper levels. They are quite active and spend most of the day moving. At night they sleep together, tails intertwined, on a large horizontal branch in the canopy. In some parts of its range (Peru) this species is largely restricted to areas where the soil has been leached to almost pure white sand and the stream waters are clear "black." In the

northern part of its range it is found on many soil types. Found only in mature, tall rainforest, in both terra firme and floodplain formations.

Geographic range. South America: east of the Andes in the western Amazon Basin lowlands of Colombia, Venezuela, Peru, and Brazil.

Status. CITES Appendix II. IUCN Red List vulnerable, *C. t. medemi.* Locally common but distribution patchy; too small to be hunted intensively except where larger game is scarce.

Local names. Japuça-de-coleira (Br); viduita, macaco caresebo (Co, Ve); cotoncillo (Ec); mono viudo, tocón (Pe).

References. Kinzey, W. G. 1981. The titi monkeys, genus *Callicebus.* In R. Mittermeier and A. F. Coimbra-Filho, eds., *Ecology and behavior of Neotropical primates,* vol. 1, 241–76. Rio de Janeiro: Academia Brasileiro de Ciências.

Defler, T. R. 1994. *Callicebus torquatus* is not a white-sand specialist. *Am. J. Primatol.* 33:149–54.

Masked Titi Monkey
Callicebus personatus
Plate 11, map 102
Identification. Measurements: HB = 325–550; T = 430–580; HF = 90–123; E = 30–35; WT = 1,050–1,650 g.

Upperparts pale brown, or frosted yellowish beige, or grizzled gray with brown lower back; fur long, thick, dull colored. Head small; face round and flat, black; wide fringe of black hair encircling face completely, or not extending to chin, or black only around ear; pale yellow crown of hair on top and side of head, or head grizzled gray; ears inconspicuous. Tail nonprehensile, thickly haired, dark orange, pale orange, yellowish brown, or dirty gray, sometimes with a pale tip. Hands and feet black. Throat and chest yellowish orange or brown, belly like back, but paler. Posture like dusky titi.

Variation. Titis from Minas Gerais and Espírito Santo are beige with sharply contrasting yellow and black crown; those from Rio de Janeiro and São Paulo are yellow-gray with black face and crown; and those from Bahia are grizzled gray with gray head and brown lower back.

Similar species. Marmosets (*Callithrix* spp.) and tamarins (*Leontopithecus rosalia*) are smaller, with relatively large heads and slender tails; all other monkeys in range are much larger.

Sounds. Duets with loud whooping calls at dawn, similar to dusky titi.

Natural history. Diurnal; arboreal; groups of 2–6. Masked titis feed mainly on fruit (77%) supplemented by leaves (17%) and tiny amounts of insects. They are quite inactive and spend about 40% of the day resting. They feed mainly in small-crowned trees. At night they sleep in a huddle, with tails intertwined, on a large, high branch in the canopy (25–40 m up). From mature lowland Atlantic coastal rainforest and inland gallery forests.

Geographic range. South America: SE Brazil, Bahia to São Paulo.

Status. Cites Appendix II. IUCN Red List critically endangered, *C. p. barbarabrownae;* vulnerable, all others. Threatened by deforestation of small geographic range.

Local names. Sahuí-guaçu, sauá, guigó.

References. Kinzey, W. G., and M. Becker. 1983. Activity pattern of the masked titi monkey, *Callicebus personatus. Primates* 24:337–43.

Müller, K.-H. 1995. Ecology and feeding behavior of masked titi monkeys. *Neotropical Primates* 3:51–52.

Common Squirrel Monkey
Saimiri sciureus
Plate 11, figure 7, map 103
Identification. Measurements: HB = 250–320; T = 340–435; HF = 78–90; E = 21–30; WT = 480–1,400 g; males heavier than females.

Upperparts gray-olivaceous; back faint to bright gold-olivaceous frosted with black; fur short. Head round, distinctly patterned, muzzle blackish, mask around eyes white; ears hairy, white with slight, pointed tufts, or almost bare; sides of neck behind ear white, crown a cap of black or gray. Females tend to have darker heads and cheeks than males. Tail nonprehensile, slightly tufted, olivaceous with black tip above, paler below. Forelimbs, hands, and feet bright gold to pale yellow-orange. Chin and throat whitish; chest and belly pale yellow to orange. Body slender and gracile.

Map 103

▦ Common squirrel monkey, *Saimiri sciureus*
◪ Central American squirrel monkey, *S. oerstedii*

Variation. Animals from Bolivia and SE Peru have black caps and bright gold limbs; those from S Pará and north of the Amazon have olivaceous caps; a dark, blackish form is found in a small region of islands at the mouth of the Rio Japurá. Authorities disagree on the number of squirrel monkey species; some recognize as many as seven, but a recent review of all types of evidence (reference below) concluded that all South American forms appear to intergrade.

Similar species. These are the only small monkeys with black muzzle and white eye mask. Marmosets (*Callithrix* spp.) and tamarins (*Saguinus* spp.) are smaller, with disheveled manes of long hair on the head, and travel in small groups; capuchins (*Cebus* spp.) are larger and often walk with tail coiled.

Sounds. Very noisy, travel with constant squeals and high-pitched birdlike whistles and chirps, crashing and disturbance of the vegetation.

Natural history. Diurnal; arboreal; large groups of 25 to more than 100. Feeds on insects, small ripe fruits such as figs, and nectar. Squirrel monkeys spend much of the day ranging through the forest searching for insects, especially caterpillars and large orthopterans. Small, slender, and agile, they favor vegetation with many thin vines and branches, like that at river and lake edges. In constant motion, they run and jump through the vegetation, usually keeping to the middle and lower levels of the forest, sometimes even using the ground. They also climb to reach fruits in the highest treetops. They

often associate with capuchin monkeys and travel with them for many hours. Troops use a large home range of from 100 to 500 ha. The mating system is promiscuous, and only females carry and care for the young. This species is most common in floodplain forests, especially on rich alluvial soils, and it adapts well to the dense vegetation of old secondary and disturbed forests. It is absent in areas far from rivers in tall, poor-soil, terra firme forests. Found in mature and disturbed lowland evergreen rainforest.

Geographic range. South America: the Amazon Basin and Guianas. To 1,500 m elevation.

Status. CITES Appendix II. IUCN Red List vulnerable, *S. s. vanzolinii* only. Widespread and often common. Too small to be hunted for food except where game is scarce, but formerly intensively exploited for the laboratory animal and pet trades.

Local names. Mono amarillo (Bo); macaco-de-cheiro, boca-prêta (Br); titi (Co); fraile, frailecito, barizo (Co, Ec, Pe); singe écureuil, sapajou blanc (FG); eekhoornaap, monki-monki (Su); tití, mono calavera (Ve).

References. Costello, R. K., C. Dickinson, A. L. Rosenberger, S. Boinski, and F. S. Szalay. 1993. Squirrel monkey (genus *Saimiri*) taxonomy: A multidisciplinary study of the biology of species. In W. H. Kimbal and L. B. Martin, eds., *Species, species concepts, and primate evolution,* 177–210. New York: Plenum Press.

Mitchell, C. L., S. Boinski, and C. P. van Shaik. 1991. Competitive regimes and female bonding in two species of squirrel monkey *(Saimiri oerstedi* and *S. sciureus). Behav. Ecol. Sociobiol.* 28:55–60.

Central American Squirrel Monkey
Saimiri oerstedii
Plate 11, map 103

Identification. Measurements: HB = 266–291; T = 362–389; HF = 76–86; E = 27; WT = 600–750 g.

Back, lower legs, hands, and feet orange-gold; shoulders and hips olivaceous; fur on back quite long at higher elevations. **Head with black cap, black muzzle, white mask;** cap and cheeks of males often paler than those of females; **ear and sides of neck white. Tail olivaceous, distal half black, tip tufted.** Chin and throat white; rest of underparts pale orange.

a   b

Figure 7. Head patterns of squirrel monkeys: *(a)* **common squirrel monkey,** hairy-eared form, *Saimiri sciureus; (b)* bare-eared form, *S. s. ustus.* (Drawings after P. Hershkovitz, 1984, *Am. J. Primatol.*)

Variation. There are two geographically disjunct subspecies. Males from the northern part of the range have olivaceous caps. Both sexes have black caps in Costa Rica. Because of its tiny, disjunct geographic range, unlike that of any other monkey, this species might have been introduced by man to Central America, but its distinctiveness from South American forms implies a long separation.
Similar species. These are the only small monkeys in Central America with black muzzle and white mask.
Sounds. Similar to common squirrel monkey.
Natural history. Diurnal; arboreal; groups of 36–65. Feeds on small ripe fruits and insects. These active monkeys spend two-thirds to three-fourths of the day foraging and have difficulty finding enough food in the late wet season. They use home ranges of 76–110 ha within dense growth of secondary, disturbed, and forest edge habitats, where they forage at low heights, often near the ground. They apparently cannot survive in tall, undisturbed, mature forest where there is little vegetation in the lower strata.
Geographic range. Central America: Costa Rica in two small areas on the Pacific coast, and perhaps Panama.
Status. CITES Appendix I, US-ESA endangered. IUCN Red List critically endangered, *S. o. citrinellus;* endangered. *S. o. oerstedii.* Apparently extirpated in Panama. Most of the former range has been deforested and only a few small fragmented populations re-

main. Severely threatened by habitat loss to development and agriculture, and by the return of protected forests to tall mature habitat inappropriate for the species.
Local names. Titi, mono ardilla (Pn, CR).
References. Boinski, S. 1987. Habitat use by squirrel monkeys *(Saimiri oerstedi)* in Costa Rica. *Folia Primatol.* 49:151–67.
———. 1994. Affiliation patterns among male Costa Rican squirrel monkeys. *Behaviour* 130:191–209.

Brown Capuchin Monkey
Cebus apella
Plate 11, map 104
Identification. Measurements: HB = 350–488; T = 375–488; HF = 107–132; E = 28–43; WT = 1.7–4.5 kg; males larger than females.
Upperparts dark yellowish to slightly reddish **brown,** darkest middorsally; **shoulders paler than back,** mustard yellow or brown; sides and thighs frosted yellowish. **Head broad, crown completely covered with black or dark brown cap that extends down cheek as a distinct dark bar in front of ear; hair of cap forming short tufts above ears,** giving **top of head** a characteristic **flat, squared-off, or eared** frontal outline; **face flat, dark brown, pink, or brown mottled pink, fringed** completely **with yellowish to white. Tail prehensile, black or brown, darkest at tip, usually carried with tip in tight, downward coil. Hands, hindlimbs, and feet black or brown, darker than body.**

Map 104

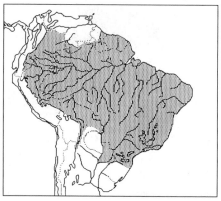

Brown capuchin monkey, *Cebus apella*

Underparts brown or chest yellow. Body stocky and robust, especially adult males.
Variation. Animals from drier regions of Paraguay, and Pernambuco and Bahia, Brazil, are pale yellow-brown with yellow shoulders, yellow or reddish chests, and brown extremities; those from SE Brazil are dark brown or blackish overall; those from the Amazon Basin have black hands, feet, caps, and tails. There is much individual variation among members of the same troop, especially in face color. Adult males are often darker than females.
Similar species. These are the only capuchin monkeys that occur with other capuchin species over large regions. They are distinguished from all others by dark bars in front of ears, brown face, flat or tufted top of head, more robust build, and no sharp, narrow, dark stripe on midbrow. No other yellowish brown, black-capped monkeys carry the tail coiled; woolly monkeys *(Lagothrix lagothricha)* have round heads, and black or dark faces with no markings or cap.
Sounds. Noisy. Emit frequent short, yipping whines (like hungry newborn puppies); alarm call a distinctive two-toned tonal clunking; birdlike rising whistles. Forage destructively by ripping apart and dropping vegetation, hammering nuts against branches, jumping noisily from tree to tree. Sounds of foraging can be heard for long distances.
Natural history. Diurnal; arboreal; groups of 5–20, usually about 10. Feeds on ripe fruits, palm nuts, arthropods and small vertebrates, and some nectar. Brown capuchins usually forage in the middle to lower levels of the forest but will feed on high-canopy fruit. These robust capuchins find animal foods by tearing apart dead vegetation; they will eat almost any small animal they find, including mouse opossums, bird's eggs or nestlings, and lizards. They also tear open well-protected buds or stems of palms and eat the young flowers or pith, or pry ripe palm fruits from a tight cluster. The stronger individuals break the hard nuts of *Astrocaryum* palms by banging them against branches (other capuchins also do this, but more rarely). They spend much foraging time in palm trees. Brown capuchins are one of the most commonly seen monkeys in Amazonia. Their noisy behavior makes them easy to detect, but where hunted they are difficult to approach and quickly give the alarm and flee silently. Squirrel monkeys often forage for many hours with a group of brown capuchins. Found in many types of forest including mature rainforest, dry forest, gallery forest, and disturbed and secondary forest.
Geographic range. South America: east of the Andes from Colombia and Venezuela south to Paraguay and N Argentina. To 1,500 m elevation.
Status. CITES Appendix II. IUCN Red List critically endangered, *C. a. margaritae, C. a. xanthosternos;* vulnerable, *C. a. robustus.* Widespread and often common. Intensively hunted for meat in most parts of its range, but has greater reproductive potential, habitat flexibility, and recuperative capacity than larger monkeys.
Local names. Cai comun (Ar); mono martín, mono, mono negro (Bo); macaco-prego (Br); mico maicero (Co); machin negro, mono negro, koshiri (Ec, Pe); macaque noire (FG); ringtail (Gu); mutsaap, bruine rolstaartaap, keskesi (Su); caí (Gua).
References. Terborgh, J. 1983. *Five New World primates.* Princeton, N.J.: Princeton University Press.

Janson, C. 1985. Aggressive competition and individual food consumption in wild brown capuchin monkeys *(Cebus apella). Behav. Ecol. Sociobiol.* 18:125–38.

Wedge-capped or Weeping Capuchin Monkey
Cebus olivaceus
Plate 11, map 105

Map 105

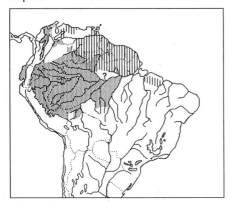

▨ White-fronted capuchin monkey, *Cebus albifrons*
▥ Wedge-capped or weeping capuchin monkey,
   C. olivaceus

Identification. Measurements: HB =
374–460; T = 400–554; HF = 120–143;
E = 35–51; WT = 2.3–4.2 kg.
**Upperparts uniform tawny brown, some-
times frosted with dirty yellow; shoulders
and upper arms sometimes silvery yellow;
back of head and neck reddish; fur quite
long, stands upright, giving the animals a
bristly, robust look. Head brownish yellow
with sharp, dark brown V-shaped cap,
point of V connected to a thin brown
stripe down center of face to nose. Face
pink, fringed to behind ears by yellowish
white to silvery white.** Tail prehensile, dark
brown, often carried with tip coiled. Hands,
wrists, and feet dark brown, often contrasting with paler forearms. Chest pale brown,
belly brown.
Variation. Animals from coastal Venezuela
(Falcón) are entirely dark brown with red-
tinged limbs and light brown shoulders and
face. The name *C. nigrivitattus* is sometimes
used for this species. We consider the recently
described *C. kaapori* to be a subspecies.
Similar species. Brown capuchins *(C.
apella)* have dark bars in front of ears and
no facial stripe; white-faced capuchins *(C.
albifrons)* have extremities and tail paler
than back; woolly monkeys *(Lagothrix lago-
tricha)* have completely dark head and face
with no markings.
Sounds. Frequent short yips and tonal caws,
barks like small dog, squeals and chatters.
Sometimes hammers or taps nuts, rips vegetation apart noisily.

Natural history. Diurnal; arboreal; groups
of 10–33. Feeds chiefly on ripe fruits, seeds,
and arthropods, with some vegetative plant
parts. Wedge-capped capuchins mainly use
the middle and lower levels of the forest,
on occasion even descending to the ground.
They spend several hours a day searching
for invertebrates, often in palms, rummaging
among leaves, breaking apart dead branches,
and searching under foliage. Found in mature
evergreen forest and gallery forest or, rarely,
deciduous forest.
Geographic range. South America: Venezuela, the Guianas, and Brazil, east of the
Rios Orinoco and Negro and north of the Rio
Amazonas, and Pará south of the Amazonas
E of the Rio Toncantins *(C. o. kaapori).*
Status. CITES Appendix II. IUCN Red List
vulnerable, *C. o.* kaapori. Common in some
parts of Venezuela, where brown capuchins
are absent, but over most of its geographic
range is much less common than brown
capuchin.
Local names. Caiarara (Br); cebus tête
blanche, macaque blanc, capuchin (FG);
ringtail (Gu); Grijze capucijneraap, bergi-
keskesi (Su); mono chuco, mono capuchino
(Ve).
References. Robinson, J. G. 1986. Seasonal
variation in use of time and space by the
wedge-capped capuchin monkey, *Cebus
olivaceus:* Implications for foraging theory.
Smithsonian Contrib. Zool., no. 431.
   Mittermeier, R. A., and M. G. M. van
Roosmalen. 1981. Preliminary observations
on habitat utilization and diet in eight Surinam monkeys. *Folia Primatol.* 36:1–39.

White-fronted Capuchin Monkey
Cebus albifrons
Plate 11, map 105
Identification. Measurements: HB =
358–460; T = 401–475; HF = 112–136;
E = 32–45; WT = 1.2–4.3 kg.
**Upperparts pale smoky gray-brown, yel-
lowish brown, or reddish brown; forearms
and hindlegs yellow or rusty reddish. Face
pink fringed with silvery white; crown
with sharp, wedge-shaped cap of dark
smoky brown, generally extending forward
as a thin stripe down center of brow. Tail
prehensile, smoky silvery yellow, usually
paler at tip than base,** often carried with
tip coiled. Underparts yellowish. A gracile,
slender, medium-sized monkey.

Variation. There is much individual color variation even within the same troop.
Similar species. Brown capuchins *(C. apella)* have dark bars in front of ears; both they and wedge-capped capuchins *(C. olivaceus)* have dark hands, feet, and tail tip and robust build; squirrel monkeys *(Saimiri* spp.) have black muzzles, white masks, and do not coil their tails; woolly monkeys *(Lagothrix lagothricha)* have dark faces with no markings.
Sounds. Calls often, with low whines and series of rising whistles. Travels and forages noisily.
Natural history. Diurnal; arboreal; groups of 7–30. Feeds on fruit, seeds, and arthropods, especially wasps. White-fronted capuchins use all levels of the forest, including occasionally the ground. They forage for hidden arthropods by manipulating the leaves and branches, but they do less destructive ripping and tearing than brown capuchins. Sometimes they hammer palm nuts to crack them. White-fronted capuchins are gracile and agile, moving more quickly and lightly than brown capuchins, and they are often more wary and difficult to approach. They form loose foraging associations with squirrel monkeys. Troops have large home ranges that overlap each other. Found in mature and disturbed rainforest.
Geographic range. South America: isolated areas in N Colombia and Venezuela and coastal Ecuador; and the middle and upper Amazon Basin of Colombia, Venezuela, Ecuador, Peru, Bolivia, and Brazil west of the Rios Negro and Tapajóz. To 2,000 m elevation.
Status. CITES Appendix II. Hunted for meat, but widespread in Amazon Basin; coastal populations may be threatened by deforestation and hunting.
Local names. Caiarara (Co, Br); silbador (Bo); machin blanco, mico, mono lanudo (Ec); machin blanco, makieri (Pe); maicero cariblanco, mico (Co); mono carablanca (Ve).
References. Terborgh, J. 1983. *Five New World primates.* Princeton, N.J.: Princeton University Press.
   Hershkovitz, P. 1949. Mammals of northern Colombia. Preliminary report no. 4: Monkeys (Primates), with taxonomic revisions of some forms. *Proc. U.S. Nat. Mus.* 98:323–427.

Map 106

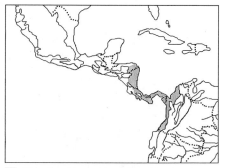

White-throated capuchin monkey, *Cebus capucinus*

White-throated Capuchin Monkey
Cebus capucinus
Plate 11, map 106
Identification. Measurements: HB = 335–453; T = 350–551; HF = 110–150; E = 21–42; WT = 1.4–3.9 kg; males larger than females.
Upperparts black except sides of neck, shoulders, and upper arms sharply contrasting pale yellow to whitish. Head yellow with black, V-shaped cap on crown; face pink with whitish hairs. Tail prehensile, black, sometimes brownish below, often carried with tip coiled. Hands and feet black. Throat and chest pale yellow; belly thinly haired, blackish.
Variation. Capuchins from Honduras have inner thighs frosted yellow.
Similar species. Squirrel monkeys *(Saimiri* spp.) have dark muzzles and do not hold their tails in a coil.
Sounds. Calls with yips, barks, whistles, and hoarse screams; forages noisily.
Natural history. Diurnal; arboreal; groups of 2–24. Feeds on ripe fruits and arthropods. White-throated capuchins use all levels of the forest, sometimes including the ground. They forage for prey by examining leaves, picking through dead litter, splitting twigs, and pulling off bark. These are active monkeys that spend much of the day traveling and foraging. They may defend territories, unlike other capuchins. Found in a wide range of habitats including mature, disturbed, and secondary evergreen and deciduous forests.
Geographic range. Central and South America: Honduras south to N Colombia, west of the Andes to N Ecuador.

Status. CITES Appendix II. IUCN Red List vulnerable, *C. c. curtus* from Isla Gorgona, Colombia. Many populations are probably threatened by deforestation.

Local names. Mico negro, carita blanca, maicero (Co); carablanca (CR, Ho); lanco, capuchino, mico (Ec); cariblanco (Pn).

References. Freese, C. H., and J. R. Oppenheimer. 1981. The capuchin monkeys, genus *Cebus.* In A. F. Coimbra-Filho and R. A. Mittermeier, eds., *Ecology and behavior of Neotropical primates,* vol. 1, 331–90. Rio de Janeiro: Academia Brasileira de Ciências.

Boinski, S. 1989. Why don't *Saimiri oerstedii* and *Cebus capucinus* form mixed-species groups?. *Int. J. Primatol.* 10: 103–14.

Guianan Saki Monkey
Pithecia pithecia
Plate 12, map 107

Identification. Measurements: HB = 285–480; T = 328–455; HF = 107–131; E = 28–40; WT = 1.65–2.35 kg; males larger than females.

Male entirely glossy pitch black except brow and sides of face pure white or rusty orange, white cheeks sometimes grading to orange or pale buff beard; **face around mouth, nose, and eyes dark, framed by white. Female entirely grizzled gray or gray-brown;** face whitish, with a diagonal white stripe from below eye, down cheek beside muzzle. **Fur long and shaggy** on upperparts, hanging in a fringe on sides over thinly haired belly, makes the body appear much larger than it is; **long hair grows forward in a mop over crown from rear of head.** Head held low, below level of shoulders. **Tail nonprehensile, exceptionally bushy,** usually hangs straight downward when saki is sitting, and is carried behind, not above, the body when traveling. Hands and feet black or gray; belly sometimes bright rusty orange.

Variation. On the N bank of the Rio Amazonas east of Manaus, males have orange faces and females have whitish crowns and orange bellies.

Similar species. Brown bearded sakis *(Chiropotes satanas)* have no white on face, are not shaggy over the shoulders, and often carry the tail raised in an arc over the back.

Sounds. Generally quiet; low grunts in alarm; clear, high-pitched whistles and trills.

Natural history. Diurnal; arboreal; groups

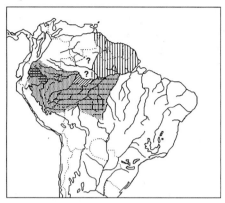

Map 107

Monk saki monkey, *Pithecia monachus*
Guianan saki monkey, *P. pithecia*
Equatorial saki monkey, *P. aequatorialis*
Buffy saki monkey, *P. albicans*
Gray monk saki monkey, *P. m. irrorata*

of 1–4, usually 2–3. Feeds on fruit and seeds (about 90% of diet) and small amounts of leaves and insects, and in the dry season, flowers. Groups consist of a monogamous pair and their young. Guianan sakis seem to prefer tall primary forest habitat, where they use the middle and lower levels most intensively. The diet is specialized on soft, nutritious seeds from fruits protected by hard husks that are broken with the large canine teeth. Because they chew and destroy the seeds, sakis are poor seed dispersers. Sakis are quiet and wary. They characteristically travel with springing leaps, sometimes hopping bipedally without using their hands, and are locally known as "flying monkeys." When a saki is sitting and apparently slightly alarmed, the plumelike tail is sometimes waggled back and forth. They are occasionally seen in apparent association with other species of monkeys. Found in a variety of lowland evergreen rainforest types, including terra firme, riverside, dense, and open, and although they can be found in disturbed or secondary vegetation, they are chiefly mature-forest monkeys.

Geographic range. South America: the Guianas, Venezuela, and Brazil north of the Rio Amazonas and east of the Rios Negro and Orinoco.

Status. CITES Appendix II. Usually uncommon or rare and occurs patchily, but widespread, and much of its geographic range is still undisturbed by man.

Local names. Parauacu (Br); flying jack, breezy monkey (Gy); maman guinan, yarké (FG); witkopap, pluimstaartap, wanakoe (Su); viuda (Ve).

References. Oliveira, J. M. S., M. G. Lima, C. Bonvíncíno, J. M. Ayres, and J. G. Fleagle. 1985. Preliminary notes on the ecology and behavior of the Guianan saki (*Pithecia pithecia*, Linnaeus 1776; Cebidae, Primate). *Acta Amazonica,* 15:249–63.

Kinzey, W. G., and M. A. Norconk. 1993. Physical and chemical properties of fruit and seeds eaten by *Pithecia* and *Chiropotes* in Surinam and Venezuela. *Int. J. Primatol.* 14:207–27.

Monk Saki Monkey
Pithecia monachus
Plate 12, map 107
Identification. Measurements: HB = 370–480; T = 404–500; HF = 116–147; E = 25–37; WT = 2.2–2.5 kg.
Upperparts black, or lightly to heavily frosted with dirty white, appears blackish or gray with black hair base showing below; **fur long and shaggy. Face** almost naked, **blackish** thinly haired with white, **white stripe diagonally down side of face from below eye; crown** almost bald, **lacking a mat of short hair; mop of long hair growing forward over brow; brow and cheeks the same color as back.** Hands and feet dirty white; legs sometimes whitish. Chin and throat naked; chest thinly haired whitish or pale buffy; belly thinly haired black with pale skin showing through. Tail, posture, and locomotion like Guianan saki. **Male and female alike.** Young almost black.
Variation. Blackish animals are from Rondônia (Brazil), Peru, Ecuador, and Colombia; those from east of the Rio Juruá and Bolivia are generally frosted with white and may have gray-rimmed faces and white tails: these have recently been recognized as a separate species, *P. irrorata.* Formerly sometimes called *P. hirsuta.*
Similar species. Equatorial sakis *(P. aequatorialis)* have contrasting white brow and cheeks, and sometimes red chest or beard; buffy sakis *(P. albicans)* have ruff, limbs, and underparts buff or orange.
Sounds. Generally quiet; calls with low-pitched whines, grunts, chirps, trills, crow-like croaking, and high-pitched whistles.

Natural history. Diurnal; arboreal; groups of 2–8, usually 3–4. Feeds on fruit, ripe and green seeds, leaves, and ants. Monk sakis use the middle and upper levels of the forest. They are shy and quiet, with behavior like that of equatorial sakis. At night they sleep on large branches in the upper canopy. There is only one breeding female in a group, and she gives birth every 2–3 years. Home ranges of groups overlap. Monk sakis are usually found in terra firme forest, where they frequent dense, leafy vegetation, but they can also be found in seasonally flooded forests. Found in lowland evergreen rainforest only.
Geographic range. South America: upper Amazon Basin of Colombia, Ecuador, Peru, Bolivia, and Brazil west of the Rio Tapajós. To 600 m elevation.
Status. CITES Appendix II. IUCN Red List vulnerable, *P. m. milleri* (Colombia). Distribution patchy, locally common, uncommon, or rare, but widespread.
Local names. Mico volador, huapo negro (Co); parauacu (Br, Ec); saki (Ec); huapo negro, oso mono, yana huapo (Pe).
References. Happel, R. E. 1982. Ecology of *Pithecia hirsuta* in Peru. *J. Hum. Evol.* 11:581–90.

Soini, P. 1986. A synecological study of a primate community in the Pacaya-Samiria National Reserve, Peru. *Primate Conservation* 7:63–71.

Equatorial Saki Monkey
Pithecia aequatorialis
Plate 12, map 107
Identification. Measurements: HB = 392–440; T = 448–474; HF = 111–131; E = 28–40; WT = 2.0–2.5 kg.
Upperparts black, sparsely frosted with dirty white; fur long, shaggy, and wavy, hangs in a long fringe on sides over nearly naked belly. **Face** nearly naked, **whitish, male with face surrounded by a broad, sharply contrasting border of short white hair,** or cheeks and beard red; **crown covered by a distinctive mat of short white or buffy hair. Female with cheeks and crown white like male, or distinctly darker,** similar to back. **Feet dirty white. Throat, chest, and belly orange or dark brown.** Tail and body hair pattern, locomotion, and posture as in Guianan saki *(P. pithecia).*
Variation. This species was until recently confused with the monk saki; older accounts of saki natural history often do not

adequately distinguish which species was studied.

Similar species. Monk sakis *(P. monachus)* have cheeks and crown the same color as upperparts or black, lack a mat of short, dense hair on crown, and are generally grayish; titis *(Callicebus* spp.) do not have bushy tails.

Sounds. Usually silent; gives low grunts in alarm, also chirps and moaning calls.

Natural history. Diurnal; arboreal; groups of 1–4. Diet probably fruit, seeds, and leaves, similar to that of other sakis. In some remote areas, equatorial sakis can often be seen from a boat, from which they look like big black balls of fur with bushy tails, sitting in riverside trees. They seem to be most common in riverside, flooded, and swamp forests in poor-soil (blackwater) regions, but they are also found in terra firme forests away from water. Equatorial sakis are stealthy and quiet and will sit absolutely motionless in dense, leafy vegetation, their presence betrayed only by long bushy tails hanging below a branch. If put to flight, they bound away acrobatically and suddenly disappear into hiding. Appear to be found mostly in lowland primary forests.

Geographic range. South America: in the upper Amazon Basin of Ecuador, Peru, and possibly Brazil.

Status. CITES Appendix II. Distribution appears patchy, rare or absent in many areas, locally common in a few.

Local names. Parauacu (Br, Ec); saki (Ec); huapo negro, oso mono (Pe).

References. Hershkovitz, P. 1987. The taxonomy of South American sakis, genus *Pithecia* (Cebidae, Platyrrhini): A preliminary report and critical review with the description of a new species and a new subspecies. *Am. J. Primatol.* 12:387–468.

Buffy Saki Monkey
Pithecia albicans
Plate 12, map 107

Identification. Measurements: HB = 365–560; T = 405–570; HF = 120–132.

Back and tail blackish; ruff, arms, legs, fringe on sides, and underparts buff, orange, or reddish. Sexes similar. Long, shaggy hair of body and tail as in other sakis. Infants under 3 weeks old are chocolate brown.

Similar species. Monk sakis *(P. monachus)* are gray or black, with a dark belly.

Natural history. Diurnal; aboreal; groups of 1–7. Feeds on fruits, young leaves, seeds of unripe fruits, and insects. Buffy sakis use the middle and upper levels of tall forest. They move with frequent strong leaps, but they spend much of the day sitting motionless, and hide from observers in the same way as equatorial sakis (above). Found in undisturbed and logged primary forests.

Geographic range. South America: Brazil, south of the Solimões between the lower Rios Purus and Juruá.

Status. CITES Appendix II. Locally common and not intensively hunted, but geographic range small.

Local names. Parauacu.

References. Johns, A. 1986. Notes on the ecology and current status of the buffy saki, *Pithecia albicans. Primate Conserv.* 7:26–29.

Brown Bearded Saki Monkey
Chiropotes satanas
Plate 12, map 108

Identification. Measurements: HB = 327–440; T = 300–465; HF = 121–122; E = 30–33; WT = 2–4 kg; males slightly larger than females.

Head, limbs, and tail glossy deep brownblack; back from neck to tail pale yellow to red-brown, palest on center of back between shoulders, darkening to brown on sides; or back rich, dark brown. Face dark, crown hair growing from a whorl forward and sideward into two large tufts with sharp part on midline; tufts much larger in males than in females and supported by enlarged mounds of muscle on temples; chin and cheeks with a long, thick, rounded beard, small in females. Tail nonprehensile, very bushy, often carried in a high arc over back when walking, waved back and forth in this position, or sometimes waggled rapidly vertically below body in excitement. Underparts blackish except genital area naked, pure white.

Variation. Animals from north of the Rio Amazonas have pale yellow or reddish backs; several colors are seen in a single troop. Those from south of the river have dark brown backs except close to the Rio Xingu, where they have pale brown backs.

Similar species. Capuchins *(Cebus* spp.) have slender prehensile tails carried in a coil; howler monkeys *(Alouatta* spp.) are either entirely red (N of Rio Amazonas) or black with brown hands (S of Rio Amazonas)

Map 108

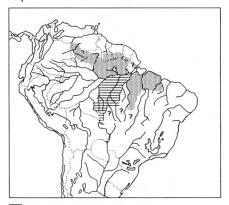

▨ Brown bearded saki monkey, *Chiropotes satanus*
▤ White-nosed bearded saki monkey, *C. albinasus*

and have slender, often coiled tails; spider monkeys (*Ateles* spp.) have long, slender limbs and tail and swing by their arms below branches.

Sounds. High-pitched whistles rising in a crescendo and abruptly cut off; other whistles.

Natural history. Diurnal; arboreal; groups of 2 to more than 30, usually more than 10. Feeds largely on the seeds of ripe and unripe fruits with hard husks and on ripe fruits (over 90% of diet), and a few leaves, flowers, and insects. Brown bearded sakis are most often seen in the canopy of tall forests. They often sit silently, but when they feed on seeds, a rain of discarded fruit parts patters to the ground, betraying their location. Sometimes they associate with other monkey species. They travel by walking on the tops of branches, tail held high, and jump across gaps with powerful springs from the hindlegs. They may hang by their hindfeet to reach down and pick fruits. Troops use large areas of several km² and travel long distances each day. Found in mature lowland rainforest or, rarely, dry forest.

Geographic range. South America: eastern Amazon Basin in Venezuela, Brazil, and the Guianas; north of the Amazon east of the Rios Negro and Orinoco, south of the Amazon east of the Rio Xingu.

Status. CITES Appendix II, US-ESA endangered. IUCN Red List endangered, *C. s. satanas*, vulnerable, *C. s. utahicki*. Generally uncommon; hunted for food and rare or absent near settlements, but much of large geographic range is uninhabited. Apparently

extirpated from former range in northern Guianas.

Local names. Cuxiú (Br); saki noire, macaque mon pé (FG); black saki, satan-aap, baard-aap, bisa (Su); mono capuchino, capuchino del Orinoco (Ve).

References. Ayres, J. M. 1981. Observaçóes sobre a ecologia e o comportamento dos cuxiús (*Chiropotes albinasus* e *Chiropotes satanas,* Cebidae: Primates). M.S. thesis, INPA, Manaus (CNPq, INPA, FUA).

Hershkovitz, P. 1985. A preliminary taxonomic review of the South American bearded saki monkeys, genus *Chiropotes* (Cebidae, Platyrrhini), with the description of a new subspecies. *Fieldiana Zool.,* n.s., no. 27.

White-nosed Bearded Saki Monkey
Chiropotes albinasus
Plate 12, map 108

Identification. Measurements: HB = 360–460; T = 355–480; HF = 115–130; E = 25–30; WT = 2.2–3.6 kg; males slightly larger than females.

Upperparts glossy pitch black, back frosted with sheen of silvery olivaceous or silver; hair of back growing laterally from a part down midline. **Head entirely black except muzzle from between eyes to upper lip a triangle of red skin overlaid with pure white hairs; crown with long, thick hair growing in two prominent tufts** divided by sharp part; hair growing in a whorl from back of crown forward, sideward, and backward; tufts much larger in males than females; **beard of thick hair** much larger on males than females. **Tail nonprehensile, thickly furred and bushy; often carried in a high arc over back,** or waggled vertically below body. Underparts thinly haired, black; genital area and scrotum naked, red. Infants have a prehensile tail. Locomotion similar to brown bearded saki.

Variation. The skin of the face and scrotum is unpigmented and pure white in specimens, but flesh colored to vermillion red in life.

Similar species. These are the only black monkeys with a red nose and bushy, nonprehensile tail; howlers (*Alouatta* spp.) are black with black faces and slender tails; spider monkeys (*Ateles* spp.) have long, slender limbs and prehensile tails; monk sakis *(Pithecia monachus)* are shaggy and grizzled with white.

Sounds. Loud, high-pitched whistles.
Natural history. Diurnal; arboreal; groups
of 2 to more than 26. Feeds on ripe fruits,
many seeds of unripe fruits, and a few flowers and insects. White-nosed sakis most often use the middle and upper levels of the
forest. A troop has a home range of about
1 km² and travels 2–5 km a day. Sometimes white-nosed bearded sakis associate
with other monkeys, especially brown capuchins. Found only in undisturbed terra
firme rainforest.
Geographic range. South America: Brazil,
south of the Rio Amazonas between the
Rios Madeira and Xingu, extending to Mato
Grosso on the Rio Guaporé.
Status. CITES Appendix I, US-ESA endangered. Apparently fairly common where not
hunted, but rare from overhunting near settlements; threatened because of restriction to
undisturbed forest.
Local names. Cuxiú, pirolucu, piroculú.
References. Van Roosmalen, M., R. M.
Mittermeier, and K. Milton. 1981. The
bearded sakis, genus *Chiropotes*. In A. F.
Coimbra-Filho and R. M. Mittermeier, eds.,
*Ecology and behavior of Neotropical pri-
mates*, 419–41. Rio de Janeiro: Academia
Brasileira de Ciências.

Red or White Uakari Monkey
Cacajao calvus
Plate 12, map 109
Identification. Measurements: HB =
360–570; T = 137–185; HF = 122–152;
E = 26–33; WT = 2.3–3.5 kg; males
larger than females.
There are two distinctive races; generally:
**face and crown naked, bright red, fore-
head of males with large bulging muscles
over temples,** giving head a square top; canines form large bulges under lips; **body
hair long, shaggy and coarse, forming a
cape over shoulders; tail short, stumpy,
one-third the length of head and body,
thickly haired,** wagged in excitement.
White uakari, *C. calvus calvus.* **Upperparts
dirty white,** yellowish or pale gray frosted
with blackish; crown frosted with black,
with short cap of hairs growing forward;
chin with fringe of blackish hairs; underparts and front of thighs buffy to red brown,
or hindleg entirely washed with faint buff.
Red uakari, *C. calvus rubicundus.* **Upperparts entirely bright orange,** or cape paler,

Map 109

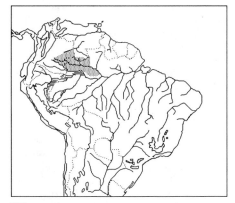

▦ Black uakari monkey, *Cacajao melanocephalus*
▤ Red and white uakari monkey, *C. calvus*

yellowish white; crown hair whitish yellow.
Variation. There are a number of local color
variations of the above forms.
Similar species. These are the only monkeys
with short tails and scarlet faces.
Sounds. Staccato series of sonorous, low-
pitched "ca-ca-ca-ca."
Natural history. Diurnal; arboreal; groups
of 10 to over 120. Large groups often split
into smaller groups of 1–10 that forage inde-
pendently. Feeds on seeds of immature fruits,
ripe fruits, leaves, nectar, and some insects,
especially caterpillars. Husks of hard fruits
are cracked open with the canines. Red and
white uakaris often inhabit permanently or
seasonally flooded forests, especially palm
swamps and near small streams (aguajales,
várzea), but they can also be found in terra
firme forests far from water, and they may
move seasonally between habitats to follow
fruit resources. Usually they use the middle
and upper levels of the forest, but in the dry
season they will descend to the ground to
feed on seedlings and fallen seeds. Uakaris
are very active and spend more of the day
moving, and move farther per day, than most
other New World monkeys. The home range
of a large group in Peru exceeded 150 km².
Often they are seen in association with other
monkeys. At night they climb into large trees
and sleep on the highest thin branches. Apparently found only in primary forests.
Geographic range. South America: central
Amazon Basin of Brazil, Colombia, and
Peru. The white form is found only on or

around the large delta island in the Rio Solimões below the mouth of the Rio Japurá; the red form is found south of the Solimões and west of the Juruá, and between the Ríos Ucayali and Yavarí in Peru. Range poorly known.
Status. CITES Appendix I, US-ESA endangered. IUCN Red List, *C. c. ucayalii* vulnerable, other subspecies endangered. Hunted for meat in Peru and for bait in Brazil (where people apparently do not eat them because of their disconcertingly human-looking faces). A habitat specialist with limited, fragmented distribution in habitat that is particularly vulnerable to logging and hunting owing to accessibility by boat. Apparently extirpated from much of its former habitat in Peru.
Local names. Uácari vermelho; macaco-ingles, cacajao (Br); huapo colorado, mono ingles, puca huapo (Pe).
References. Ayres, J. M. 1986. Uakaris and Amazonian flooded forest. Ph.D. dissertation, University of Cambridge, Cambridge.
    Aquino, R. 1995. Conservacíon de *Cacajao calvus ucayalii* en la Amazonia Peruana. *Neotropical Primates* 3:40–42.

Black Uakari Monkey
Cacajao melanocephalus
Plate 12, map 109
Identification. Measurements: HB = 300–500; T = 125–210; HF = 111–143; E = 19–36; WT = 2.4–4 kg; males slightly larger than females.
Head, forearms, hindfeet, and belly black; shoulders dark brown at top **with cape of long hair, yellowish or black, grading to dark red hindquarters and tail.** Face and crown thinly haired, black with surrounding fringe of hair growing forward; male with large bulge of muscle on each temple. **Tail short, one-third the length of head and body, thickly haired.** Chest with red skin showing below black hairs.
Variation. Animals from Venezuela have black capes; those from Brazil and Colombia have yellow capes.
Similar species. No other monkey in range has a short tail or similar color.
Natural history. Diurnal; arboreal; groups of 15 to 100 or more. Feeds on seeds and fruits. Black uakaris use permanently or seasonally flooded forests, especially near blackwater streams, but they can also be found in a number of terra firme forest

types. They use the canopy layer of tall forests. They have a patchy distribution in mature lowland rainforest.
Geographic range. South America: the upper Amazon Basin north of the river in SE Colombia, S Venezuela, and adjacent Brazil. Range poorly known.
Status. CITES Appendix I, US-ESA endangered. Said to be common in some remote parts of its range, but generally rare; hunted for food in Colombia.
Local names. Cauirí, uacari-preto (Br); chucuto, ichaca (Co, Ve).
References. Hershkovitz, P. 1987. Uakaries, New World monkeys of the genus *Cacajao* (Cebidae, Platyrrhini): A preliminary taxonomic review with the description of a new subspecies. *Am. J. Primatol.* 12:1–58.
    Boubli, J. P. 1994. The black uakari monkey in the Pico da Neblina National Park. *Neotropical Primates* 2:11–12.

Red Howler Monkey
Alouatta seniculus
Plate 13, map 110
Identification. Measurements: HB = 439–690; T = 540–790; HF = 128–162; E = 30–48; WT = 3.6–11.1 kg; males larger than females.
Head, shoulders, limbs, tail, and usually underparts dark red to purplish red; back and sides paler, bright orange to gold. Head large, throat swollen; face naked, black; **chin with** forward-growing **beard, longest in males. Shoulders robust, hindquarters disproportionately small and weak-looking; tail prehensile, often carried in a coil.** Adult males often have blackish beard, limbs, and tail.
Variation. Some animals from Venezuela are almost entirely purplish red; those from Peru have head, limbs, and tail red-orange, back gold. *A. s. sara,* from near Santa Cruz de la Sierra, Bolivia, is considered by some to be a separate species.
Similar species. These are the only large, red and gold, prehensile-tailed monkeys. Some white-bellied spider monkeys *(Ateles belzebuth)* are pale brown, with paler bellies; other spider monkeys in range have black backs.
Sounds. Loud choruses of roaring or howling, usually lasting many minutes, especially at dawn, late afternoon, and during rainstorms. Howling starts as an accelerating

Map 110

▦	Red howler monkey, *Alouatta seniculus*
▤	Red-handed howler monkey, *A. belzebul*
⦀	Black howler monkey, *A. caraya*
▬	Brown howler monkey, *A. fusca*

series of deep grunts by a male, which grades into long, deep, roars; the females join in with higher-pitched roars. In alarm, one monkey in the group gives a few soft grunts. *Natural history.* Diurnal; arboreal; groups of 3–9, usually 5–7. Feeds on fruits and many leaves. Red howlers use the middle and upper levels of mature and disturbed forests. They are quiet, slow-moving monkeys that have small home ranges and move little during the day. Their inactivity makes them inconspicuous, but their recent presence is shown by copious droppings with a strong, pungent, stablelike smell. When alarmed, they may position themselves to defecate and urinate on a person below. These are one of the monkeys most easily seen from a boat, as they sit high in riverside trees. In the dry season they come to the ground at the river's edge to drink or eat earth. They use salt licks. Because their home range needs are small, howlers can persist in small forest fragments or gallery forests. In habitats where there are few other monkey species, they can reach densities of over 100/km². Found in mature and disturbed evergreen to deciduous forest and gallery forest.
Geographic range. South America: east of the Andes in Colombia, Venezuela, Trinidad, the Guianas, and Brazil north of the Amazon, and Ecuador, Peru, Bolivia, and Brazil west of the Purus. To 1,200 m elevation. SE limits of range poorly known.

Status. CITES Appendix II. IUCN Red List vulnerable, *A. s. insulanus* (Trinidad) only. Widespread and locally common but intensively hunted for meat in many areas. Rare or extinct from hunting near human settlements in the upper Amazon Basin, where its flesh is tasty; in French Guiana and some other areas, its meat is bitter and it is hunted only as a last resort.
Local names. Maneche colorado (Bo); guariba (Br); mono colorado, cotudo, roncador, araguato, berreador (Co, Ve); coto mono, coto rojo, aullador (Ec); babouin (FG); coto mono, keníri (Pe); brulaap, baboen (Su).
References. Crockett, C. M., and J. F. Eisenberg. 1987. Howlers: Variations in group size and demography. In *Vertebrate ecology in the northern Neotropics,* 107–26. Washington, D.C.: Smithsonian Institution Press.
Rudran, R. 1979. The demography and social mobility of a red howler *(Alouatta seniculus)* population in Venezuela. In J. F. Eisenberg, ed., *Vertebrate ecology in the northern Neotropics,* 107–26. Washington, D.C.: Smithsonian Institution Press.

Red-handed Howler Monkey
Alouatta belzebul
Plate 13, map 110
Identification. Measurements: HB = 374–560; T = 580–700; HF = 125–164; E = 30–45; WT = 4.8–8 kg; males larger than females.
Entirely black except hands, feet, tail tip, and sometimes forehead or back dull rusty red. Scrotal area of males rust-red. Body shape like red howler.
Variation. At one locality near the Rio Madeira, all animals are completely black. Some males from between the Rios Tapajós and Madeira and from the east bank of the Xingu are entirely black, but some have red feet and tail tip, like females. In this region males may tend to be entirely black and females to have red extremities.
Similar species. Sympatric spider monkeys (*Ateles* spp.) are entirely black, some with red faces or white bands on the forehead; they have small pointed heads, no beard, and long, slender arms and legs. Bearded sakis (*Chiropotes* spp.) have bushy, nonprehensile tails often carried up over the back, never in a coil.
Sounds. Roars like red howler, but higher-pitched.

Natural history. Diurnal; arboreal; groups of 2–8. Probably similar to red howler. Uses the upper levels of terra firme and flooded forests, including disturbed forests.
Geographic range. South America: Brazil south of the Amazon east of the Rio Purus to Alagoas.
Status. CITES Appendix II. IUCN Red List critically endangered, *A. b. ululata* only. Locally common, widespread.
Local names. Guariba.
References. Branch, L. C. 1983. Seasonal and habitat differences in the abundance of primates in the Amazon (Tapajós) National Park, Brazil. *Primates* 24:424–31.

Black Howler Monkey
Alouatta caraya
Plate 13, map 110
Identification. Measurements: HB = 420–550; T = 530–650; HF = 126–155; E = 30–38; WT = 3.5–7.3 kg; males larger than females.
Males entirely black or dark blackish brown; scrotum rust-red. Females pale yellow-brown or dirty yellow; lower back sometimes brown darker than shoulders. Underparts sometimes tinged orange. Body shape like red howler. Young males to about 4.5 years colored like females.
Variation. Males from Bahia and Goiás are black; those from Mato Grosso and Paraná have brown back and hindparts, black head and forequarters. Those from São Paulo and Minas Gerais are brown-black with yellowish hands, feet, belly, and tail tip; these parts vary from pale yellow to almost entirely brown-black.
Similar species. These are the only monkeys in their range with strikingly different blackish males and yellowish females; they are the only large monkeys in most of their range. See brown howler *(Alouatta fusca).*
Sounds. Like red howler.
Natural history. Diurnal; arboreal; groups of 3–19, usually 7–9. Probably similar to red howler. Gallery forests and dry forests in pantanal, cerrado, and chaco habitats. The southernmost howler monkey.
Geographic range. South America: southern Amazon Basin and Paraguay Basin from E Bolivia, W Paraguay, and N Argentina, to SE Brazil from Bahia to Rio Grande do Sul, in the latter region in the western higher elevations away from the coast.

Status. CITES Appendix II. Geographic range large, but suitable habitat highly fragmented in some regions. Locally common, little hunted in Paraguay because its meat is bitter.
Local names. Manechi, manechi negro (Bo); Guariba, bugio, carajá (Br); mono aullador negro (Ar); carayá (TGua).
References. Rumiz, D. I. 1990. *Alouatta caraya:* Population density and demography in Northern Argentina. *Am. J. Primatol.* 21:279–94.

Brown Howler Monkey
Alouatta fusca
Plate 13, map 110
Identification. Measurements: HB = 440–570; T = 510–610; HF = 129–140; E = 30–40; WT = 4.1–7.2 kg.
Males with upperparts brown, red-brown, or orange-red; back brown frosted with yellow or gold, rump red or orange. Beard dark red to blackish. Tail dark red-brown to dark red or orange. Underparts dark brown, red, or blackish. **Females usually paler overall,** yellowish brown to dark brown (but see below). Body shape like red howler.
Variation. Males from Minas Gerais are brown, with orange-brown rumps; females are yellow-brown. Males from São Paulo are orange-red to red-brown, with red bellies. Males from Santa Catarina and Rio Grande do Sul are bright red-orange with dark brown feet; females are dark brown or blackish or with back frosted with orange or yellow-brown. There is also much individual variation within populations; the largest males seem reddest. The name *A. guariba* is also applied to this species.
Similar species. Muriquis *(Brachyteles arachnoides)* are entirely beige, with prominent, furry ears; black howlers *(A. caraya)* have blackish males and yellow-brown females.
Sounds. Probably like other howlers.
Natural history. Diurnal; arboreal; groups of 2–11. Feeds on leaves and fruit. Brown howlers chiefly use the upper levels of the forest. Like other howlers, they are slow-moving and sedentary. This species has been associated with the endemic Paraná pine *(Araucaria angustifolia).* It favors these trees for resting and sleeping, and its brown color matches that of clusters of dead leaves

remaining on the tree. Found in mature and secondary Atlantic coastal forests.

Geographic range. South America: Brazil, coastal forests from Bahia to Rio Grande do Sul; and Misiones, Argentina.

Status. CITES Appendix II. IUCN Red List critically endangered, *A. f. fusca;* vulnerable, *A. f. clamitans.* Said to be drastically declining; much of the original habitat has been deforested, the species has been exterminated by hunting for meat in some of the remaining forests, and yellow fever epidemics have decimated populations.

Local names. Bugio, ruivo, guariba, barbado (Br); mono aullador rufo (Ar).

References. Cordeiro da Silva, E., Jr., 1981. A preliminary survey of brown howler monkeys *(Alouatta fusca)* at the Cantareira reserve (São Paulo, Brazil). *Rev. Brasil. Biol.* 41:897–909.

Di Bitetti, M. S., G. Placci, A. D. Brown, and D. I. Rode. 1994. Conservation and population status of the brown howling monkey *(Alouatta fusca clamitans)* in Argentina. *Neotropical Primates* 2:1–3.

Mantled Howler Monkey
Alouatta palliata
Plate 13, map 111

Identification. Measurements: HB = 405–555; T = 585–710; HF = 124–154; E = 25–39; WT = 4.8–7.7 kg; males larger than females.

Entirely black except sides with a fringe (mantle) of long, pale hairs that are yellow, gold, pale brown, or buff; pale color sometimes extending as a saddle across entire lower back. Scrotum of adult males white. Hair of head, limbs, and tail relatively short. Subadult males indistinguishable from females in the field. Body shape like red howler.

Variation. Animals from the Azuero peninsula and Coiba Island are sometimes considered a separate species, *A. coibensis.*

Similar species. Mexican black howler monkeys *(A. pigra)* are entirely black and live in small groups; Central American spider monkeys *(Ateles geoffroyi)* are black, brown, or red, with a paler belly and face often pale around eyes; Colombian black spider monkeys *(A. g. fusciceps)* are entirely black or have a brown head; spider monkeys have small heads, no beard, long arms and tail, and make whinnying calls.

Map 111

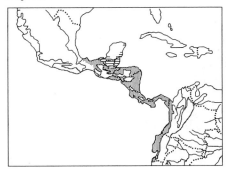

Mantled howler monkey, *Alouatta palliata*
Mexican black howler monkey, *A. pigra*

Sounds. Roars as described for red howler, but often with a series of single loud grunts emitted by a single animal.

Natural history. Diurnal; arboreal; groups of 2–45, usually 10–18. Feeds on fruit and leaves. Mantled howlers use the upper levels of the forest. They are slow-moving and sedentary, often inconspicuous. Like other howlers, their presence is shown by strong-smelling droppings on the ground below them, and they are prone to defecate on the heads of people below. They have small home ranges, so they can survive in small fragments of forest. In protected areas populations can be extremely high. These are the monkeys most often seen in parks and around ruins in archaeological sites. Found in mature and old secondary evergreen, semi-deciduous, and riverine forests.

Geographic range. Central and South America: E Mexico south through Panama, except the Yucatán Peninsula (see following species); west of the Andes from Colombia to N Peru. To 1,500 m elevation.

Status. CITES Appendix I, US-ESA endangered. IUCN Red List critically endangered, *A. p. coibensis;* vulnerable, *A. p. mexicana.* Threatened by deforestation in many parts of its range, which is now fragmented; hunted for food in some areas.

Local names. Mono negro (Co, Pn); mono congo (CR); mono chongo, chongón (Co), coto negro, mongón, aullador (Ec, Pe); olingo (Ho); saraguato, mono aullador (Me).

References. Carpenter, C. R. 1934. A field study of the behavior and social relations of howling monkeys *(Alouatta palliata). Comp. Psychol. Monogr.* 10:1–168.

Milton, K. 1980. *The foraging strategy of howler monkeys.* New York: Columbia University Press.

Mexican Black Howler Monkey
Alouatta pigra
Plate 13, map 111
Identification. Measurements: HB = 521–639; T = 590–690; HF = 133–165.
Entirely black; hair long. Scrotum of males over 3–4 months old white and prominent. Body shape like red howler.
Similar species. Mantled howlers *(A. palliata)* have a pale saddle or fringe on sides; Central American spider monkeys *(Ateles geoffroyi)* in same area are brown or silvery with pale bellies and pale masks around eyes.
Sounds. Like mantled howler. Usually howls in late afternoon and early morning.
Natural history. Diurnal; arboreal; groups of 2–10, usually 4–6. Feeds on fruit and leaves. These monkeys can often be seen in forests around archaeological sites. Their behavior apparently is similar to that of other howlers. Found in evergreen and riverine lowland forests.
Geographic range. Central America: Yucatán Peninsula of Mexico, Guatemala, and Belize.
Status. CITES Appendix I, US-ESA endangered. Threatened by deforestation and hunting for meat, and some populations may have been decimated by yellow fever epidemics.
Local names. Mono aullador (Span); baboon (Be); saraguate (Me); batz (May).
References. Horwich, R. H., and C. Gebhard. 1983. Roaring rhythms in black howler monkeys *(Alouatta pigra)* of Belize. *Primates* 24:290–96.

Schlichte, H.-J. 1978. A preliminary report on the habitat utilization of a group of howler monkeys *(Alouatta villosa pigra)* in the National Park of Tikal, Guatemala. In G. G. Montgomery, ed., *The ecology of arboreal folivores,* 551–59. Washington, D.C.: Smithsonian Institution Press.

Common Woolly Monkey
Lagothrix lagothricha
Plate 14, map 112
Identification. Measurements: HB = 390–580; T = 550–800; HF = 118–158; E = 25–37; WT = 3.6–10.2 kg.

Map 112

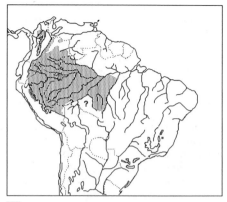

▨ Common woolly monkey, *Lagothrix lagothricha*
▤ Yellow-tailed woolly monkey, *L. flavicauda*

Upperparts dark brown, pale smoky brown, dark smoky gray, pale gray, red-brown, or olivaceous; limbs and tail and/or head sometimes darker than back, brown or reddish; but in general monkey **looks uniformly colored in the field;** fur soft and woolly, short and even, except old animals have a fringe of longer hair on rear of arms and legs and under belly. **Head conspicuously round, face black;** crown with short, even hair; **ears inconspicuous. Tail strongly prehensile, thick and muscular at base, tapering to thinner tip. Limbs and body robust and muscular; pot belly rounded and protruding. Underparts dark.** Newborns straw-colored.
Variation. There is much individual and geographic variation: in Ecuador and Colombia animals of different colors can be seen in the same troop; gray or almost black animals are found along the base of the Andes in Colombia; olivaceous animals with dark heads south of the Amazon in Brazil and Bolivia; brown animals in Peru and north of the Amazon in Brazil. There are reports of gray *(L. l. cana)* and brown *(L. l. poepiggii)* animals existing segregated in the same region, perhaps evidence that there is more than one species.
Similar species. Howler monkeys *(Alouatta* spp.) are red-orange or black, with bearded chins; spider monkeys *(Ateles* spp.) are black, or black or brown with pale underparts, with small pointed heads and long,

lanky arms, legs, and tail; capuchins (*Cebus* spp.) are smaller, yellow-brown with pink or brown faces and black caps.

Sounds. Loud, descending musical trills, barks, screams, and other calls; alarm a chorus of "yoohk-yoohk."

Natural history. Diurnal; arboreal; groups of 2–70. Larger groups subdivide into two to three subunits that reunite from time to time. Groups are not territorial, and each overlaps several others. Occasionally, neighboring troops will unite temporarily into a large group of 40 to 60 or more monkeys. Common woolly monkeys feed on ripe fruits, new leaves, and some seeds and arthropods. They favor fruits with adherent pulp and large seeds, which they generally swallow whole and thus disperse. Daily foraging paths are long (2–3 km), and home ranges are large (700–900 ha). They use the upper and middle levels of the forest, and seem to prefer tall, terra firme habitats. Like many cebids, male woollys threaten humans by shaking branches and defecating. Common woolly monkeys seem to reproduce slowly, with females first giving birth at 6–8 years old and every other year thereafter. Found in mature terra firme and seasonally flooded rainforest; apparently not in secondary forests, but this may be because humans hunt them to extinction wherever they cut the forest.

Geographic range. South America: upper Amazon Basin of Colombia, Ecuador, Peru, and Brazil west of the Rios Negro and Tapajós, and isolated populations in Córdoba and Bolívar Departments of Colombia. To at least 1,800 m elevation.

Status. CITES Appendix II. IUCN Red List critically endangered, *L. l. lugens;* vulnerable, *L. l. cana* and *L. l. poeppigii.* The most intensively hunted monkey; its meat is considered better than that of other large species. Extirpated or naturally absent from many areas. This species is unable to maintain its populations under much hunting pressure and is usually the first species to disappear where subsistence hunting is intense. Intolerance to disturbance of vegetation and a low reproductive rate make this species vulnerable to local extirpation, but its geographic range is large.

Local names. Macaco barrigudo (Br); choyo, churuco, chuluco (Co); choro, barrigudo (Ec, Co, Pe); chorongo (Ec).

References. Defler, T. R. 1989. Recorrido y uso del espacio en un grupo de *Lagothrix lagothricha* (Primates: Cebidae) mono lanudo churuco en la Amazonia Colombiana. *Trianea* 3:183–205.

Peres, C. A. 1994. Diet and feeding ecology of gray woolly monkeys *(Lagothrix lagothricha cana)* in Central Amazonia: Comparisons with other Atelines. *Int. J. Primatol.* 15:333–72.

Yellow-tailed Woolly Monkey
Lagothrix flavicauda
Plate 14, map 112

Identification. Measurements: HB = 515–535; T = 560–610.

Upperparts glossy dark red-brown; hair long and thick, especially on legs, giving **body a muscular, robust look. Face brown with pale yellow triangular patch covering mouth and nose, point of triangle between eyes. Tail prehensile, robust, dark brown above, yellow below for distal half.** Underparts dark brown except genital area with a tuft of yellow or reddish hairs, prominent only in males.

Similar species. White-bellied spider monkeys *(Ateles belzebuth)* are black or brown, with distinctly pale underparts; red howler monkeys *(Alouatta seniculus)* are orange-red and have a beard; capuchins *(Cebus* spp.) are yellow-brown with black caps.

Sounds. A sharp, puppylike bark, like the bark of spider monkeys.

Natural history. Diurnal; arboreal; groups of 6–12. Feeds on fruits, flowers, roots of epiphytes, and leaf petioles. The largest male in a group may threaten an observer by shaking and dropping branches and by urinating and defecating. Yellow-tailed woolly monkeys are found only in steep, humid premontane and montane forests.

Geographic range. South America: NE Peru in a small region in the Departments of Amazonas, San Martín, and La Libertad, on the Andean slopes at 1,800–2,500 m elevation.

Status. CITES Appendix I, US-ESA endangered, IUCN Red List critically endangered. Apparently rare; the geographic range is small and fragmented by valleys, although steep and inaccessible. The species is partly protected in Río Abiseo National Park. Locally hunted for meat. This monkey was discovered in 1802 by the explorer Von Humboldt, but it was not recorded again until

1925–26, when five specimens were taken. No more were recorded for 50 years, and the species was thought to be extinct, but it was rediscovered in 1974 by an expedition launched to search for it.

Local names. Paccorrunto, tupa, chú, choba, mono barroso, quillirunto, mono choro.

References. Leo Luna, M. 1980. First field study of the yellow-tailed woolly monkey. *Oryx* 15:386–89.

Macedo-Ruiz, H. de, and R. A. Mittermeier. 1979. Redescubrimiento del primate peruano *Lagothrix flavicauda* (Humboldt 1812) y primeras observaciones sobre su biología. *Rev. Cienc. Univ. Nac. San Marcos* 71:78–92.

Black Spider Monkey
Ateles paniscus
Plate 14, map 113

Identification. Measurements: HB = 430–620; T = 710–900; HF = 175–220; E = 30–40; WT = 7.5–13.5 kg.

Entirely black. Hair long, giving the body a robust, apelike appearance. Tail thickly haired for two-thirds of its length, tapering sharply at prehensile tip. Head small, narrow at crown; **face red; pointed tufts of hair protrude sideways from in front of ear. Size large; arms, legs, and prehensile tail extremely long;** hands with only four functional digits, thumb rudimentary. **Often swings by arms below branches and hangs supported by tail.**

Variation. Black-colored spider monkeys *(Ateles belzebuth chamek)* from the upper Amazon Basin south of the Amazon of Peru, Bolivia, and Brazil were formerly placed in this species. Following the reference below, we consider those a subspecies of *A. belzebuth.*

Similar species. North of the Amazon these are the only large, prehensile-tailed, completely black monkeys with a red face; bearded sakis (*Chiropotes* spp.) have bushy, nonprehensile tails. Spider monkeys are the only monkeys in their geographic range that regularly travel by swinging by their arms from branch to branch.

Sounds. Loud, prolonged screams that can be heard for long distances; short quavering whinnies heard at close range; in alarm, bark like a pack of dogs. Noisy travelers that make long, crashing leaps through the canopy.

Map 113

▨ White-bellied spider monkey, *Ateles belzebuth*
▤ Black spider monkey, *A. paniscus*
▦ Muriqui, *Brachyteles arachnoides*

Natural history. Diurnal; arboreal; large social groups of about 20, but members of a group usually feed in small units of 1–5, and the whole social group is rarely seen together. Feeds on ripe fruits and some new leaves and flowers. Spider monkeys use the upper levels of the trees, usually in tall, open forest. They often sprawl quietly on a branch or hang motionless and soundless and may be difficult to spot. When disturbed they may threaten a human by standing on a branch and stamping and shaking the vegetation, breaking and dropping dead branches, and growling. Black spider monkeys have a low reproductive rate: females first give birth at 4–5 years old and have young only every 3–4 years thereafter. Found mainly in tall, undisturbed evergreen rainforest.

Geographic range. South America: Brazil and the Guianas east of the Rio Negro and north of the Amazon.

Status. CITES Appendix II. Common where not hunted, locally extinct, or threatened from overhunting for meat in inhabited areas, but widespread, and parts of range are uninhabited.

Local names. Coatá, macaco-aranha (Br); kwata (FG, Su); spinaap, slingeraap (Su).

References. Mittermeier, R. A., and M. G. M. van Roosmalen. 1981. Preliminary observations on habitat utilization and diet in eight Surinam monkeys. *Folia Primatol.* 36: 1–39.

Froehlich, J. W., J. Supriatna, and P. H. Froehlich. 1991. Morphometric analyses of *Ateles:* Systematic and biogeographic implications. *Am. J. Primatol.* 25 : 1–22.

White-bellied Spider Monkey

Ateles belzebuth

Plate 14, map 113

Identification. Measurements: HB = 416–582; T = 680–899; HF = 176–217; E = 31–46; WT = 5.9–10.4 kg.

Upperparts black, pale brown, dark brown, or reddish brown. Underparts black, or sharply contrasting paler white, yellowish, or pale brown. Head entirely black or with white or pale brow, triangular patch on brow, or white fringe completely surrounding face; face black or red. Hands, feet, and forearms like upperparts. Tail dark like upperparts, or dark above, pale below. Shape and posture like black spider monkey.

Variation. Animals from Amazonian Colombia and Venezuela are black, with yellow to gold underparts, legs and tail gold frosted with black, and little or no white on face. *A. b. marginatus* in Pará is black, with a black belly, white brow patch, and red mask; *A. b. chamek* from S Peru, Bolivia, and W Brazil is completely black. In some populations there is much individual variation, and recent work suggests that neighboring subspecies intergrade to form a ring of populations around the Amazon River.

Similar species. These are the only large, pale-bellied, prehensile-tailed monkeys; in the range of the black-bellied form *(A. b. marginatus)* south of the Amazon, red-handed howlers *(Alouatta belzebuth)* have large, bearded heads and robust limbs; woolly monkeys *(Lagothrix lagothricha)* are uniform brown or gray; bearded sakis *(Chiropotes* spp.) have bushy, nonprehensile tails.

Sounds. Screams, barks, and whinnies like black spider monkey.

Natural history. Diurnal; arboreal; groups of 20–40, members of groups often split into small parties of 1–9. Feeds on ripe fruits and some dead wood and leaves. Party size decreases when fruit becomes scarce, but the whole group may unite when fruit is abundant. Members of a group share a home range of 150–250 ha, which may overlap slightly with those of neighboring groups. There are

three times as many adult females as adult males in a group. Spider monkeys use the upper levels of tall forest. They spend much of the day resting and feed most intensively in early morning and late afternoon. They sometimes feed on bright moonlit nights (as do brown capuchins), and they have a distinctive, moderately bright whitish eyeshine. They are nervous at night and sometimes bark for many minutes. Their reproduction, postures, and locomotion are like those of black spider monkeys. Found in mature rainforest and deciduous forest.

Geographic range. South America: east of the Andes from N Colombia and Venezuela; the upper Amazon Basin of Peru, Bolivia, and Brazil south of the Rios Amazonas and Solimões. To 1,800 m elevation.

Status. CITES Appendix II. IUCN Red List endangered, *A. b. brunneus, A. b. marginatus;* vulnerable, *A. b. belzebuth.* Threatened by deforestation in northern part of range; Amazonian populations are subject to similar pressures as black spider monkeys.

Local names. Macaco-aranha, coatá (Br), marimono, mono araña (Bo); braceador, choiba, coatá (Co); marimonda (Co, Ec, Ve); maquisapa (Pe); ca'í (Gua).

References. Symington, M. McFarland. 1988. Demography, ranging patterns, and activity budgets of black spider monkeys *(Ateles paniscus chamek)* in the Manu National Park, Peru. *Am. J. Primatol.* 15 : 45–67.

———. 1988. Food competition and foraging party size in the black spider monkey *(Ateles paniscus chamek). Behaviour* 105 : 117–34.

Central American Spider Monkey

Ateles geoffroyi

Plate 14, map 114

Identification. Measurements: HB = 305–630; T = 635–855; HF = 135–196; E = 30–45; WT = 6.6–9.0 kg.

Upperparts black, brown, or reddish. Face often with "mask" of pale, unpigmented skin around eyes and muzzle. Forearms, lower legs, and feet usually black or dark. Underparts usually paler than back: white, pale brown, reddish or buff, but black in some populations. **Body shape and posture like black spider monkey.**

Variation. Highly variable both individually and geographically; animals from W Mexico,

Map 114

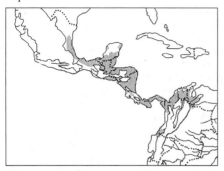

Central American spider monkey, *Ateles geoffroyi*

Guatemala, Honduras, and El Salvador are mostly blackish above; those from the Yucatán Peninsula, Nicaragua, and Costa Rica are brown to silvery above; those from Panama are red-brown with a reddish belly, except for one grayish population. Those from the west coast of Colombia and Ecuador *(A. g. fusciceps)* are entirely black, sometimes with brown heads. Those from N Colombia and Zulia, Venezuela *(A. g. hybridus),* formerly considered a subspecies of *A. belzebuth,* are pale brown above with a darker brown head and white brow triangle, and whitish underparts; some have blue eyes.

Similar species. Howler monkeys (*Alouatta* spp.) are black or red, with dark faces, large, bearded heads, and heavy shoulders; they do not travel by swinging by their arms from branch to branch. There are no other large monkeys in Central America.

Sounds. Barks, whinnies, and screams like black spider monkey.

Natural history. Diurnal; arboreal; groups of 1–35; members of large groups frequently split into smaller groups to forage. Feeds on ripe fruits, with a few leaves and flowers. Spider monkeys use the upper levels of the forest. The locomotion, reproduction, and behavior of Central American spider monkeys are similar to those of black spider monkeys, except that a female may have young every 2–4 years. Found in mature rainforest and deciduous forests.

Geographic range. Central America: both coasts of S Mexico south to Colombia; Darién, Panama, west of the Andes south to central Ecuador; and the northern coast

of Colombia and Venezuela to the gulf of Maracaibo and the Cauca and Magdalena valleys. To 1,600 m elevation.

Status. CITES Appendix I. US-ESA endangered (Pn, CR, Ni). IUCN Red List critically endangered, *A. g. azuerensis* and *A. g. fusciceps;* endangered, *A. g. grisescens, A. g. panamensis,* and *A. g. hybridus;* vulnerable, *A. g. frontatus, A. g. yucatanensis,* and *A. g. ornatus.* Threatened by deforestation and fragmentation of range and by hunting for meat; common in a few parks, but locally extirpated in many areas. May also suffer from yellow fever epidemics. Spider monkeys seem to need large areas of unbroken forest for survival.

Local names. Monkey (Be); mono araña, mico (Span); mono negro (Co); marimonda (Co, Ec, Ve); brasílargo (Ec); mono colorado, yerré (Pn, CR); maax (May).

References. Carpenter, C. R. 1935. Behavior of red spider monkeys in Panama. *J. Mammal.* 16:171–80.

Richard, A. 1970. A comparative study of the activity patterns and behavior of *Alouatta villosa* and *Ateles geoffroyi. Folia Primatol.* 12:241–63.

Muriqui or Woolly Spider Monkey
Brachyteles arachnoides
Plate 14, map 113
Identification. Measurements: HB = 462–630; T = 650–800; WT = 12–15 kg.
Upperparts and underparts pale, smoky, grayish gold-beige; rump, basal half of tail, and sometimes rear of ankle pale orange; fur dense and woolly. **Head round, face black or pale and speckled,** fringed with pale ring; crown slightly darker grayish brown than body; **ears thickly haired, standing out from head** like teddy bear, dark brown at center. **Arms, legs, and prehensile tail long and thin; belly large and protruding.** Thumb vestigial. The largest Neotropical monkey. Swings below branches by arms and tail.

Similar species. Brown howler monkeys *(Alouatta fusca)* are dark brown to dark red, or yellow-brown with darker legs and head, with large head, prominent beard, and naked ears; brown capuchins *(Cebus apella)* are smaller, dark brown, with black limbs and tail; there are no other large monkeys in range.

Sounds. Loud, piercing screams, whinnying or neighing sounds, barks in alarm, staccato chutters.
Natural history. Diurnal; arboreal; groups of 8–42. Groups include similar numbers of adults of both sexes, but where populations have been decimated, skewed sex ratios can result. Feeds chiefly on fruit and leaves, with some flowers. Muriquis use the upper levels of tall forest. They are the most specialized New World leaf-eating monkeys, but they eat more fruit than leaves when it is available, and will linger at a fruit tree until the fruit is gone. They are generally inactive: two-thirds of the day is spent resting, and they do not begin activity until well after dawn. Like spider monkeys, they may threaten an observer by jumping and shaking branches and defecating. Found in mature evergreen to deciduous lowland Atlantic forest.

Geographic range. South America: Brazil, SE coastal forests from Bahia to São Paulo.
Status. CITES Appendix I, US-ESA endangered, IUCN Red List endangered. Only 700–1,000 individuals are thought to remain, scattered widely in about 25 small fragments of forest. Only 2–4% of the original forest habitat of this species has not been deforested. Formerly intensively hunted for meat; now close to extinction owing to habitat loss and continued hunting.
Local names. Muriqui, mono-carvoeiro.
References. Strier, K. B. 1992. *Faces in the forest: The endangered muriqui monkeys of Brazil.* New York: Oxford University Press.

Strier, K. B., F. D. C. Mendes, J. Rímoli, and A. O. Rímoli. 1993. Demography and social structure of one group of muriquis (*Brachyteles arachnoides*). *Int. J. Primatol.* 14:513–26.

Carnivores (Carnivora)

All of the five families of New World carnivores (Canidae, Ursidae, Procyonidae, Mustelidae, and Felidae) are found in the rainforest. In general, carnivores are adapted to find, catch, and kill animal prey, although they also have other ecological roles. The skull, muscles, and teeth are fitted for giving a powerful puncturing, crushing, or slicing bite, but do relatively little grinding of the food. The canines are usually large, and the premolars are often developed into blades with a scissorlike shear. The range of diets is wide. While some species are truly carnivorous or meat-eating, others feed mainly on insects, fruits, or even leaves. Mammalian carnivores do not usually specialize on particular species of prey; most are opportunists that will eat whatever they encounter that can be caught and killed without danger. Perhaps for this reason, the meat-eating Carnivora usually have geographic ranges much larger than those of any single species of their prey (e.g., puma eat different species of deer in different parts of their range). Purely meat-eating species generally live at low densities; those that also eat insects and fruit are often numerous in good habitat. Carnivores have an important and beneficial role in ecosystems. We follow the taxonomy of W. C. Wozencraft, "Order Carnivora" (in D. E. Wilson and D. M. Reeder, eds., *Mammal species of the world*, 279–348. Washington, D.C.: Smithsonian Institution Press, 1993).

Dog Family (Canidae)

Dental formula usually I3/3, C1/1, P4/4, M2/3 = 42; the bush dog is unique, with M1/2 = 38. There are four weight-bearing toes on each foot; a few species have a fifth "dewclaw" that does not bear weight or appear in tracks. Canids are long-legged and digitigrade (walk on the toes, the heel not touching the ground), with long, pointed snouts, straight backs, bushy tails, and wet noses. Their senses of smell and hearing are acute, and most also see well. They can travel long distances and run economically, and many species track prey by following a scent trail. Canids are carnivores or scavengers that feed on vertebrate and/or invertebrate prey, but most species will eat fruit if prey is scarce. They do not have a "killing bite," and they kill their mammal prey by shaking it to break the back if it is small or by disabling it with bites to the legs or nose and tearing and worrying messily at vital organs if it is large. In a number of species, several individuals join in a pack to immobilize and kill prey larger than themselves. This requires complex social organization, cooperation, and strategy. Even species that forage solitarily for small prey often live in monogamous pairs or groups. They are in general highly intelligent (good at learning, remembering, and solving problems). Canids usually have large litters of pups that are born helpless and with eyes and ears closed (altricial). Pups are raised in a burrow or den and are usually fed prey by both parents (or all members of a pack) as they are weaned. The world's canid species are mostly occupants of savanna woodlands, grasslands, and woodlands and forests that have an open canopy or are interspersed with open meadows. The only two deep tropical rainforest dogs in the world are the two described in this book, which are of particular interest because of their unique habitat. Unfortunately, both are rare, and little is known about their behavior in the wild. Other canids occur on all the fringes of the entire rainforest zone, including coyotes *(Canis latrans),* maned wolves *(Chrysocyon brachyurus),* and several foxes *(Lycalopex* spp., *Urocyon* spp.). There are about 36 species in about 11 genera worldwide, 2 genera each with one species in the Neotropical rainforest, and a third species (included here) from the forest edge.

Short-eared Dog
Atelocynus microtis
Plate 15, map 115
Identification. Measurements: HB = 583–1,000; T = 260–350; HF = 140–150; E = 34–65; SH = 350–356; WT = 6.5–9 kg.

Upperparts grizzled blackish gray; hair short, dense, stiff and oily. Head large, grizzled more brownish than back; **muzzle with a marked black streak slanting from nose to beneath the eye;** neck long and thick; **ears small but protruding well above**

Map 115

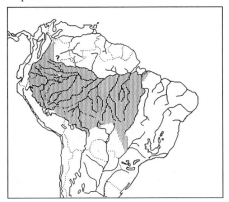

▨ Short-eared dog, *Atelocynus microtis*

Geographic range. South America: east of the Andes in Colombia, Ecuador, Peru, and Bolivia; Brazil south of the Amazon from the Rio Toncantins to Mato Grosso, in the Rio Paraguay Basin. To 1,000 m elevation.
Status. Appears to be rare throughout its range, but there are regular sightings on the Peru-Bolivian border and in Pando, Bolivia, where they may be relatively common.
Local names. Cachorro do Mato (Br); zorro ojizarco (Co); zorro negro, zorro de monte (Span).
References. Defler, T. R., and A. Santacruz. 1994. A capture of and some notes on *Atelocynus microtis* (Sclater, 1883) (Carnivora: Canidae) in the Colombian Amazon. *Trianea* 5:417–19.
    Berta, A. 1986. *Atelocynus microtis.* Mammalian Species, no. 256.

crown, rounded, contrasting pale brown against black head; eyes blue or hazel; eye-shine bright pale green; tips of upper canines visible when mouth is closed. **Tail bushy, long enough to touch the ground, black** except pale beneath base. Legs slender, dark brown or black, feet small, webbed between toes. Underparts grizzled brown. Young like adults.
Variation. Placed by some in the genus *Dusicyon* or *Canis.*
Similar species. Bush dogs *(Speothos venaticus)* are smaller, with a pale tan head, short legs, and short tail; tayras *(Eira barbara)* are smaller, with inconspicuous ears and usually a contrasting gray head and yellow throat patch; crab-eating foxes *(Cerdocyon thous)* are grizzled tawny, with tail shorter than legs; jaguarundis *(Herpailurus yagouaroundi)* have a small head and slender tail.
Sounds. Growls when threatened.
Natural history. Terrestrial; probably diurnal and crepuscular; solitary. The habits of short-eared dogs are little known. They have been observed preying on a frog and pursuing a spiny rat, and are said to also chase other small rodents, agoutis, and pacas, and to eat fruit. There is an anecdotal record of a den with two pups in a hollow log. These dogs move with catlike grace. When excited, males emit a strong musky odor. Most records are from lowland tropical rainforest, but they may inhabit dry forests in Bolivia and S Pará, Brazil.

Crab-eating Fox
Cerdocyon thous
Figure 8, map 116
Identification. Measurements: HB = 590–765; T = 243–335; HF = 132–165; E = 66–80; WT = 3.6–7.9 kg.
Upperparts grizzled tawny gray, lightly to heavily streaked with black; midback usually with black streak from neck to tail; muzzle darker brown than cheeks; ears medium-sized, thinly haired, without thick pale hair on inner rim near tips, **dark brown behind at tips;** neck behind ears usually tawny orange. **Tail** moderately bushy, shorter than hindleg, **black at tip and usually black above, continuous with black midline on back, pale below near base. Feet and lower legs usually brown, darker than body, often dark brown.** Underparts cream to buff. Young entirely dark brown when tiny, dark gray with black tail and dark brown feet when half-grown.
Variation. Amount of dark color variable; an animal from Paraguay has pale feet and no back stripe, but foxes near rainforest seem to have dark extremities.
Similar species. Gray foxes *(Urocyon cinereoargenteus)* of northern coastal savannas have a bright fulvous collar around neck from behind ears. Short-eared dogs *(Atelocynus microtis)* are entirely dark brown and blackish. Bush dogs *(Speotheos venaticus)* have short legs and tail and have pale brown forequarters and dark hindquarters. Other foxes of southern cerrado and pampa

Map 116

Crab-eating fox, *Cerdocyon thous*

Map 117

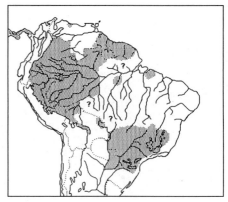

Bush dog, *Speothos venaticus*

(*Lycalopex* spp.) have pale feet and thick whitish hair inside large ears; *L. culpaeus* has a white chin and no dark spot on reddish thigh; *L. gymnocercus* has a dark chin and dark spot on rear of thigh.
Sounds. "Sings" with a tonal cry.
Natural history. Nocturnal and crepuscular; solitary and in pairs. Feeds on small vertebrates (rodents, frogs, reptiles), invertebrates including crabs and insects, and fruit. A male and female pair occupies a common territory of at least 45–100 ha. Members of pairs often travel and rest together, but each captures its own food. These foxes use llanos, savanna, cerrado, and dry forest habitats, but they enter evergreen rainforests when their ranges abut them. They are widespread even within savannas now isolated in forest, such as in S Suriname, and in dry or semi-deciduous forests.
Geographic range. South America: encircling the Amazon Basin in Colombia, Venezuela, the Guianas, Brazil, Bolivia, Paraguay, Argentina, and Uruguay. To 2,000 m elevation.
Status. Widespread and often common.
Local names. Zorro (Span); graxaim, raposa (Br); zorro perruno, zorro perro (Co); zorro de monte (Pe).
References. Sunquist, M. E., F. Sunquist, and D. E. Daneke. 1989. Ecological separation in a Venezuelan llanos carnivore community. In K. H. Redford and J. F. Eisenberg, eds., *Advances in Neotropical mammalogy,* 197–232. Gainesville: Sandhill Crane Press.

Berta, A. 1982. *Cerdocyon thous.* Mammalian Species no. 186.

Bush Dog
Speothos venaticus
Plate 15, map 117
Identification. Measurements: HB = 610–750; T = 110–130; HF = 110–120; E = 40–51; WT = 5–7 kg.
Head and neck to top of shoulders pale brown to tawny yellow, darkening gradually to black or dark brown hindquarters; fur long and soft. Muzzle short; ears short and rounded; eyes brown. **Tail short and stumpy, black,** thickly furred. **Legs very short, black** or dark brown; feet webbed. Underparts dark brown or black, sometimes chest with white spot. Back long, body cylindrical, thick. Young gray-black.
Variation. Some bush dogs from Brazil have completely pale brown backs.
Similar species. Tayras *(Eira barbara)* have longer tails, yellow throat patches, and small ears, and they climb trees; short-eared dogs *(Atelocynus microtis)* are grizzled blackish gray, with a bushy tail long enough to touch the ground, longer legs, and short fur.
Sounds. Groups communicate with high-pitched whining (as do coatis and several monkeys); they yap while chasing prey.
Natural history. Diurnal; terrestrial; usually in small groups of four to seven or more, but sometimes alone. They have been reported attacking pacas and, anecdotally, agoutis, acouchys, and mice. Captives swim well,

Figure 8. Crab-eating fox, *Cerdocyon thous.*

and in an anecdotal report, two swam underwater in pursuit of a swimming paca. They may hunt in packs, yapping like puppies as they pursue prey. The teeth of bush dogs are more specialized for meat-eating than those of other New World canids, and they may be strongly carnivorous. Bush dogs den in burrows, the whole group together. They seem rare everywhere, and few forest Indians report ever having seen one. Most information on wild bush dogs consists of undocumented anecdotes. Found in a range of habitats from rainforest to wooded savanna, but the most records seem to be from dry forests and rainforests near savannas.

Geographic range. Central and South America: W Panama south to Bolivia, Paraguay, and N Argentina. To 1,500 m elevation.

Status. CITES Appendix I. Extremely rare in a large geographic range.

Local names. Zorro vinagre (Ar, Pe); cachorro-do-mata (Br); perrito venadero (Co); perro de monte (Co, Ec, Pn, Pe); guanfando (Ec); boshond, busdagoe (Su);.

References. Peres, C. A. 1991. Observations on hunting by small-eared *(Atelocynus microtis)* and bush dogs *(Speothos venaticus)* in central-western Amazonia. *Mammalia* 55:635–39.

Bear Family (Ursidae)

Dental formula I2–3/3, C1/1, P4/4, M2/3 = 40–42, but first three premolars are small and one or more may be missing. Bears include the largest living terrestrial carnivores. They have massive bodies, stout legs, broad plantigrade feet with five forward-pointing toes armed with long claws, and tiny, inconspicuous tails. The fur is dense and coarse. The head is broad, with small ears, a long snout, and large, conical canine teeth. Most bears are omnivores that eat both meat and plant material, but polar bears eat little but meat, and giant pandas feed almost entirely on plants (bamboo). Despite their size, bears are good climbers that can ascend trees to rest, escape danger, or forage on localized foods such as fruit or honey. They are also powerful diggers and can unearth tubers and bulbs, or rodents from their burrows. They have poor, myopic vision, but a good sense of smell. Apart from polar bears, they are not specialized hunters, and take easily captured, vulnerable prey, which they may kill with bites and blows from the forefeet. They also scavenge carrion. Bears give birth in a den to 1–3 minuscule, altricial young (200–700 g), which are raised by the mother alone. Young stay with the mother for 2–4 years, until her next litter is born. The time between litters depends upon the food supply. The best-known species seem to have a difficult life and produce few young. This may be why they readily take to artificial food supplies such as garbage dumps and cornfields. Worldwide, there are only six genera and eight species of bears, one genus and species in the Neotropics.

Andean Bear
Tremarctos ornatus
Fig. 9, map 118
Identification. Measurements: HB = 1,120–2,200; T = 70–75; HF = 195; E = 93; WT = 80–175 kg; males larger than females.
Upperparts entirely pure black or with dark, red-brown tones; head marked with a broad white to yellow band across brow, down cheeks and muzzle, encircling eyes and joining a large triangular patch on chest; these markings variable from complete to partial or absent; muzzle dark brown. Fur dense, long and coarse. Tail small and inconspicuous.
Variation. There is much individual variation in the extent of facial markings and general color. In Peru, local people often believe there are two species of bears, a carnivorous one that kills livestock and a herbivorous one. Color variation encourages such folklore.
Similar species. None, but the presence of bears is usually determined by signs of feeding or claw marks, and it should be noted that cloud forest *Cebus* monkeys also rip apart arboreal and cliff-hanging bromeliads to eat the hearts, and big cats can also leave scratch marks on trees.
Natural history. Nocturnal and diurnal; terrestrial and partly arboreal; solitary. Andean bears are primarily vegetarian, occasionally eating meat of rodents and large ungulates, which may be scavenged. Their dominant food is the hearts of bromeliads, supplemented by fruit, bulbs, palm leaf petioles, bamboo, and other vegetation. They often raid cornfields. They climb to reach arboreal bromeliads and fruit, and may build a platform of branches, resembling a nest, to support them while they reach for food, and later to rest on. They make tunnels through dense vegetation such as bamboo thickets. These bears are montane and generally live in forested habitats between 1,800 m and

Map 118

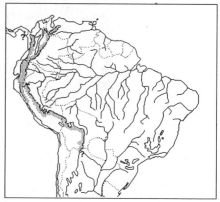

Andean bear, *Tremarctos ornatus*

3,300 m elevation, but when food is scarce above, they will descend to as low as 500 m to feed in maize fields.
Geographic range. South America: the Andes and their outliers from Colombia and Venezuela to N Argentina.
Status. CITES Appendix I. These bears seem present throughout most of their theoretical range, which is both geographically large and difficult of access. They are hunted for parts for use in folk medicine, and killed when they destroy crops. Northern and southern regions of the range are severely deforested.
Local names. Oso de anteojos (Span); uco (Ar); ucumari, jucumari, frontino, tomasito (Bo); oso careto, oso congo (Co); ucumari, ucucu, meéni (Pe).
References. Peyton, B. 1980. Ecology, distribution, and food habits of spectacled bears, *Tremarctos ornatus,* in Peru. *J. Mammal.* 61:639–52.
  Salazar, J., and S. Anderson. 1990. Informe sobre el estado actual del conocimiento del oso andino en Bolivia. *Ecologia en Bolivia* 15:3–23.

Figure 9. Andean bear, *Tremarctos ornatus.*

Raccoon Family (Procyonidae)

Dental formula: I3/3, C1/1, P4/4, M2/2 = 40, except kinkajou with P3/3. Five well-developed toes on all feet. Procyonids are medium-sized; have pointed muzzles; broad, crushing molars; quite forward-set eyes; dense, dull fur; short legs; and plantigrade feet (their heels touch the ground when they are standing). Their hands are mobile, and they are able to dig and rummage for prey or hold and manipulate food items. They are omnivores that feed on invertebrates such as insects and crabs, small vertebrates such as frogs, snakes, and nestling birds, fruit, and flower nectar. Members of this family are all good tree climbers, although some forage mainly on the ground. Several species are entirely arboreal, but even the most terrestrial species climb trees to escape danger when alarmed and to sleep during inactivity. All species raise their young in arboreal nests (although northern raccoons on the prairies may den on the ground). The young are tiny at birth. Kinkajous have a single young; the others a litter of three to seven. Weanlings follow the mother to forage. There are 6 genera and about 13–16 species, all in the New World, and all with ranges entering the Neotropics. Pandas were once thought to belong to this family, but are now placed with the bears. Five genera and 7–9 species of procyonids are found in the rainforest region.

Crab-eating Raccoon
Procyon cancrivorus
Plate 15, map 119
Identification. Measurements: HB = 543–650; T = 252–380; HF = 130–150; E = 50–60; WT = 3.1–7.7 kg.
Upperparts blackish grizzled with gray or rusty brown; sides grizzled rusty brown or gray; fur short and stiff or medium length, usually with no underfur; hair on back of neck "reversed," slanting forward. Head foxlike, broad, with pointed muzzle; **face grizzled whitish traversed by broad black "mask" around eyes, mask usually fading** out just behind eyes; **ears relatively short, white** inside and out. **Tail** about 50% as long as head and body, moderately bushy, **with prominent, wide, black and pale rings. Legs and feet dark brown.** Underparts deep rust to whitish; chin and throat grizzled whitish like face; band across throat, if present at all, faint, dirty white or grizzled tan. Body thick; legs and feet slender, with long, dexterous fingers and no webs between toes; rump much higher than forequarters, tail carried low; strong, distinctive odor. Eyeshine bright orange; eyes medium-sized and far apart.

Map 119

Crab-eating raccoon, *Procyon cancrivorus*

Variation. Grizzled coloring is due to banded hairs; the tips of the hairs are black and the basal parts of the back hair, belly hair, and paler rings on the tail vary from dark rust-red to whitish. Raccoons from Argentina and E Brazil have underfur and white bellies and more closely resemble northern raccoons.
Similar species. Northern raccoons *(P. lotor)* have whitish forelegs and usually feet; normal, backward-slanting hair on neck; and prominent brown band across throat extending from mask that reaches well behind eyes. Coatis *(Nasua* spp.) have no mask and a long snout and carry their slender tails vertically. Olingos *(Bassaricyon* spp.) are small, with long backs and tails and no mask, and tail rings are faint. Cacomistles *(Bassariscus* spp.) are small and slender, with long tails. No other mammals in region except spotted cats have black and pale rings on tail.
Natural history. Nocturnal; terrestrial but climbs well; solitary. Feeds mainly on molluscs, fish, and crabs, with some amphibians and insects, and probably fruits. Crabs can be found throughout the lowland rainforest in large and small, permanent and temporary bodies of water. Crab-eating raccoons seem to be restricted to waterside habitats such as swamps, rivers, streams, and beaches. During the day they den in hollow trees. This species is rarely seen deep in rainforest, although it is found throughout the region. It is also found in llanos and in deciduous as well as evergreen forests. In the zone of geo-

graphic overlap with the northern raccoon in Central America, the northern raccoon is found in mangrove swamps, while the crab-eating raccoon is found on inland rivers, but both have been found on the same river.
Geographic range. Central and South America: E Costa Rica and Panama south to Uruguay and NE Argentina.
Status. Unknown, but widespread and probably rarely hunted.
Local names. Mapache, osito lavador (Span); zorrino (Bo); guaxinim, mao-pelada (Br); manipelado, zorra patona, cangrejera (Co); tejón (Ec); ratón laveur, chien crabier (FG); osito cangrejero (Pe); gato manglatero (Pn); wasbeer, krabdagoe (Su); aguará-popé, goá-xiní (Gua); mayuato (Qui).
References. Bisbal, F. J. 1986. Food habits of some Neotropical carnivores in Venezuela (Mammalia, Carnivora). *Mammalia* 50:329–39.

Northern Raccoon
Procyon lotor
Plate 15, map 120
Identification. Measurements (Central American animals only): HB = 440–625; T = 270–360; HF = 102–142; E = 60–75; WT = 2.7–6.4 kg.
Upperparts grizzled blackish gray; fur long and soft, on neck "normal," slanting backward. Head broad, with pointed muzzle; **face white with broad black mask reaching across and well behind eyes to nearly below ear and extending below as sharply contrasting brown bar across throat;** nose black; ears white inside and out, relatively large; eyes medium-sized, with bright greenish white eyeshine at night. **Tail about 60% as long as head and body,** moderately bushy, **with broad, brightly contrasting black and gray or buff rings.** Underparts whitish, covered with long, white hairs, **soft brown underfur showing below. Forelegs whitish,** hindlegs and feet usually whitish, sometimes brown. Shape and postures like crab-eating raccoon.
Variation. Five Neotropical island species of raccoons have been described: *P. gloveralleni* (Barbados), *P. insularis* (Maria Madre, Maria Magdalena, Mexico), *P. maynardi* (Bahamas), *P. minor* (Guadeloupe), and *P. pygmaeus* (Cozumel). All are close to *P. lotor.*

Map 120

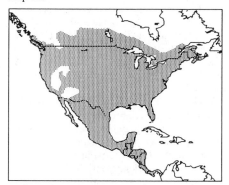

▨ Northern raccoon, *Procyon lotor*

Map 121

▨ South American coati, *Nasua nasua*
▦ White-nosed coati, *N. narica*

Similar species. See crab-eating raccoon.
Sounds. A variety of whines, yips, and
growls; not usually heard in field.
Natural history. Nocturnal; terrestrial and
arboreal; solitary except females with young
or congregations at food sources. Feeds on
fruits and small animals, especially aquatic
ones such as crayfish, crabs, and fish; also
raids garbage bins and dumps and is a pest
when it eats young ears of corn. Northern
raccoons are better tree climbers than are
crab-eating raccoons, but nevertheless do
most of their foraging on the ground. When
alarmed, they will often escape by running
up the nearest tree into the canopy. Before
eating, they rub and manipulate their food
with their hands. During the day they usually den in hollow trees. Northern raccoons
are common on the beaches of both coasts
of Central America and are also found inland
in mangrove swamps and near rivers. They
adapt well to man and thrive in towns and
cities.
Geographic range. North and Central
America: S Canada south to Chiriquí,
Panama.
Status. Widespread and common.
Local names. Mapache, osito lavador
(Span); zorra manglera (Co); pizote solo
(Ho); tzil (May).
References. Sanderson, G. C. 1983. *Procyon lotor.* In D. N. Janzen, ed., *Costa Rican
natural history,* 485–88. Chicago: University of Chicago Press.
    Lotze, J.-H., and S. Anderson. 1979. *Pro-
cyon lotor.* Mammalian Species, no. 119.

South American Coati
Nasua nasua
Plate 15, map 121
Identification. Measurements: HB =
470–580; T = 420–550; HF = 84–100;
E = 38–45; WT = 3–7.2 kg; males larger
than females.
**Upperparts dark brown, dark or bright
rust red-orange,** or gray; fur thick and dull.
**Head narrow, snout long, narrow, mobile;
nose slightly upturned,** wet, black; cheeks
often grizzled gray; muzzle brown; **prominent pale spots usually above, below, and
behind eye; ear small,** rounded, **fringed
with white** on inner rim, dark brown on
back and base; canine teeth large, especially
lowers. **Tail** long, well-haired, **tapering** to
point at tip, **blackish brown with bright or
faint pale yellow, brown, or buff rings; often carried vertically upright.** Chin and
throat grizzled white, underparts whitish to
rich buff. Feet dark brown except in orange
individuals; **forefeet with long claws,** hindfeet with strong, curved claws. Small young
darker than adults, with faint markings.
Variation. Color highly variable, from almost black, with gray face without white
markings (Peru), to entirely bright red-
orange (Pará, Brazil; parts of Bolivia) or
gray (Minas Gerais, Brazil). Facial markings and pale rings on tail range from
sharply contrasting to faint or absent.
Similar species. Raccoons *(Procyon cancrivorus)* have black masks, short tails, and
normal snouts; olingos *(Bassaricyon* spp.)
are much smaller, with no facial markings;

tayras *(Eira barbara)* have short snouts and no rings on the tail. At high elevations in the Andes (>2,000 m), Andean coatis *(Nasuella olivacea)* are externally similar to South American coatis but smaller, with small canine teeth. The tails of spotted cats may be held vertically and resemble those of coatis in the undergrowth: cat tails are not tapered and are irregularly spotted with pure black and white.

Sounds. Members of active groups emit constant soft whining sounds; alarm calls include explosive woofs and clicks.

Natural history. Diurnal; terrestrial and arboreal; solitary and in groups of up to 30. Omnivorous, feeds on fruit, invertebrates, and other small animals. Coatis can be seen feeding on fruit high in the canopy or searching the forest floor for animal prey by poking their long noses into crevices, turning over rocks, or ripping apart dead logs with their claws. Males are often solitary, but females and young travel in conspicuous troops; they are often seen as a group of vertical tail tips waving among the shrubbery. When one coati spots danger and gives the alarm, all run partway up trees to look. After a few moments they drop to the ground and disperse rapidly through the undergrowth. At night coatis sleep in a treetop. Females leave the group when they give birth to young, which are raised in a nest in a tree. Found in forested habitats ranging from tropical rainforest and gallery forest to chaco, cerrado, and dry scrub.

Geographic range. South America: east of the Andes in all countries from Colombia and Venezuela south to Argentina and Uruguay.

Status. CITES Appendix III (Ur). Generally uncommon, but ubiquitous in rainforest, perhaps most numerous in dry forests; widespread. Hunted moderately for meat and the live animal trade.

Local names. Coatí (Span, Gua); cusumbo, guache (Co); quati, quatimundé (Br); tejón (Bo, Co, Ec); coachí (FG); achuni, sehuaro (Pe); neusbeer, kwaskwasi (Su).

References. See white-nosed coati.

White-nosed Coati
Nasua narica
Plate 15, map 121
Identification. Measurements: HB = 463–690; T = 490–620; HF = 93–146; E = 34–42; WT = 4.5 kg.

Upperparts dark brown, shoulders or entire forequarters grizzled gray. Head sometimes rusty brown; **muzzle and chin white, extending as a white stripe above eye; other white spots above and behind eye;** ear short, lined with whitish fur. **Tail** 75–100% as long as head and body, thickly furred but not bushy, tapering at the tip; **dark brown with paler rings usually inconspicuous to invisible.** Feet dark brown, forefeet often contrasting with gray shoulder. Throat and chest frosted white, belly dark brown or frosted white. **Body shape and postures like South American coati.** Young dark brown with pale facial markings like adult.

Variation. The above description applies to coatis from the rainforest region; animals from more arid habitats may be paler, tawny brown. Coatis from Cozumel Island, Quintana Roo, Mexico, are sometimes considered a distinct species *(N. nelsoni).* They are smaller and have silkier fur than the mainland form. Some consider the white-nosed coati to be a subspecies of the South American coati *(N. nasua).*

Similar species. See South American coati.

Sounds and natural history. Like South American coati. Occupies a wide range of habitats from rainforest to Sonoran desert.

Geographic range. North, Central, and South America: Arizona and Texas south through all of Central America to the west coasts of Colombia and Ecuador.

Status. CITES Appendix III (Ho). Locally common to rare.

Local names. Quash (Be); cusumbo, cusumbo solo, gato solo (Co); pizote, pizote solo (CR, Ho); tejón, cuchucho, andasolo (Ec, Me); cuzumbo (Co); gato solo (Pa); chic, sis (May).

References. Kaufmann, J. H. 1962. Ecology and social behavior of the coati, *Nasua narica,* on Barro Colorado Island, Panama. *Univ. Calif. Pub. Zool.* 60:95–222.

———. 1983. *Nasua narica.* In D. H. Janzen, ed., *Costa Rican natural history,* 478–80. Chicago: University of Chicago Press.

Olingos
Bassaricyon spp.
Plate 15, map 122
Identification. Measurements: HB = 360–411; T = 370–520; HF = 62–92; E = 30–44; WT = 1.1–1.4 kg.

Map 122

Olingos, *Bassaricyon* spp.
Cacomistle, *Bassariscus sumichrasti*

Upperparts drab **brown,** sometimes darker on midline; fur soft, dense, quite long (1.5 cm at midback). Crown often dark brown; **face grizzled gray; muzzle pointed;** ears rounded, set low on head, brown with fringe of white hairs on rim; eyes large, round, brown; eyeshine bright. **Tail slightly longer than head and body; nonprehensile; well-haired, tip almost tufted** with longer hair than base (2–4 cm); **brown with indistinct dark rings,** tip usually darker than base, or dirty white; carried straight out behind during travel. Underparts cream or buff. Tiny young gray-brown, sides with faint dark stripes. A slender-bodied, short-legged, monkeylike animal.
Variation. Olingos from higher elevations in Panama are frosted with gray; many from Panama have white tail tips. Six species of olingos from different regions are sometimes recognized: *B. gabbii,* from Central America; *B. alleni,* from the western Amazon Basin; *B. beddardi,* from northern South America; *B. pauli,* from one locality in Panama; and *B. lasius,* from one locality in Costa Rica. The differences between these forms are minor, and they probably are all conspecific *(B. gabbii).*
Similar species. Kinkajous *(Potos flavus)* are similar and often confused with olingos: they are about twice the weight, usually reddish, and have short fur and a tapered, prehensile tail that is often curled under branches during travel. Cacomistles *(Bassariscus sumichrasti)* have the tail prominently ringed. Night monkeys *(Aotus* spp.) have a white-rimmed face with dark stripes

on crown, large, close-set eyes, flat face, and inconspicuous ears. Opossums have small eyes and prehensile tails either naked at the tip or strongly tapered.
Sounds. Usual alarm call a two-toned "wake-up," less often a sneezing. A noisy traveler that jumps from tree to tree in the canopy, but more lightly than kinkajou.
Natural history. Nocturnal; arboreal; solitary. Feeds on fruits and some invertebrates; in the dry season drinks the nectar of flowers such as balsa. Olingos are agile and active and usually travel high in the canopy, lightly leaping and running through the branches, feeding quickly, and running on. They often feed on fruits, especially figs, in the same trees with kinkajous and night monkeys, and in the dry season feed on nectar in the same trees as these animals and opossums. They are restricted to humid forests. They are generally less common than kinkajous and do not seem to adapt as readily to disturbed or secondary forests, or plantations and gardens. Found in mature and disturbed rainforest.
Geographic range. Central and South America: Nicaragua south, west of the Andes to N Ecuador; east of the Andes from Venezuela and Guyana to Bolivia, in the western half of the Amazon Basin only. To 1,600 m elevation.
Status. CITES Appendix III (CR). Locally common, not hunted, and widespread; but in Central America may be locally threatened by destruction of mature forests.
Local names. Usually the same as kinkajou: jupará (Br); macoperro, leoncillo (Co); olingo (Ec, Pn); ocate, chosna pericote (Pe).
References. Decker, D. M., and W. C. Wozencraft. 1991. Phylogenetic analysis of recent procyonid genera. *J. Mammal.* 72:42–55.

Kinkajou
Potos flavus
Plate 15, map 123
Identification. Measurements: HB = 390–547; T = 400–570; HF = 70–108; E = 30–55; WT = 2–3.2 kg.
Upperparts reddish brown to smoky gray-brown, often with a dark brown stripe on midback; fur dense, soft, and short (about 1 cm at midback). Head and face red-brown to blackish; **head round, muzzle short, pointed;** nose brown; **eyes brown, large and round, set wide apart;** eyeshine bright

Map 123

Kinkajou, *Potos flavus*

pale yellow-white to greenish white; **ears brown, thinly haired, set low on sides of head;** tongue long and highly extensible. Tail slightly longer than head and body, prehensile, tapered toward tip, brown or darkening to black at tip. Feet colored like back, all with five long, curved claws. **Underparts contrasting yellow to pale orange buff.** Juveniles gray with dark stripes behind shoulders. An agile, muscular, short-legged, long-backed, monkeylike animal.
Variation. Occasional animals have a small white tail tip; some animals from Central America are gray-brown.
Similar species. Olingos (*Bassaricyon* spp.) are similar but smaller, with a nonprehensile tail, slightly bushy at the tip. In the field at night, the belly of olingos looks dark, that of kinkajous pale. Cacomistles *(Bassariscus sumichrasti)* have a brightly ringed tail and white markings on face; night monkeys (*Aotus* spp.) have a flat face, white brow, three black stripes on crown, close-set eyes, and long, nonprehensile, black-tipped tail with no taper.
Sounds. Highly vocal: when alarmed emits sneezing whistles; jumps noisily from tree to tree at night.
Natural history. Nocturnal; arboreal; solitary, in pairs, or several may congregate in a fruit tree. Feeds on fruit, primarily figs (*Ficus* spp.), and insects, especially ants, and in the dry season drinks flower nectar. Kinkajous are animals of the forest canopy, and they are the most commonly seen large, nocturnal, arboreal mammal. They are agile and can travel quickly, running and jumping

noisily from tree to tree. When seen at night in the canopy, they can be distinguished from olingos and night monkeys (which may be feeding in the same tree) by the tapered, prehensile tail, which is frequently curled partially under or around a branch during travel or feeding. During the day they den in tree hollows. They have home ranges of about 8 to 50 ha; a group of several males with large ranges may overlap a female with a smaller range, but the social system is not yet clear. Found in mature, disturbed, and secondary rainforest, gardens and plantations, gallery forest, and occasionally in deciduous forest.
Geographic range. Central and South America: S Mexico south to Bolivia and Mato Grosso, Brazil. Rarely above 500 m, occasionally to 1,750 m elevation.
Status. CITES Appendix III (Ho). Widespread and often common; hunted for meat and for the pet trade.
Local names. Mono michi (Bo); jupará, macaco-de-noite (Br); night walker (Be); perro de monte, oso mielero, leoncillo, micoleón (Co); marta, martucha (Co, Me); martilla (CR); martica, tutamono, chuche, cuchicuchi, cusumbo (Ec); singe de nuit (FG); chosna, martucha, chuchumli (Pe); meti-keskesi (Su); cusumbi (Pn). Sometimes not locally distinguished by name from night monkeys; note the occurrence of "monkey" in many names.
References. Bisbal, E. F. J. 1986. Food habits of some Neotropical carnivores in Venezuela (Mammalia, Carnivora). *Mammalia* 50:329–39.

Kays, R. W., and J. L. Gittleman. 1995. Home range size and social behavior of kinkajous *(Potos flavus)* in the Republic of Panama. *Biotropica* 27: 530–34.

Cacomistle
Bassariscus sumichrasti
Plate 15, map 122
Identification. Measurements: HB = 389–468; T = 400–545; HF = 77–94; E = 42–55; WT = 900 g.
Upperparts tawny brown; midline of back blackish. Crown blackish; **eyes narrowly ringed with black, bordered by prominent pale, dirty yellow rings encircling outside of eye;** eyeshine bright yellow-white; **ears large and rounded, standing above crown,** rims white inside and out, inside of pinnae pale. **Tail longer than head**

and body, thickly furred; **strongly banded with black and whitish or tawny rings,** pale rings disappearing distally, **tip entirely black** for about terminal quarter. **Feet and toes dark** brown or blackish. Underparts yellowish white, grayish around midsection. A small, slender, long-bodied animal with a relatively small head and long neck.

Similar species. Ringtails (*B. astutus,* in dry habitats) are paler, with pale feet, and pale rings to near tip of tail; olingos (*Bassaricyon* spp.) are similar in shape, with no facial markings and indistinct rings on tail.

Sounds. Frequently calls with loud, two-syllable barks.

Natural history. Nocturnal; arboreal; solitary. Feeds on fruit, insects, and probably small vertebrates. Cacomistles use the middle and upper levels of the forest. Found in mature and secondary lowland and montane rainforest and dry forest.

Geographic range. Central America: S Mexico to W Panama. To about 2,000 m elevation.

Status. CITES Appendix III (CR). Common in remnant forests of Veracruz; rare in Panama.

Local names. Olingo, cacomistle (CR, Pn); guía de león, guayanoche (Gu); uayuc (Ho); cacomixtle (Me).

References. Estrada, A., and R. Coates-Estrada. 1985. A preliminary study of resource overlap between howling monkeys *(Alouatta palliata)* and other arboreal mammals in the tropical rainforest of Los Tuxtlas, Mexico. *Am. J. Primatol.* 9:27–37.

Weasel Family (Mustelidae)

Dental formula: I3/3, C1/1, P3 or 4/3 or 4; M1/1 or 2 = 32–40. Five toes on all feet. Mustelids have shearing premolars and crushing molars. They are plantigrade, with broad heads, small ears and eyes, tails shorter than the head and body, and usually long slinky bodies and short legs, convenient for using burrows and holes. The back is humped when the animal is standing. Most have soft, dense, glossy fur, valuable in the fur trade. The New World rainforest species all forage terrestrially or in the water, but at least one species also climbs to forage. Mustelids have an extremely powerful bite for their size (legendary in the case of the wolverine), and some of them are among the few carnivores in the world that can single-handedly kill prey much larger than themselves. They do this with a strong killing bite to the head or neck. The family includes pure carnivores, some that feed on specific prey types (e.g., black-footed ferrets); worm eaters (some badgers); omnivores (tayras, skunks); and fish eaters (otters). The senses of smell and hearing are acute, but most species do not seem to see particularly well, and a motionless observer can often watch them undetected. Most species have large anal glands that produce strong-smelling musk: this has been carried to an extreme in the skunks, zorillas, and stink badgers, which spray vile-smelling fluid at their enemies. Litter size is 2–18 altricially born young. Most species sleep and raise their young in burrows or hollow logs. The family includes about 26 genera and 67 species worldwide; 7 genera and 8 species occur in Neotropical rainforest.

Amazon Weasel

Mustela africana

Plate 16, map 124

Identification. Measurements: HB = 260–333; T = 170–234; HF = 50–56; E = 22.

Upperparts entirely glossy chestnut brown. Head broad, narrowing sharply to blunt snout; ear short, broad at base; eyes small; upper lip, chin, and lower cheek to below ear yellowish white. Tail shorter than head and body, well furred. **Throat and belly pale buff or yellowish with a sharp dark brown stripe down midline** from throat or chest to belly. A small, slender mammal with long body and neck, strongly humped back, and very short legs.

Similar species. These are the only true weasels in their range and the smallest lowland carnivores in the region; no other mammals are of similar color and shape; long-tailed weasels and Don Felipe's weasels *(M. frenata* and *M. felipei)* have no dark ventral stripe.

Natural history. Unknown. Amazon weasels probably feed on rodents and other small mammals, like other members of the genus. Several were found denning in a hollow tree

Map 124

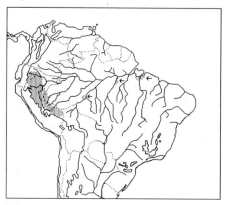

Map 125

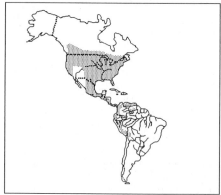

▦ Amazon weasel, *Mustela africana*

▦ Long-tailed weasel, *Mustela frenata*

stump. A group of four adult-sized unidentified weasels was seen traveling in the lower forest canopy in the daytime, in terra firme forest near the Rio Madeira, where no species other than this is suspected. Found in lowland rainforest.

Geographic range. South America: east of the Andes in the lowland Amazon basins of Peru, Ecuador, and Brazil.

Status. This is one of the rarest carnivores in South America; known from fewer than 50 specimens from widely scattered localities.

Local names. Comadreja (all weasels, Span); furão (Br); vasóabejero (Co); chucuri (Ec).

References. Izor, R. J., and L. de la Torre. 1978. A new species of weasel (*Mustela*) from the highlands of Colombia, with comments on the evolution and distribution of South American weasels. *J. Mammal.* 59: 92–102.

Ferrari, S. F., and M. A. Lopes. 1992. A note on the behaviour of the weasel *Mustela* cf. *africana* (Carnivora, Mustelidae), from Amazonas, Brazil. *Mammalia* 56:482–83.

Long-tailed Weasel

Mustela frenata
Plate 16, map 125

Identification. Measurements: HB = 215–320; T = 115–207; HF = 35–55; E = 10–20; WT = 85–340 g; males usually much larger than females.

Upperparts entirely rich to dark, glossy chocolate brown. Head with or without following facial markings: broad white stripe from nose over eye and down neck in front of ear, or a white spot or spots between eyes, and a diagonal white stripe in front of ear. Tail short, about 60% as long as head and body, thickly furred, with slight tuft, **black at tip. Underparts entirely creamy white, or chin and throat creamy white, grading to pale orange on belly.** Shape like Amazon weasel.

Variation. Much geographic variation in size and color: the white facial stripe is most highly developed in Mexico and gradually fades southward, from a broad, solid stripe in Mexico, to striking white spots in Nicaragua, to faint traces in Colombia, to head entirely brown above, with no white spots, in Peru; but some Venezuelan populations have prominent white spots. Body color varies from almost black (Colombia) to paler brown (Central America).

Similar species. See Amazon weasel.

Natural history. Diurnal and nocturnal; terrestrial; solitary. Feeds chiefly on small mammals such as rabbits and rodents, but will also take birds and reptiles. These weasels have been called snakes with legs: their short legs and long neck and back, fitted for running down the narrow burrows of their prey, give them a graceful, sinuous bounding gait. They are active and nervous, and it is rare to get more than a brief glimpse of one. When alert, they briefly stand on their hindlegs and crane their necks. They den in burrows or holes under roots or among rocks. Long-tailed weasels are highly adaptable, but they are not a rainforest species. In general they are found either in drier cleared or open country, agricultural land, or montane forest at high elevations, but they are occasionally found below 1,000 m in rainforest in Central America. In South America they are chiefly found at high elevations in the Andes or at low elevations on the west coasts of Colombia, Ecuador, and Peru.

Geographic range. North, Central, and South America: Canada to Bolivia. To 4,000 m elevation.

Status. Widespread and locally common.

Local names. Comadreja (Span); chucurí (Co, Ec); chucuro (Co); oncilla (Me); lince (Pa); tolompeo (Pe).

References. Hall, E. R. 1951. American weasels. *Univ. Kans. Pub. Mus. Nat. Hist.* 4:1–466.

Grison
Galictis vittata
Plate 16, map 126
Identification. Measurements: HB = 465–552; T = 135–195; HF = 75–97; E = 20–30; WT = 1.5–2 kg.
Upperparts grizzled pale gray mixed with dark brown. Head tricolored: crown gray; forehead with broad white band above eyes, across ears, and down side of neck; muzzle to eyes black, continuous with black chin, throat, and chest. Ears small and inconspicuous, white; eyes small, black. **Tail short and stumpy,** one-third as long as head and body, well-haired with fur grizzled gray like back. Legs and feet black; toes joined by partial webs that reach to base of last joint. Underparts black anteriorly, grizzled gray posteriorly. **A big weasel with very short legs and tail, long neck and back.** Young like adults.
Similar species. No other animals in region are gray with a broad white band across face and black muzzle and throat. Similar grisons (*G. cuja*) live in south temperate latitudes and at higher elevations; hog-nosed skunks (*Conepatus* spp.) are black with a pure white cape over crown, neck, and back; tayras (*Eira barbara*) have black bodies and long legs and tails.
Sounds. Snorts in alarm; growls in defense.
Natural history. Nocturnal, with some diurnal activity in the morning; terrestrial; solitary and in pairs that travel together or females with young. Carnivorous; feeds on small vertebrates including rodents, birds,

Map 126

Grison, *Galictis vittata*

reptiles, amphibians, and fish. Grisons are often seen near rivers or streams, and local anecdotes report that they are good swimmers that can hunt in water. An adult female had a home range of about 4 km². Found in rainforest and savanna.

Geographic range. Central and South America: S Mexico south to Peru and Bolivia.

Status. CITES Appendix III (CR). Widespread but apparently uncommon or rare everywhere.

Local names. Bushdog (Be); furão, furax (Br); hurón (Bo, Co, Ec); mapuro (Co); grisón (CR, FG, Ho, Me); perro de agua (Ec); lobo gallinero, tigrillo rosillo (Pn); weti-aira (Su); yaguagumbé, yuguapé (Gua); zabin (May).

References. Sunquist, M. E., F. Sunquist, and D. E. Daneke. 1989. Ecological separation in a Venezuelan llanos carnivore community. In K. H. Redford and J. F. Eisenberg, eds., *Advances in Neotropical mammalogy*, 197–232. Gainesville: Sandhill Crane Press.

Tayra
Eira barbara
Plate 16, map 127
Identification. Measurements: HB = 559–712; T = 365–460; HF = 90–123; E = 30–42; WT = 2.7–7 kg.
Upperparts glossy dark brown to black, including legs, feet, and tail; fur of head short and stiff; fur of body and tail long and glossy, underfur brown. **Head and neck sharply contrasting grizzled tan, gray-brown, or yellowish,** rarely dark brown slightly paler

Map 127

Tayra, *Eira barbara*

Sounds. Snorts in alarm; growls and squeals when cornered.

Natural history. Diurnal except near human habitations, where also crepuscular; terrestrial and arboreal; solitary or in pairs that travel together. Feeds on small vertebrates, especially rodents, insects, fruit, and honey. Tayras have a broad habitat range and diet and usually forage terrestrially, but they will climb trees to eat fruit, rob birds' nests, or raid beehives. They are generally seen as they travel quickly through the forest on the ground during the day. They often travel along the tops of fallen logs, where they leave seed-filled scats. When alarmed at close quarters they may climb a tree and growl and spit in threat or run away through the branches. If disturbed from a distance, they gallop away noisily on the ground. Tayras den in hollow trees or holes in the ground. The available field data shows that they travel long distances of 2 to 8 km each day, within enormous home ranges of 10–24 km^2. They are wary and not often seen, but are much commoner than any of the similar species. They are fond of papayas, and if undisturbed become tame when they raid fruit near houses. They inhabit mature and secondary rainforests, dry forests, gallery forests, cloud forests, and gardens and plantations.

Geographic range. Central and South America: S Mexico south to N Argentina. To elevations of 2,400 m.

Status. CITES Appendix II (Ho). One of the most common and widespread carnivores, tayras can live in disturbed habitats near man.

Local names. Bushdog (Be); melero, taira (Bo); irara, papa-mel (Br); comadreja, mama, umba, melero, tayra, zorro palmichera (Co); tolomuco (CR); tejón, manco, perro de monte (Pe, Ec); cabeza de mate (Ec); martre (FG); lepasil, cadejo (Ho); gato eirá (Pa); comadreja grande, gato negro, gato cutarra (Pn); cabeza de viejo (Ho, Me); guache, guanico (Ve); eirá (Gua); sacol (May).

References. Konecny, M. J. 1989. Movement patterns and food habits of four sympatric carnivore species in Belize, Central America. In K. H. Redford and J. F. Eisenberg, eds., *Advances in Neotropical Mammalogy,* 243–64. Gainesville: Sandhill Crane Press.

than back, or same color as back. **Ears small and round,** same color as head, **not protruding above crown.** Underparts completely black or dark brown, except for a **bright pale yellow to orange spot, often triangular, on chest and throat.** Tail bushy, two-thirds as long as head and body. Toes joined by partial webs that reach to base of last joint. A large, muscular weasel, **much like a small dog** with a long, slightly humped back and long tail. Young entirely black, sometimes with white throat patch and/or white head.

Variation. Some tayras from Costa Rica and Panama are completely blackish, with dark brown head and no throat patch; pale phases, almost pure pale yellow, occur in savannas of Guyana and Bolivia in mixed populations with typical individuals; occasional animals from northern South America have a large bright yellow spot on the shoulders, or even a complete yellow collar joining with the throat patch.

Similar species. Bush dogs *(Speothos venaticus)* are similar in size and color, but have pale brown shoulders and back, short tail, ears that project above crown, straight back, and thick, cylindrical body; short-eared dogs *(Atelocynus microtis)* are larger, uniformly grizzled dark gray, with ears that protrude above crown; grisons *(Galictis vittata)* are gray, with white band across head and very short legs and a stumpy tail; jaguarundis *(Herpailurus yaguarondi)* are uniformly brown or reddish, with a slender, feline tail and small head.

Map 128

▓ Striped hog-nosed skunk, *Conepatus semistriatus*
≡ Andean hog-nosed skunk, *C. chinga*

Striped Hog-nosed Skunk
Conepatus semistriatus
Plate 16, map 128
Identification. Measurements: HB =
332–500; T = 166–317; HF = 70–102;
E = 26–35; WT = 1.4–3.4 kg.
**Entirely dark brown or black with a broad
white band from crown of head over neck;
band splits at shoulder into two parallel
stripes down top of back** with a narrow
black stripe between, stripes ending short
of rump; fur coarse and thick. **Head conical,
snout long and naked;** ears short, not protruding above crown, black with a few white
hairs at base; eyes black, eyeshine bright
green. **Tail bushy, entirely white distally,
black near base, often carried high.** Forefeet with very long, black claws. Underparts
dark brown or blackish. In defense squirts
vile-smelling fluid from its anal glands.
Variation. Details of color pattern (extent
and width of white stripes, white on tail)
are highly variable geographically, and three
South American species have been recognized based mainly on pelage differences.
These need systematic review.
Similar species. Grisons (*Galictis* spp.) have
a grizzled gray body and tail. Several other
skunks are found in open habitats fringing
the rainforest: other species of hog-nosed
skunks in dryer areas of Mexico have a solid
white back and tail; hooded skunks (*Mephi-
tis* spp.) have black tails; spotted skunks
(*Spilogale putorius*) are small, with short
stripes and spots all over the body. A similar

species, *C. chinga,* is found on the southern
borders of rainforest of Brazil, Bolivia, Paraguay, and Argentina.
Natural history. Nocturnal; terrestrial; solitary. Feeds mainly on insects and other invertebrates, and probably small vertebrates
and occasional fruit. When alarmed at close
range, this and other skunks aim their rear
at the enemy, raise the tail, and spray. Well
protected, skunks move slowly, ambling
along snuffling the ground, scratching and
rooting for their prey. By day they rest in
or under logs or in burrows. Skunks are not
truly rainforest animals, but this species
occasionally is found in Central American
rainforest, especially in secondary or disturbed forests near the clearings and gardens that are its more usual habitat. Occurs
in pastures, clearings, roadsides, and other
cultivated areas, in rainforest, dry forest, and
scrub. In montane, temperate, and savanna
habits; around its fringes but not in Amazonian rainforest.
Geographic range. Central and South
America: Veracruz south through Central
America, absent from eastern Panama; in
South America in coastal Peru and Ecuador,
N Colombia and Venezuela, and at high elevations in the Andes, and in E Central Brazil
in caatinga and cerrado. To at least 4,100 m
elevation.
Status. Widespread; adaptable and rarely
hunted.
Local names. Polecat (Be); zorrino común,
anatuya (Bo); jaritataca, zorilho (Br); mofeta,
mapuro, mapurito (Co, Ec); zorrillo, zorrillo
pijón (Me, Co); gambá zorrino, anas (Pe);
gato cañero (Pa); yaguaré (Gua).
References. Kipp, H. 1965. Beitrag zur
Kenntnis der Gattung *Conepatus* Molina,
1782. *Z. Saügetierk.* 30:193–232.

Neotropical Otter
Lontra longicaudis
Plate 16, map 129
Identification. Measurements: HB =
532–809; T = 360–570; HF = 80–130;
E = 19–23; WT = 5–14.75 kg; males larger
than females.
Upperparts entirely glossy dark brown;
fur short and dense. **Upper lip, lower cheek,
throat, and belly silvery whitish to yellowish, or brown. Head small, flat; muzzle
broad; nosepad completely or partially**

Map 129

Neotropical otter, *Lontra longicaudis*

but amount of current hunting and population status are unknown.

Local names. Nutria, lobito de rio (Span); guaiao (Ar); water dog (Be); lontra, cachorro-d'agua, nútria (Br); tig d'eau (FG); perro de agua (Ho, Me); gato de agua (Pn); watradagoe (Su).

References. Van Zyll de Jong, C. G. 1972. A systematic review of the Nearctic and Neotropical river otters (genus *Lutra*, Mustelidae, Carnivora). Roy. Ontario Mus. Life Sci. Contrib., no. 80.

Foster-Turly, P., S. Macdonald, and C. Mason, eds. 1990. *Otters: An action plan for their conservation.* IUCN/SSC Otter Specialist Group.

naked; neck thicker than head; eyes small; **ears short and rounded. Tail long, thick at base and tapering.** Legs short and stout, **feet webbed.** On land, head and tail are carried low and back humped high. Young like adults.

Variation. Animals from the northern part of the range have pale bellies; those from Bolivia have dark underparts.

Similar species. Giant otters *(Pteronura brasiliensis)* are much larger, with throat irregularly spotted brown on pale, and tail strongly flattened.

Sounds. Probably whistles, hums, and screeches like other river otters.

Natural history. Probably diurnal and nocturnal; semiaquatic; solitary or mothers with young. Feeds chiefly on fish and crustaceans, supplemented by other aquatic animals. Neotropical otters are always in or near water and are graceful swimmers and divers. On land they are awkward and move with a humping gallop or waddling walk. These little-known otters appear to favor clear, fast-flowing rivers and streams, and they may be rare in or absent from sluggish, silt-laden lowland rivers. Found in riverine habitats in both deciduous and evergreen forests, savannas, llanos, and pantanal, and in warm and cool climates.

Geographic range. Central and South America: N Mexico south to Uruguay. To 3,000 m elevation.

Status. CITES Appendix I, US-ESA endangered. Widespread but rarely seen; formerly intensively hunted for the fur trade,

Giant Otter
Pteronura brasiliensis
Plate 16, map 130

Identification. Measurements: HB = 1,000 – 1,200; T = 530 – 700; HF = 176 – 207; WT = 24 – 34 kg.

Upperparts rich brown, almost black when wet; fur short, dense, velvety. Head round, muzzle blunt, **nosepad completely hairy,** with fine hairs; whiskers stout, ears small, set low on sides of head. Upper lip and **throat with irregular splotches of cream and brown. Underparts apart from throat** about **the same color as back. Tail thick at base tapering to dorsoventrally flattened tip.** Legs short and thick; feet large, toes completely webbed to tips of digits. Young like adults.

Variation. Each individual has a different pattern of spots on throat.

Similar species. Neotropical otters *(Lontra longicaudis)* are smaller, with no spots on throat and tail cylindrical.

Sounds. Alarm call a loud explosive snort; group members constantly interact with loud, high-pitched hums and whining squeals and screeches.

Natural history. Diurnal; semiaquatic; usually in groups of five to nine, rarely solitary. Feeds primarily on large fish, but will kill and eat other vertebrates such as snakes and small caiman. Groups are territorial and consist of an adult pair or harem and their offspring from several years. Giant otters are always found in or near large bodies of water. When hunting fish, they often travel swimming in a phalanx underwater together,

Map 130

Giant otter, *Pteronura brasiliensis*

alternately surfacing to breathe. When alarmed, all members of the group surface and crane their necks high out of water with a chorus of loud snorts. An otter will climb onto the bank or a half-submerged log to eat a large fish, but fish are often eaten while the otter swims on its back holding the prey in its forepaws. Group members constantly squabble over food, growling and squealing. Their territories are marked by large trampled places on the waterside, usually around a log projecting into the water. These have a strong, unpleasant fishy odor and much residue of scats and fish scales. At night the group sleeps in a large burrow in a bank, with the entrance above a worn slide into the water. Giant otters fiercely defend their young,

attacking in a group, and can even drive a jaguar away. They occupy lowland forest rivers and lakes of many types, from large, silt-laden, fish-rich waterways to clearwaters, blackwaters, and flooded forest in the rainy season.

Geographic range. South America: east of the Andes from S Venezuela and Colombia south to N Argentina. There is a historical record from Uruguay, but probably none remain there.

Status. CITES Appendix I, US-ESA endangered. Quite common where undisturbed, but rare or extirpated in much of its former range and can be seen only in remote and protected areas. Decimated by overhunting for the skin trade and killed because of competition with man for fish. However, populations in some areas are now recovering as a direct result of implementation of the CITES convention ban on trade. Giant otters are much more endangered than spotted cats because their waterside habitat is limited and accessible and their behavior is highly conspicuous.

Local names. Londra (Bo); ariranha (Br); perro de agua, lobo, colón, ariraña (Co); lobo del rio, arirai (Pe, Ec); grote waterhond, watradagoe (Su); araraí (Gua).

References. Duplaix, N. 1980. Observations on the ecology and behavior of the giant river otter *Pteronura brasiliensis* in Suriname. *Rev. Ecol.* 34:496–620.

Munn, M. B., and C. A. Munn. 1988. The Amazon's gregarious giant otters. *Animal Kingdom,* Sept.-Oct., 34–41.

The Cat Family (Felidae)

Dental formula: I3/3, C1/1, P2 or 3/2, M1/1 = 28–30. Four weight-bearing toes on all feet and a fifth, non-weight-bearing claw on the forefoot (the first digit); forefeet with retractile claws (except in cheetahs). Cats have teeth that are highly specialized for killing and meat eating, and their sharp, retractile claws and strong shoulders allow them to single-handedly grasp and drag down large prey, which they kill either by a powerful bite to the head or neck or by throttling with a bite to the throat or muzzle; however, most species eat small prey. Cats see and hear well; their vision is binocular, and they see colors. They are purely carnivorous. Cats are major predators in tropical rainforests worldwide. Most species are solitary hunters that capture their prey by surprise, either by stealthy approach or by patient waiting in ambush, followed by a short charge or pounce. They prey on almost anything they encounter that is not too large, including mammals, birds, snakes, turtles, caiman, fish, and even large insects. Some are territorial; others appear to share their home ranges but avoid an area where another cat is hunting. They communicate their presence to one another with marking behavior that includes spraying urine, scratching the ground and trees, and leaving their scats in prominent places. The Neotropical cats scratch fallen logs, and logs with scratch marks can be seen wherever cats are common. The litter size is about one to four. The kittens are sheltered in a den and raised by the mother alone, who brings them prey as they are weaned. Mothers must leave their young and hunt for many hours to find food; healthy kittens found alone have not been abandoned and should not be touched. Cats can be active at any time of day or night, but rainforest species are most active nocturnally; the best time to see them is after nightfall, when they walk on trails and their bright eyeshine makes them visible. There are about 36 species worldwide, with 4 genera and 6 species in Neotropical rainforest. There is entrenched disagreement about the classification of cat genera: the species are lumped into as few as 5 or split into as many as 18 genera. All cat species are listed under CITES Appendix II.

Ocelot
Leopardus pardalis
Plate 17, map 131
Identification. Measurements: HB = 710–875; T = 320–410; HF = 140–170; E = 50–65; WT = 8–14.5 kg; adult males are usually 11–12 kg, females 8–9 kg.
Upperparts tawny yellowish to buff with black spots and lines in longitudinal rows, many spots in open rosettes; fur usually short, smooth and slightly stiff, rarely soft and woolly. **Neck with heavy black stripes dorsally, fur "reversed," slanting forward.** Eyes large, eyeshine very bright pale yellow; muzzle profile slightly convex. **Tail distinctly shorter than hindleg, banded and spotted with black.** Underparts white with black spots. Feet large; forefeet broader than hindfeet. Kittens spotted. A long-legged, gracile cat the size of a medium-sized dog.
Variation. Each individual has a different spot pattern; ground color varies from bright buff to gray (in arid regions of Mexico).
Similar species. Jaguar *(Panthera onca)* are much larger, with spots, not stripes, on neck; margays *(L. wiedii)* and oncillas *(L. tigrinus)* are smaller, with tails longer than hindleg.
Sounds. None usually heard in field.
Natural history. Nocturnal and diurnal; terrestrial; solitary. Entirely carnivorous, feeds chiefly on rodents, supplemented by birds, snakes, lizards, and other small vertebrates. Ocelots hunt and capture their prey on the ground; they rarely climb trees, but will climb to cross over a stream on a branch or sometimes to rest. They are mainly active at night, when they spend many hours walking, often on man-made trails. When active by day they tend to keep hidden in dense brush. At rest they shelter under treefalls or between the closed buttress roots of large trees. Ocelots are the most commonly seen spotted cats, and their tracks can be found on muddy spots in trails and on riverbanks. They leave fine scratch marks on horizontal fallen logs. Found in habitats with good cover from rainforest to riverine scrub in deserts. Where they are not hunted for their skins, ocelots adapt well to disturbed habitats around villages, where they sometimes kill poultry.
Geographic range. North, Central, and South America: S Texas south to N Argentina. To 1,800 m elevation.
Status. CITES Appendix I, US-ESA endangered (two subspecies). Formerly intensively hunted for the skin trade, much less so currently. Widespread, common in some areas, naturally or artificially rare in others.
Local names. Tigrillo, ocelote (Span); tigrecillo, gato montés (Bo); gato-maracajá (Br, Pa); maracajá-açu, gato mourisco (Br); tiger cat (Be); manigordo (Co, CR, Pn), gato solo, canaguaro, tigrillo podenco (Co); chat tig (FG); gato tigre, tigre chico (Pn); onsa, yaguareté-í (Gua); tigri-kati (Sar).
References. Emmons, L. H. 1988. A field study of ocelots in Peru. *Rev. Ecol.* 42: 133–57.

Margay
Leopardus wiedii
Plate 17, map 132
Identification. Measurements: HB = 501–720 (few > 600); T = 351–490; HF = 107–137; E = 45–60; WT = 3–9 kg; males larger than females.
Upperparts tawny yellowish to grayish brown, with rows of black spots and lines in longitudinal rows, some spots in open rosettes; neck with heavy black stripes, hair "reversed," slanting forward; fur often soft, parted here and there into clumps, more rarely stiff and smooth. Eyes very large, eyeshine bright; **whiskers long, muzzle area at base of whiskers sometimes bulging. Tail longer than hindleg,**

Map 131

Ocelot, *Leopardus pardalis*

Map 132

Margay, *Leopardus wiedii*

spotted and banded with black. Feet large, fore- and hindfeet about the same width. This is the only New World cat with ankle joints that can rotate sufficiently for it to climb headfirst down vertical trees, with the hindfeet turned facing the trunk (like a squirrel). Kittens spotted. Somewhat larger than a large house cat.

Variation. Much individual variation in ground color and spot pattern.

Similar species. Ocelots *(L. pardalis)* have a shorter tail and are usually much larger, but the largest margay males can be as large as the smallest female ocelots. Oncillas *(L. tigrinus)* have the nape hair growing in normal slant and are usually smaller, with smaller, solid dotlike spots and body and head shaped like a house cat's, but they may not be distinguishable from margays with certainty in the field.

Natural history. Nocturnal; arboreal and terrestrial; solitary. Feeds on arboreal and terrestrial small mammals, birds, and reptiles, but also some insects and fruit. The diet in Belize consists largely of scansorial rodents, mouse opossums, and birds, implying that margays capture prey in the trees. When inactive during the day, they rest 7–10 m above the ground in a vine tangle or tree. Unlike ocelots, they cover their scats with leaves or dirt, and deposit them in inconspicuous spots. When traveling, they walk on the ground. Margays may not adapt well to human disturbance of the habitat, despite intensive use of old second-growth forest. Found in mature and secondary evergreen and deciduous forests.

Geographic range. Central and South America: Mexico south to Uruguay and Argentina. To 900 m elevation.

Status. CITES Appendix I, US-ESA endangered. Widespread, but always seems rarer than ocelot; status unknown. Fur has less value than that of ocelot, but margays are caught in traps set for ocelots. Threatened by deforestation in Central America.

Local names. Tigrillo, gato tigre (Span); gato pintado, gato brasilieño (Ar); maracajá-peludo, gato-do-Mato (Br); tiger cat (Be); gato Brasil, gato montés (Bo); maracaya, tigrillo peludo (Co); caucel (CR, Ho); pichigueta (Cent Am); burricón (Ec); chat tig (FG); mbaracayá (Gua); chulul (May); kuichua (Gy).

References. Tewes, M. E., and D. J. Schmidly. 1987. The Neotropical felids: Jaguar, ocelot, margay and jaguarundi. In M. Novak, ed., *Wild furbearer management and conservation in North America,* 696–711. Ontario: Ontario Trappers Association.

Konecny, M. J. 1989. Movement patterns and food habits of four sympatric carnivore species in Belize, Central America. In K. H. Redford and J. F. Eisenberg, eds., *Advances in Neotropical Mammalogy,* 243–64. Gainesville: Sandhill Crane Press.

Oncilla
Leopardus tigrinus
Plate 17, map 133
Identification. Measurements: HB = 452–648; T = 255–330; HF = 96–145;

Map 133

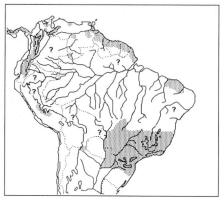

Oncilla, *Leopardus tigrinus*

E = 39–78; WT = 1.5–3 kg; males larger than females.
Similar to margay, but spots often solid small dots, sometimes rosettes; or body entirely black; neck hair slants backward in "normal" direction; fur soft or coarse. **Head shaped like house cat's; muzzle at base of whiskers not bulging. Feet small, like house cat's.** Kittens spotted. **Body size and shape of slender house cat.**
Variation. Color highly variable, from margaylike to quite distinct tawny brown, gray, or blackish with rows of tiny solid spots; black individuals not uncommon.
Similar species. See margay *(L. wiedii);* other species of small spotted cats, *Oncifelis geoffroyi* and *O. colocolo,* occur in parts of range outside the rainforest region; the small spotted cats are difficult to tell apart in the field and are often confused. Found at higher elevations than margay or ocelot.
Natural history. Feeds on small mammals and birds. Found in evergreen and deciduous forest and brush; in lowlands in some regions and in cloud forest and high elevation thickets in others.
Geographic range. Central and South America: Costa Rica south to S Brazil and N Argentina; Atlantic coastal forests of SE Brazil. Range poorly known. To 3,200 m elevation.
Status. CITES Appendix I (Central America), US-ESA endangered. Apparently always rare.
Local names. Tigrillo (Span); gato tigre, chivi (Ar); gato-do-Mato (Br); tigre gallinero, tigrillo peludo (Co); caucel (CR);

tigrillo chico (Ec); chat-tigre (FG); ocelot-cat, tigricati (Su).
References. Gardner, A. L. 1971. Notes on the little spotted cat, *Felis tigrina oncilla* Thomas, in Costa Rica. *J. Mammal.* 52: 464–65.

Jaguarundi
Herpailurus yaguarondi
Plate 17, map 134
Identification. Measurements: HB = 505–645; T = 330–609; HF = 120–152; E = 25–40; SH ≈ 350; WT = 4.5–9 kg.
Upperparts usually uniform grizzled gray or brown, or red, black, or tawny yellow; no spots. Head small; ears small, rounded. Underparts the same as back or slightly paler. Feet small and doglike. A long-backed, slender-bodied cat with a long, slender tail, long neck, and short legs. At least some kittens are unspotted; some are reported to be dark-spotted.
Variation. Probably the most variable in color of all wild cats: red and gray animals are found in the same populations; rainforest animals tend to be dark brown or blackish, sometimes grizzled gray on head with black bodies; specimens from dry habitats are usually paler, gray, yellowish brown, or red; some from N Venezuela are tawny yellow like puma.
Similar species. Tayras *(Eira barbara)* usually have a pale spot on throat, inconspicuous ears, and bushy tail; bush dogs *(Speothos venaticus)* have thick bodies and short tails; puma *(P. concolor)* are much larger, with dark tail tips and pale muzzles; in dry habitats could be mistaken for a fox.
Sounds. None usually heard in field; makes birdlike chirps.
Natural history. Diurnal and nocturnal; terrestrial but can climb trees; solitary and in pairs. Feeds on small mammals, birds, and reptiles. Jaguarundis travel widely (about 7 km per 24 hours) in a huge home range (13–100 km^2). They den in hollow logs, treefalls, and thickets. Found in many habitats from rainforest to fields, savannas, and dense thickets in scrub, and can live in secondary vegetation near villages, where they may raid poultry. They may be more common in dry forest, secondary forest, and savanna than in rainforest.
Geographic range. North, Central, and South America: Texas south to S Brazil and Paraguay. To 2,200 m elevation.

Map 134

Jaguarundi, *Herpailurus yaguarondi*

Map 135

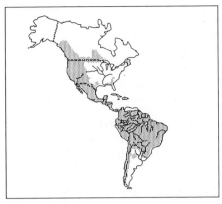

Puma, *Puma concolor*

Status. CITES Appendix I, US-ESA endangered (Central and North America only). Widespread and not hunted for the fur trade but apparently uncommon or rare everywhere.

Local names. Yaguarundi (Span); halari (Be); gato gris (Bo); maracajá-prêto, gato-prêto, gato mourisco (Br); gato montés, gato pardo, zorrogato, gato servante, onza (Co); león breñero (CR); gato pardo, gato moro (Co); gato cerban (Ho); tigrillo congo, tigrillo negro (Pn); anushi-puma (Pe); weasel-cat, boesikati (Su); mbaracaya-eirá, yaguarundí (Gua); ekmuch (May).

References. See margay *(L. wiedii).*

Puma

Puma concolor
Plate 17, map 135

Identification. Measurements: HB = 860–1,540; T = 630–960; HF = 230–290; E = 83–102; SH = 530–790; WT = 29–120 kg; males larger than females.

Upperparts uniform tawny yellow-brown to dark reddish. Head relatively small; facial markings prominent: muzzle around mouth white, patch at base of whiskers blackish; throat whitish. Eyeshine bright pale yellow. Tail darkening to blackish tip. Underparts paler than back. Young spotted with dark brown. Large, long-legged, rangy cats, slightly swaybacked when standing.

Variation. Puma from dry habitats are pale gray or yellow; those from rainforest range from yellow-brown to dark red-brown.

Similar species. Puma are the only large, uniformly colored cats in the region; jaguarundi *(H. yaguarondi)* are much smaller, with no contrasting dark tail tip.

Sounds. None usually heard in field, but young emit a shrill whistle and females in heat call with loud screams.

Natural history. Nocturnal and diurnal; terrestrial; solitary. Feeds chiefly on medium-sized and large mammals such as deer, agoutis, and pacas, but also eats smaller prey such as snakes and rats. Puma are found throughout the rainforest, where they seem to keep mostly to dry ground. They are lovers of wilderness, normally shy and wary of man, and are rarely seen even where common. They can sometimes be encountered walking down a rainforest trail. They mark their paths about every 200 m by scraping up a patch of dirt with their hindfeet and urinating or spraying on it (jaguar and collared peccaries also scrape), and like other cats, they scratch fallen logs. One to three females and one adult male seem to share overlapping home ranges. Puma will attack livestock where their habitat has been converted to pasture and their natural prey reduced, and many populations have been extirpated by ranchers. In centuries of historical association with man, only a few attacks on humans have ever been recorded. Puma sometimes follow humans discreetly from apparent curiosity, but they will flee if faced and approached boldly. Puma are the most adaptable of the world's cats and are found in many climates from boreal to tropical, desert to rainforest, and lowland to montane, in forest, woodland, and scrub habitats with abundant game.

168   Species Accounts

Geographic range. North, Central, and South America: Alaska to S Argentina and Chile to at least 51° S. To 4,500 m elevation.
Status. CITES Appendix I. US-ESA endangered (CR and SE United States subspecies only). Widespread but always uncommon or rare; locally hunted as a predator of livestock, extinct over much of former range, and locally threatened by deforestation and overhunting of its prey.
Local names. León, león colorado, león de montaña, puma (Span); red tiger (Be); onça-vermelha (Br, Co); tig rouge (FG); poema, redi-tigri (Su); guasura, yaguá-pytá (Gua); cabcoh (May).
References. Anderson, A. E. 1983. *A critical review of literature on puma* (Felis concolor). Colorado Div. Wildl. Special Rep., no. 54.

Seidensticker, J. C. IV, M. G. Hornocker, W. V. Wiles, and J. P. Messick. 1973. *Mountain lion social organization in the Idaho Primitive Area.* Wildl. Monographs, no. 35.

Jaguar
Panthera onca
Plate 17, map 136
Identification. Measurements: HB = 1,100–1,850; T = 44–56; HF = 22–25; E = 64–88; SH = 64–76; WT = 31–158 kg; males larger than females.
Upperparts tawny yellow with black spots, many on back and sides in open circles or rosettes; neck spotted above and below, not striped; fur short and smooth. **Head very large, canine teeth long and stout; eyeshine bright greenish yellow; ears rounded, white inside, black behind tips.** Tail long, spotted or banded with black. **Underparts white with black spots.** Kittens spotted. **Very large and heavy, built for power, not speed, with a short back, thick body, robust, short legs, and large feet.**
Variation. There is much size variation: jaguar from open habitats seem on average larger than those of rainforest. Dark brown or black jaguar (with spots always faintly visible at certain angles to light) occur rarely in all populations; Indians give these a separate name and often consider them a separate species, claiming different habits than tawny jaguar.
Similar species. Jaguar are the only very large spotted cats in the New World; ocelots *(Leopardus pardalis)* are much smaller, with stripes on the neck.

Map 136

Jaguar, *Panthera onca*

Sounds. Occasionally roar, day or night, with a pulsed series of single, deep, resonant, hoarse grunts that can be heard for several hundred meters.
Natural history. Nocturnal and diurnal; terrestrial; solitary. Feeds chiefly on large mammals such as capybaras, peccaries, and deer; also on turtles, tortoises, caiman, birds, fish, and smaller mammals such as sloths and agoutis. Jaguar hunt at any time of the day or night. They are fond of walking on man-made trails at night (as are other cats). Jaguar often use wet or waterside habitats, where they hunt capybaras, turtles, caiman, and fish. Large cat tracks on river beaches are usually those of jaguar. In remote areas free from hunting, jaguar are most often seen when they lie stretched out on a log over the water in the morning sun. They rest, and carry their kills to feed, in dense vegetation. Jaguar may kill livestock where their natural habitat has been destroyed and their prey replaced by cattle. Several individuals of both sexes have overlapping home ranges. Although most jaguar flee quickly from man and attacks are very rare, these cats are potentially dangerous. Never run away from a big cat, since that may cause it to give chase; face it or walk toward it making loud noises (e.g., shouting and clapping hands). Found in a wide range of habitats from rainforest to wet grasslands and arid scrub.
Geographic range. North, Central, and South America: Mexico to Argentina. Formerly in SW United States and Uruguay, where now extirpated. To 2,000 m elevation.
Status. CITES Appendix I, US-ESA endangered. It is difficult to estimate populations, but jaguar are rare or absent in many parts

of their former range due to overhunting for the fur trade, loss of habitat by deforestation, persecution by ranchers, and probably loss of their prey. Only a few hundred are thought to remain in all of Mesoamerica, but they are still widespread and can be locally common in Amazonia.

Local names. Tigre, jaguar (Span); tiger (Be); onça, onça pintada, jaguareté (Br); tigre real, tigre mariposo (Co); tig (FG); oto-rongo (Pe); penitigri (Su); yaguareté (Gua); zacbolay (May).

References. Crawshaw, P. G., Jr., and H. B. Quigley. 1991. Jaguar spacing, activity, and habitat use in a seasonally flooded environment in Brazil. *J. Zool.* (Lond.) 223:357–70.

Emmons, L. H. 1987. Comparative feeding ecology of felids in a Neotropical rainforest. *Behav. Ecol. Sociobiol.* 20:271–83.

Dolphins (Cetacea)

River Dolphins and Dolphins (Cetacea: Platanistidae, Delphinidae)

The dolphins are toothed whales (Odontoceti), with many conical teeth. They have long fusiform bodies, long, narrow beaks, pectoral flippers, a dorsal "fin," and a horizontally flattened tail. They breathe through a blowhole on top of the head. Dolphins are fast and agile predators that feed mainly on fish. They can detect objects and find prey by echolocation (sonar), and they communicate with one another by sound. The river dolphins (Platanistidae) are a primitive group of four genera and five species worldwide that mostly live in muddy river systems; one genus and species is endemic to South America. They have degenerate eyes, highly developed echolocation, and flexible necks. Little is known of their behavior; they are often solitary. The single young is apparently weaned before 8–9 months of age. The dolphins (Delphinidae) are an advanced group of largely marine species that have large brains, good vision, acrobatic movements, and complex social behavior. The single young may nurse for up to 18 months. There are 17 genera and 32 species worldwide, one in Neotropical rainforest rivers.

Boto or Pink River Dolphin
Inia geoffrensis
Figure 10, map 137
Identification. Measurements: Total length to 1.8–2.7 m, weight to 160 kg. Males much larger than females.
Dorsal "fin" a long, low hump with inconspicuous point. Head with long, narrow beak and large bulge on forehead (the melon); eyes small. **Flippers broad at base, long.** Color variable, often pale, sometimes with a pinkish tinge, especially on paler underparts. **When they surface to breathe, only a small portion of the back emerges from the water.** Young usually dark gray.
Variation. Animals from the Orinoco are somewhat smaller than those from the Amazon; those from the upper Rio Madeira in Bolivia are considered by some to be a distinct species *(I. boliviensis)*.
Similar species. Tucuxis *(Sotalia fluviatilis)* are smaller, with pointed, triangular dorsal fins; their backs rise well out of the water when they surface.
Sounds. Explosive snorts or sighs when breathing; a variety of underwater calls can be heard when one is swimming near them.
Natural history. Diurnal and nocturnal; aquatic; often solitary, more rarely in groups of two to four. Feeds on fish, including many bottom-feeding species such as catfish, and sometimes other animals such as crabs and turtles. Botos have flexible bodies and can

Map 137

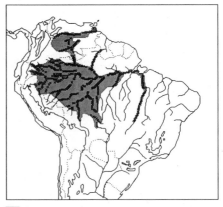

■ Boto, *Inia geoffrensis*

turn their heads and twist their tails, but they rarely leap above the water surface. They are curious and will approach swimmers but will not attack. They are easy to see almost anywhere along the main Amazon and in many tributaries, especially near the mouths of rivers and streams. Found in fresh water only, in both turbid whitewaters and clear blackwaters.
Geographic range. South America: rivers of the Amazon and Orinoco drainages, below significant waterfalls except for an isolated population in the upper Madeira drainage in Bolivia.
Status. Often common; widespread. River

Figure 10. Dolphins: *(a)* **boto,** *Inia geoffrensis; (b)* **tucuxi,** *Sotalia fluviatilis.*

dolphins are the object of many superstitions and myths, and they rarely used to be killed by local fishermen or hunters. However, where large-scale mechanized commercial fishing is carried out, or poisons or dynamite are extensively used, their food supply is threatened and accidental deaths may occur. They have recently vanished from parts of their range (such as the Río Napo in Ecuador). Their body parts are used in folk medicines, and deliberate hunting may be on the increase.
Local names. Bufeo colorado, tonina (Span); boto vermelho (Br); uyara (Ec) uynia (TGua).
References. Best, R. C. 1984. The aquatic mammals and reptiles of the Amazon Basin. In H. Sioli, ed., *The Amazon,* 371–412. Dordrecht: W. Junk.

Meade, R. H., and L. Koehnken. 1991. Distribution of the river dolphin, tonina *Inia geoffrensis,* in the Orinoco river basin of Venezuela and Colombia. *Interciencia* 16:300–312.

Tucuxi or Gray Dolphin
Sotalia fluviatilis
Figure 10, map 138
Identification. Measurements: Total length

= 1.3–1.5 m; weight to 53 kg (for riverine animals).
Dorsal fin triangular and prominent. Head with short beak, hump on forehead small. Bicolored gray, brown, or bluish above, paler gray, whitish, or pinkish below. Flippers quite short, narrow at junction with body. **When surfacing, back and often top of head emerge high out of water.**
Variation. Coastal forms are sometimes considered a distinct species.
Similar species. See boto *(Inia geoffrensis).*
Sounds. Snorts or puffs when surfacing to breathe; many underwater calls.
Natural history. Apparently diurnal; aquatic; usually in groups of two to nine, sometimes solitary. Feeds on fish, chiefly pelagic schooling species in the Amazon, armored catfish in Suriname. These little dolphins are more acrobatic than the larger botos and will more often leap clear of the water. They are easy to see in many areas, especially near the mouths of rivers and streams. Found in both fresh and salt water, in rivers below significant falls, and in estuaries and nearshore coastal waters.
Geographic range. South America: rivers draining into the Atlantic and Caribbean, and coastal waters to Panama.

Map 138

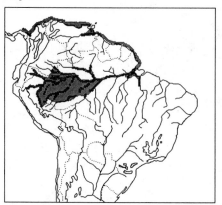

Tucuxi, *Sotalia fluviatilis*

Status. CITES Appendix I. Widespread and often common; status like boto.

Local names. Bufeo negro, bufeo plomo, boto (Span); tucuxi (Br); tonina, delfín de río (Co); profossoe (Su).

References. Magnusson, W. E., R. C. Best, and V. M. F. da Silva. 1980. Numbers and behaviour of Amazonian dolphins, *Inia geoffrensis* and *Sotalia fluviatilis fluviatilis,* in the Rio Solimões, Brasil. *Aquatic Mammals* 8:27–32.

Tapirs (Perissodactyla)

Tapiridae

Dental formula: I3/3, C1/1, P4/3, M3/3 = 42. Forefeet with four toes; hindfeet with three. Most of the body weight is borne on the third toe, which is the largest and centrally located, but the toes on each side are also functional; the small outer digit of the forefoot (actually the fifth digit) touches the ground only when the animal sinks into a soft substrate. There are hooves on all the toes. Tapirs are the only extant native New World odd-toed ungulates (Perissodactyla). They are the largest terrestrial mammals in the region. They have large, rotund bodies; thick, muscular necks; an upper lip elongated into a proboscis; and short stumpy tails. Members of this order, which includes horses and rhinoceroses, are large herbivores. Mammals do not themselves possess the enzymes with which to digest the cellulose that makes up the structural material of plants. Herbivorous mammals have evolved a variety of systems to house and maintain mutualist microorganisms that digest ("ferment") the plant material for them. In the Perissodactyla the stomach is simple, and the cecum is enlarged to form a chamber in which microorganisms live and digest plant cellulose. This system is not extremely efficient, so these animals must eat large amounts of forage each day, and spend most of the day eating in order to obtain enough energy from the leaves they eat. They produce correspondingly large amounts of droppings, which are full of undigested, chopped fibers and seeds of fruits. The large teeth of tapirs are fitted for grinding up plants. The prehensile proboscis is used to reach out and sweep browse into the mouth. Tapirs have single, precocial young. There is one genus with four species worldwide: one in Asia and three in the New World.

Brazilian Tapir
Tapirus terrestris
Plate 18, map 139
Identification. Measurements: HB = 1,700–2,010; T = 46–100; HF = 330–350; E = 120–137; SH = 800–1,100; WT = 227–250 kg.
Body, head, and legs uniform blackish brown; hair short and smooth, skin gray, sometimes not completely covered with hair; short, stiff mane or crest of longer, dark hair in narrow strip from forehead to shoulder. **Upper lip elongated in a down-curved proboscis; forehead rises in a distinct convex hump under base of mane in front of ears; ears brown, tips white,** base with small white spot; eyes small, eyeshine bright pale yellow. **Tail a short stump. Hindfeet with three toes, forefeet with three large toes** that appear in tracks **and a fourth, smaller toe** that appears as a small crescent in tracks on soft ground. Throat and sometimes chest grayish or white; belly brown, paler than back. Young with bright yellow or white longitudinal stripes alternating with lines of spots, legs and cheeks heavily spotted, belly white. **A robust animal the size of a small pony, with a cylindrical body, thick neck, and convex back highest above kidneys.**
Variation. Color varies from tan to blackish or reddish; undersides more or less white.

Map 139

Brazilian tapir, *Tapirus terrestris*

There seems to be much size variation, but few measurements or weights of wild adult specimens are available.
Similar species. Baird's tapirs *(T. bairdii)* are found with Brazilian tapirs only in N Colombia; they have a flat crown between the ears and a poorly developed crest; capybaras *(Hydrochaeris hydrochaeris)* are much smaller, with a blunt muzzle and no proboscis.
Sounds. Usually silent but may snort and stamp in alarm; communicates with a loud whistle (they answer imitations of this call by hunters, who then easily find and shoot them.)

173

Natural history. Mostly nocturnal, partly diurnal; terrestrial; solitary, but several use the same area. Feeds on terrestrial and swamp browse, grass, and fruits. Tapirs favor waterside habitats with some herbaceous vegetation, such as river edges, swamps, and lush stream bottoms, but they are all-terrain animals that travel widely through the forest, including terra firme areas far from water, and they negotiate almost vertical slippery hillsides. By day they rest in thick vegetation, especially in swamps. Their tracks are frequent where they occur, but they are shy, silent, and rarely seen. The dung is a large heap of fibers, seeds, and leaf fragments often (but not always) deposited in water. When strongly alarmed they run for the nearest water, plunge in, and swim beneath the surface; they can flatten everything in their path in a desperate rush. Tapirs use salt licks, along with deer and peccaries, and they are easiest to see near these sites. Found in rainforest, gallery forest, dry forest, chaco, and more open grassy habitats with water and dense vegetation for refuge.

Geographic range. South America: east of the Andes from N Colombia to S Brazil and N Argentina and Paraguay. To at least 2,200 m elevation.

Status. CITES Appendix II, US-ESA endangered. Locally common, but scarce or extirpated in overhunted regions; tapir meat is much prized, and tapirs are easy to locate with dogs or calls and are vulnerable to local extinction.

Local names. Tapir, anta (Span); anta, marebis (Br); danta, gran bestia (Co, Ec); sacha vaca (Co); maïpouri (FG); sacha vaca, huagra kemari (Pe); tapií, mborebí (Gua); boskoe, bosfroe (Su); maypouri (Qui).

References. Hershkovitz, P. 1954. Mammals of northern Colombia, preliminary report no. 7: Tapirs (genus *Tapirus*), with a systematic review of American species. *Proc. U.S. Nat. Mus.* 103:465–96.

Bodmer, R. E. 1990. Fruit patch size and frugivory in the lowland tapir *(Tapirus terrestris). J. Zool.* (Lond.) 222:121–28.

Baird's Tapir

Tapirus bairdii
Plate 18, map 140
Identification. Measurements: HB = 1,930–2,000; T = 70–100; E = 130–140; HF = 372–380; WT = 150–300 kg.

Map 140

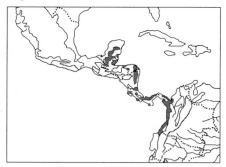

Baird's tapir, *Tapirus bairdii*

Similar to Brazilian tapir, except: head with a flat crown; mane often poorly developed or absent; proboscis slightly longer; lower cheeks and chest usually more white or gray; body more often reddish. Small young chestnut-red with white stripes and spots, older young browner.

Similar species. See Brazilian tapir.

Sounds. When surprised may stamp foot loudly; whistles like Brazilian tapir.

Natural history. Nocturnal and diurnal; terrestrial; usually solitary, family groups seem to use a small area. The habits of Baird's tapirs seem similar to those of Brazilian tapirs. They spend almost 90% of their waking time feeding on browse, grass, and fruits. Sometimes they sleep in the water. Found in rainforest and montane forest, swamps, and flooded grasslands.

Geographic range. Central and South America: S Mexico south to Panama; west of the Andes from N Colombia to the Gulf of Guayaquil, Ecuador.

Status. CITES Appendix I, US-ESA endangered. Threatened by hunting and deforestation.

Local names. Tapir, anta, danta (Span); mountain cow (Be); macho de monte (Co, CR); danto (Ho); antiburro (Me); tzimin (May).

References. Terwilliger, V. J. 1978. Natural history of Baird's tapir on Barro Colorado Island, Panama Canal Zone. *Biotropica* 10:211–20.

Piñera N., E. J. 1995. Hábitos de alimentacíon del tapir *(Tapirus bairdii)* en un bosque tropical húmedo de Costa Rica. *Vida Silvestre Neotrop.* 4:32–37.

Peccaries and Deer
(Artiodactyla)

There are two families of Artiodactyla, or even-toed ungulates, in Neotropical rainforest, the peccaries (Tayassuidae) and deer (Cervidae). In the Artiodactyla, the body weight is borne evenly on the third and fourth toes, with the center of gravity falling between them. The hooves on the tips of these toes form the familiar, two-toed "cloven hoof," but there is often one or a pair of smaller toes behind them that normally do not touch the ground.

Peccaries (Tayassuidae)

Dental formula: I2/3, C1/1, P3/3, M3/3 = 38. Forefeet with two large weight-bearing toes and two small rear toes that do not touch the ground except on soft substrates; hindfeet with two large toes and one small toe; all toes with hooves. Peccaries are large, chunky animals with slender legs, large heads, thick necks, and virtually no tails. They have large, sharp canines; the upper ones point downward (instead of out or up as in pigs). Their snouts are piglike, with nostrils opening in the center of a mobile disk that is used for rooting. They have simple digestive tracts and are omnivores that feed on fruit, seeds, some browse, and animal matter. The peccaries' nearest relatives are the pigs, but they are in a separate family and differ from true pigs in a number of features. Unlike pigs (which have large litters of altricial young that are raised in a nest), peccaries have small litters of one or two precocial young that can walk and follow their mother shortly after birth. There are two genera and three species, all in the New World.

Collared Peccary
Tayassu tajacu
Plate 18, map 141
Identification. Measurements: HB = 800–980; T = 25–45; HF = 170–200; E= 70–90; SH = 300–500; WT = 17–35 kg.
Upperparts uniformly grizzled gray-black, faint but distinct collar or stripe of pale yellow hairs from top of shoulder forward to lower cheek; midback from head to rump with crest of long hairs, raised in excitement; large scent gland along the spine; hair sparse, coarse bristles, banded black and white or yellow. **Head large, sharply tapering from large jowls to narrow nose; nostrils in a naked, mobile disk; disk small, protruding little beyond rostrum; canines large, form distinct lumps under lip,** do not protrude; eyes small, eyeshine weak, reddish; ears small, with short hair. **Crest of long hair on head starts between ears,** hair short above eyes. Tail tiny, not visible. Forefeet with two large toes and two smaller rear toes that do not touch the ground (do not appear in tracks); hindfeet with two large toes and one smaller toe. Young grizzled reddish brown. **Piglike, with a stout body, thick neck, and thin, delicate legs.**

Map 141

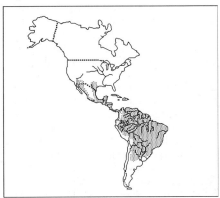

Collared peccary, *Tayassu tajacu*

Variation. Collared peccaries from dry, open habitats are small and gray; rainforest animals are large and blackish.
Similar species. White-lipped peccaries *(T. pecari)* are black with a white chin and no collar, have a crest of long hair starting above eyes, and have a large nasal disk. In the chaco, chacoan peccaries *(Catagonus wagneri)* are much larger, with long hair and

175

prominent, hairy ears. Capybaras *(Hydrochaeris hydrochaeris)* are tan or dark redbrown, with large, square muzzles, robust legs, and no markings.

Sounds. Usually quiet; if surprised at close range may give very loud doglike barks, "whoof," while in flight; sharp clacks of the teeth and grunts (perhaps threats) are heard rarely; chews nuts and snails with loud cracking.

Natural history. Diurnal in rainforest; terrestrial; groups of 1–20, usually 6–9; loose groups often change composition, males sometimes solitary. In rainforest feeds on fruit, palm nuts, browse, snails, other invertebrates, and small vertebrates. Collared peccaries travel in single file along small paths and disperse in the undergrowth to feed. They release a strong odor like cheese or chicken soup, especially when alarmed, and they frequently mark trails by scraping the ground in front of a pole with their hooves, defecating in the scrape, and rubbing their back gland on the pole. They regularly use mud wallows and salt licks in favored, traditional spots. Collared peccaries are wary and surprisingly quiet and difficult to approach; they stampede in panic when they detect humans, and they are not dangerous (but those raised in captivity or habituated to man may be very aggressive, and inflict serious bite wounds). In Amazonian rainforest, groups sleep at night in deep burrows, often under the roots of trees. These are the most commonly seen peccaries, and they are resident throughout the rainforest. Found in habitats from dry, shrubby Sonoran desert and chaco to deep rainforest. Note that the behavior is different in hot, dry habitats, where peccaries feed on cactus, are active at night, and do not use burrows.

Geographic range. North, Central, and South America: SW United States to Argentina.

Status. CITES Appendix II. Widespread and locally common, but hunted intensively for meat, sport, and hides; rare or absent near many settlements but not generally threatened.

Local names. Sajino, saíno (Span); chancho de monte (Ar, Bo); peccari (Be); taitetú (Bo); caitetú (Br); jabalí (Co); tatabro (Ec); pakira (FG, Su); quequeo (Ho); báquiro de collar, chácharo (Ve).

References. Castellanos, H. G. 1983. Aspectos de la organización social del báquiro de collar, *Tayassu tajacu* L., en el Estado Guarico-Venezuela. *Acta Biol. Venez.* 11: 127–43.

Bodmer, R. E. 1989. Frugivory in Amazonian Artiodactyla: Evidence for the evolution of the ruminant stomach. *J. Zool.* (Lond.) 219:457–67.

White-lipped Peccary
Tayassu pecari
Plate 18, map 142

Identification. Measurements: HB = 950–1,100; T = 28–56; HF = 210–230; E = 80–90; SH = 500–600; WT = 25–45 kg.

Upperparts uniform black, sometimes brownish; hair very long and coarse, few or no pale bands on individual hairs. **Crest of long hair on head starts from above eye, well in front of ear. Chin and area around corner of mouth and lower cheek white,** cheek below eye sometimes with pale band; ears small, with short hair. **Nasal disk large, protruding prominently above rostrum.** Chest, lower legs, and belly sometimes white. Young grizzled gray, reddish, or dark tan with or without white throat and dark midback stripe. Similar in shape to collared peccary, but larger.

Variation. Usually black in rainforest, sometimes brownish or reddish in sunny habitats (e.g., captives); amount of white on jaw and body varies from faint sprinkling of white hairs to striking snow white.

Similar species. Collared peccaries *(T. tajacu)* are smaller, grizzled gray-black with pale collar; chacoan peccaries *(Catogonus wagneri)* have a collar and hairy ears; capybaras *(Hydrochaeris hydrochaeris)* are tan with a square muzzle.

Sounds. Large active herds produce a continuous racket of bellowing, screaming, and loud tooth-clacking, which can be heard for several hundred meters. Young in the herd call with loud retching sounds. Small groups seem to be quiet.

Natural history. Mostly diurnal; terrestrial; large herds of 50–300 or more. Feeds mostly on fruit, palm nuts, browse, snails, and other invertebrates. White-lipped peccaries travel long distances, periodically visiting each area for a few hours or a day or two and then moving on. Their presence in a given area is episodic and unpredictable. They forage with much rooting; when they have passed, the ground is pocked and churned. Wide bands of densely packed tracks on beaches mark

Map 142

White-lipped peccary, *Tayassu pecari*

their river crossings. They have a strong
odor distinct from that of collared peccaries.
Their reputation for aggressiveness has been
greatly exaggerated and may be partly due
to their impressive appearance, numbers,
and acoustic performance. A herd generally runs away from a detected human, but
their vision seems poor, and they may walk
close past a stationary person without notic-
ing. Much less commonly seen in rainforest
than are collared peccaries. Found with extremely patchy distribution in rainforest, dry
forest, and chaco scrub: in Central America only in humid tropical forests, but at
the southern end of its range occurs in dry
scrub.

Geographic range. Central and South
America: Mexico south to N Argentina.
Status. CITES Appendix II. Poorly known;
probably threatened. Extensively hunted for
meat and hides; many animals from a herd
are often killed at a time. Extirpated by hunting near many human settlements and naturally absent or extinct from many rainforest
areas within its theoretical range. Anecdotes
suggest they are most common on the northern and southern central edges of the Amazon forest (Rio Branco and Mato Grosso in
Brazil and Chaco in Paraguay, where they
outnumber collared peccaries). In Central
America threatened by habitat destruction.
In areas where they seem to be disappearing,
small herds of fewer than 10 are seen. This
species has undergone recent, region-wide,
abrupt declines, with recovery after a dozen
years. Epidemic disease caught from livestock may be the cause.
Local names. Wari (Be); tropero, chancho
de tropa (Bo); queixada, porco-do-mato
(Br); huangana (Co, Ec, Pe); cochon-bois
(FG); pingo (FG, Su); pecari labiado (Co,
Ar); jaguilla (Ho); puerco de monte, chancho de monte (Pn, CR); báquiro (Ve).
References. Kiltie, R. A., and J. Terborgh.
1983. Observations on the behavior of rain
forest peccaries in Perú: Why do white-
lipped peccaries form herds? *Z. Tierpsychol.*
62:241–55.
   Sowls, L. K. 1984. *The peccaries.* Tucson:
University of Arizona Press.

Deer (Cervidae)

Dental formula: I0/3, C0 or 1/1, P3/3, M3/3 = 32–34. All feet with two large, weight-
bearing toes and two small rear toes that do not touch the ground. These are large, graceful
animals with long, slender legs, long necks, and short tails. Males have antlers that are shed
and regrown each year; females are antlered only in caribou and reindeer. Deer are true rumi-
nants with complex stomachs in which plant material is fermented. After feeding, deer must
take time to rest and regurgitate and finely ruminate (chew cuds) forage that has been partly
digested and stored temporarily in part of the stomach. Members of the family are mostly
browsers and grazers that feed on leaves, twigs, and grasses, with occasional fruits, but the
rainforest brockets are largely frugivorous. Brocket deer tend not to use trails but to wander
throughout the forest, so that a fusiform shape adapted for pushing through thickets is strongly
developed. Deer droppings consist of small, smooth, oval pellets deposited in groups. Deer
give birth to one or two well-developed young that for the first weeks of life lie in hiding,
well camouflaged by their spotted coats, while the mother feeds elsewhere. Fawns found lying alone have not been abandoned; they should be left undisturbed. Little is known of the
behavior of rainforest species in the wild. There are about 36 species of deer in 16 genera
worldwide, with 2 genera and at least 4 species in the Neotropical rainforest region. Several
other genera and species occupy montane, scrub, and grassland habitats around the fringes
of the rainforest. The brockets have been overlumped, and there is evidence that at least two
more species should be resurrected from recent synonymy.

Red Brocket Deer

Mazama americana
Plate 18, map 143
Identification. Measurements: HB = 1,050–1,440; T = 120–150; HF = 313–340; E = 84–106; SH = 670–800; WT = 24–48 kg.

Body and legs chestnut-red; head and neck gray-brown. Face without prominent markings; forehead with tuft of long, dark-tipped hairs; ears short, thinly haired, rim at inner base with thin fringe of white hair, inner surface naked, pinkish; eyes large, eyeshine brilliant yellow-white; **antlers on males only, short, straight, unbranched, directed backward.** Legs long and slender, as dark as midback or darker. Tail red above, white below; raised vertically in alarm to show undersurface. Throat at angle of jaw whitish; **belly chestnut, slightly paler but not demarcated from sides;** hind thighs white on inner surface under tail. Rump large, higher than shoulders, **back slightly humped in profile; while walking head carried low, level with back.** Young red with prominent or faint white spots, belly reddish.

Variation. Some animals from Central America are deep, rich red with dark brown or black faces; others are grayish, as are some from northern Colombia.

Similar species. Gray brockets *(M. gouazoubira)* are smaller, gray-brown, usually with prominent ridges inside ear, white or gray bellies; white-tailed deer *(Odocoileus virginianus)* are red or gray-brown with white bellies, large ears, prominent facial markings, branched antlers, and a straight back.

Sounds. Alarm a soft to very loud explosive, whistling snort; when less alarmed, stamps the ground sharply with a forefoot.

Natural history. Diurnal and nocturnal; terrestrial; solitary. Feeds on fruit, fungi, browse, and fallen flowers; browse is mainly eaten when fruits are scarce in the dry season. These deer favor the dense vegetation with abundant herbaceous understory of platanillos, swampy areas, riversides, and old plantations, but they forage throughout the forest. They are adapted for forest life; with their low forequarters and simple antlers they can slip easily through dense vegetation. When inactive they lie down in a fairly sheltered spot. Found in mature and sec-

Map 143

Red brocket deer, *Mazama americana*

ondary rainforest, dry forest, gallery forest, forest edges, gardens and plantations, and savannas near the forest edge.

Geographic range. Central and South America: S Mexico south to N Argentina. To 2,000 m elevation.

Status. CITES Appendix III (Gu). Widespread and often common. Much hunted for meat and scarce in some areas, but seems to persist where other large mammals have been exterminated, probably because of solitary habits and liking for dense vegetation.

Local names. Venado colorado (or rojo; Span); corzuela roja (Ar); antelope (Be); guazo (Bo); veado pardo, veado mateiro (Br); soche colorado (Co, Ec); cabro de monte (CR); biche (FG); güitsizil (Gu); tilopo, antilope (Ho); corzo (Pn); temazate (Me); locho (Ve); guazú-pytá (Gua); grootboshert, prasaradia, redidia (Su); chan yuc (May).

References. Branan, W. V., M. C. M. Werkhoven, and R. L. Marchinton. 1985. Food habits of white-tailed deer and red brocket deer in Suriname. *J. Wildl. Mgmt.* 49: 972–76.

Gray (Brown) Brocket Deer

Mazama gouazoubira
Plate 18, map 144
Identification. Measurements: HB = 900–1,250; T = 100–150; HF = 600–700; E = 90; SH = 350–650; WT = 11–25 kg.

Head and body gray-brown, darkest at midback, paling to sides. Face without prominent markings; forehead with tuft of dark hairs; ears thinly haired, with prominent pale interior ridges, rim at inner base

Map 144

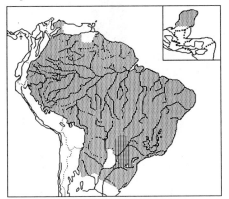

Gray brocket deer, *Mazama gouazoubira*

Bororó *M. nana*

with small tuft of white hairs; eyes large, eyeshine brilliant yellow-white. Throat whitish; neck gray, paler below; belly paler than back, white or gray. Legs dark as center of back. Tail white beneath, raised vertically in alarm. Young brown with white spots, white belly. Body shape and antlers like red brocket, but more slender and gracile.
Variation. The belly seems to be white in northern animals and gray in southern ones; a disjunct population of gray animals from the Yucatán Peninsula is considered a subspecies.
Similar species. Red brockets *(M. americana)* are larger, red with red bellies; white-tailed deer *(Odocoileus virginianus)* are much larger, with facial markings, branched antlers, and a straight back; pampas deer *(Ozotoceros bezoarticus)* are large, tawny yellowish, with prominent pale eye rings.
Sounds. Snorts and stamps forefoot in alarm; may bound away stamping hindfeet loudly.
Natural history. Mostly diurnal; terrestrial; solitary. Feeds on browse and fallen fruits and flowers. In rainforest, where they are always rarer than red brockets, gray brockets seem to favor areas of dense vegetation such as streamsides. At night they lie down in thickets or other sheltered spots. Found in rainforest and in more open, drier habitats such as cerrado and chaco. They are the more common or only brocket in dry habitats of Bolivia, Paraguay, and Uruguay.
Geographic range. Central and South

America: in Central America known only from the N Yucatán peninsula and San José Island, Panama; east of the Andes from Colombia and Venezuela south to Uruguay and N Argentina.
Status. Uncommon but widespread in rainforest.
Local names. Venado plomo, venado gris (Span); corzuela parda, virá (Ar); urina (Bo); veado-birá, veado-catingueiro (Br); soche gris (Co, Ec); cariacou (FG); matacán (Co, Vz); guazú-birá (Gua); koeriakoe (Su).
References. Stallings, J. R. 1984. Notes on the feeding habits of *Mazama gouazoubira* in the Chaco Boreal of Paraguay. *Biotropica* 16:155–57.

Bororó
Mazama nana
Figure 11, map 144
Identification. Measurements: HB = 600–1,000; T = 70–97; HF = 595–635; E = 80–97; SH = 45–50; WT = 15–20 kg.
Upperparts dark red-brown; neck gray, red-brown dorsally; head, crown, and nape gray-brown, muzzle and lower legs blackish brown. Head conspicuously short, with short ears; pale spot below nose. Antlers short, unbranched, and pointed backward. Tail white below; underparts yellowish red, not much paler than sides. Young dotted with lines of small spots on back; throat white, belly dark. **A small, short-legged, red deer; males heavy but short.**
Variation. This deer is sometimes considered a subspecies of *M. rufina,* an Andean species.
Similar species. Red brockets *(M. americana)* are much larger, and have a gray-brown head and neck; gray brockets *(M. gouazoubira)* are gray, with gray head and neck.
Natural history. Diurnal and nocturnal. Said to occupy dense vegetation such as bamboo thickets. An early report described bororós as living in rocky highlands. Small stature could be an accommodation for traveling through almost impenetrable montane thickets, as do the pygmy Andean deer.
Geographic range. South America: in the small corner where Brazil, Paraguay, and Missiones, Argentina, meet.
Status. Apparently rare in a small geographic range.
Local names. Venadito, mbororó, pororoca (Ar); veado (Br).

Figure 11. Bororó, *Mazama nana.*

References. Czernay, S. 1987. *Speis-shirsche und Pudus.* Wittenberg Lutherstadt: A. Ziemsen Verlag.

White-tailed Deer
Odocoileus virginianus
Plate 18, map 145
Identification. Measurements: HB = 1,130–2,260; T = 130–140; HF = 350–380; E = 110–130; WT of males about 50 kg, females 30 kg.
Upperparts gray-brown to light brown with a reddish tinge grading to chestnut on rear of rump, or all chestnut. **Head gray-brown with distinct markings: tip of chin white, nose and muzzle dark brown bordered by a pale band from the corner of the mouth upward; dark spot below corner of mouth; eye surrounded by pale ring, especially prominent above;** eyeshine bright yellow-white; ears large, lined with thick white hair on inner edges. Neck with white spot at angle of throat. **Antlers branched in mature males, simple spikes in yearling males,** none in females. Tail white below, long white hairs protruding as a white bordering fringe, raised and fanned in alarm to show white underside. **Belly white. Legs slightly paler than back.** Young red with white spots, belly white. Large deer with straight, horizontal back when head is

raised; head carried higher than back when walking.
Variation. Color varies from gray-brown "winter" coat to red, fine-haired "summer" coat, depending on both habitat and season.
Similar species. For lowland rainforest only: red brockets *(Mazama americana)* are red, with no facial markings, small ears, spike antlers, and red belly; gray brockets *(M. gouazoubira)* are much smaller, with spike antlers and no facial markings.
Sounds. Whistling bleats or explosive, whistling snorts in alarm; in milder alarm, stamps forefoot.
Natural history. Diurnal and nocturnal; terrestrial; small groups or solitary. Feeds on browse and grass, with some fruit and fallen flowers. White-tailed deer are not rainforest animals. Their branched antlers, tall stature, and high head and shoulder carriage are ill suited to running through dense, viny forest. Whitetails are found in savannas and secondary habitats bordering the rainforest, old secondary forest, montane habitats in upper elevations and cleared mid-elevations of the Andes, and open woodlands and fields, from which they may sometimes enter rainforest habitats in Central America.
Geographic range. North, Central, and South America (S Canada to South America): Andes from Venezuela and Colombia

Map 145

White-tailed deer, *Odocoileus virginianus*

south to N Bolivia; west of the Andes in appropriate lowland habitats from Panama to S Peru, and east of the Andes along the northern coast and in savannas from Colombia to Pernambuco in Brazil.

Status. CITES Appendix III (Gu). Widespread and often common. An important game animal for meat and sport; some local populations in South and Central America are at risk from overhunting.

Local names. Venado (Span); deer (Be); venado llanero, venado blanco, venado de cornamenta (Co); veado galeiro, cariacu (Br); biche des palétuviers (FG); cariacú (Su, Gy); lluichu (Qu); venado cola blanca (Cent Am, Ec).

References. Halls, L. K., ed. 1984. *White-tailed deer.* Harrisburg, Pa.: Stackpole Press.

Manatees (Sirenia)

Trichechidae

Dental formula: 6/6 molarlike teeth usually present at one time; vestigial incisors and deciduous premolars are lost before adulthood. There are an indefinite number of molars that erupt in sequence throughout the life of the animal; as the anterior teeth are worn down, they fall out and are replaced by a new tooth at the back. Manatees are very large, hairless, cigar-shaped mammals, with forelimbs modified to flippers, no free hindlimbs, and the rear of the body in the form of a horizontal paddle. They are entirely aquatic and never leave the water. They are gentle herbivores that browse on aquatic vegetation. They give birth to a single young, which accompanies its mother. There are two families, two genera, and four species worldwide, with one genus and two species in the New World.

Amazonian Manatee
Trichechus inunguis
Figure 12, map 146
Identification. Measurements: Length to
2.8 m; WT = 350–500 kg.
**Body large, cylindrical; hind end a single
flat, rounded, horizontal paddle;** color gray.
**Head small, upper lip modified into a large
bristly surface; eye tiny; no external ear.
Forelimbs short, rounded flippers with
no nails on tips.** Chest and **abdomen with
large irregular white patch.**
Similar species. West Indian manatees
(*T. manatus*) have nails on the flippers and
no white patch on underparts; swimming
otters (*Lontra, Pteronura* spp.), tapirs (*Tapirus* spp.), and capybaras (*Hydrochaeris hydrochaeris*) eventually surface and swim
with top of head and eyes out of water;
dolphins (*Inia, Sotalia* spp.) roll forward
and expose top of head and part of back to
breathe.
Sounds. None usually heard; may snort or
puff when coming up to breathe; makes underwater calls.
Natural history. Nocturnal and diurnal;
aquatic; solitary or females with young.
Feeds on aquatic vegetation such as grasses,
water hyacinths, and water lettuce (*Pistia*).
Manatees live entirely under the water; only
their nostrils break the surface when they rise
to breathe. They favor areas of dense aquatic
vegetation and are therefore extremely difficult to see. Their presence below is betrayed
by their droppings, which are balls of macerated fibers, like horse droppings, that float
to the surface. Manatees produce prodigious
amounts of these, since they can eat up to
8% of their body weight in vegetation per

Map 146

Amazonian manatee, *Trichechis inunguis*
West Indian manatee, *T. manatus*

day. Amazonian manatees do most of their
feeding during the wet season, when they
eat new vegetation in seasonally flooded
backwaters. When the rivers shrink in the
dry season, the manatees return to the main
channels, where they may fast for weeks for
lack of available food plants. Found in rivers
and their associated lakes below any major
rapids.
Geographic range. South America: the
Amazon and lower reaches of its tributaries
from Ecuador and N Peru to the estuaries at
its mouth; and isolated populations on the
Rupununi and Essiquibo rivers (Guyana).
Status. CITES Appendix I, US-ESA endangered. Amazonia once teemed with
these huge animals, but they were ruthlessly
hunted almost to extinction for their meat,
oil, and hides. Few populations are known
to remain; despite legal protection, they are

Figure 12. Manatees: *(a)* **Amazonian manatee**, *Trichechis inunguis; (b)* **West Indian manatee,** *Trichechis manatus.*

still killed for meat in the areas where they survive.
Local names. Peixe-boi (Br); vaca marina, manatí (Span).
References. Best, R. C. 1984. The aquatic mammals and reptiles of the Amazon. In H. Sioli, ed., *The Amazon,* 371–412. Dordrecht: W. Junk.
  Timm, R. M., and L. Albuja. 1986. Ecology, distribution, harvest, and conservation of the Amazonian manatee *Trichechus inunguis* in Ecuador. *Biotropica* 18:150–56.

West Indian Manatee
Trichechus manatus
Figure 12, map 146
Identification. Measurements: Length = 2.5–4.5 m; WT = 200–600 kg.
Similar to Amazonian manatee except larger; flippers with large, flat nails on their tips; underparts gray or with pink blotches, without white patches.
Similar species. See Amazonian manatee.
Sounds. Underwater squeals, whines, and grunts; puffs when breathing.
Natural history. Diurnal and nocturnal; aquatic; solitary or females with young, but temporary congregations form in favored spots and during the mating season. West Indian manatees feed on both bottom and floating aquatic vegetation. Like Amazonian manatees, they spend their lives almost completely submerged. They expose only the nostrils to breathe two to four times in rapid succession about every four minutes. Found in both fresh and salt water.
Geographic range. North, Central, and South America and Caribbean islands: the coasts of Georgia and Florida, United States; Mexico and Central America; the north coast of South America from Colombia to the mouth of the Amazon in Brazil; and the drainages of the Rios Cauca and Magdalena in Colombia and the Orinoco of Venezuela.
Status. CITES Appendix I, US-ESA endangered. Rare or extinct over much of its former range. It has been overhunted for meat in South America and the Caribbean; in the United States many animals are killed or maimed by boat propellers.
Local names. Vaca marina (Span); peixe-boi (Br); sea cow (Be); sekoe (Su).
References. Husar, S. L. 1978. *Trichechus manatus.* Mammalian Species, no. 93.

Rodents (Rodentia)

The rodents are most easily characterized by their teeth: they have a single large pair of chisel-like, ever-growing incisors in the front of each jaw, no other incisors or canines, and three to five cheek teeth (or rarely, fewer) on each side in the rear of the mouth, separated from the incisors by a large gap (diastema). The incisors have hard enamel only on their front surfaces; the posterior part is softer dentine that wears down more quickly. This arrangement ensures that as the teeth are used, they continually keep a sharp, bladelike edge at the front surface. These incisors are remarkably versatile tools: they can be used to cut, pry, slice, gouge, dig, stab, or delicately hold like a pair of tweezers; they can cut grass, open nuts, kill animal prey, dig tunnels, and fell large trees. To do these things, rodents have evolved several different complex systems of jaw muscles and structures of the skull that support these muscles. These differences are used to classify the rodents into two suborders: the Sciurognathi, with squirrel-like jaws, including the Old World families such as squirrels, pocket gophers, kangaroo rats, and mice; and the Hystricognathi, with porcupine-like jaws, including the native New World families such as capybaras, chinchillas, and spiny rats, as well as a few Old World families. Most of the world's rodents are small ($<$ 1 kg) ratlike animals. This body type is extremely versatile, and slight modifications allow rats to be terrestrial, arboreal, semi-aquatic, or fossorial. A number of large species, however, chiefly those in the group of porcupine-like rodents, are not at all ratlike and have body shapes convergent on those of ungulates. The wide scope of activities and diets made possible by the rodent anatomy has made this by far the most diverse order of living mammals: there are about 2,050 species of rodents worldwide, or close to as many as all other species of mammals combined (roughly 2,550), and almost twice as many as the next largest order (bats). Several new species are found yearly, and many certainly remain undiscovered.

The tails of ratlike rodents are covered with skin that is formed into patterns of tiny scales. These scales may be organized into even, prominent rings around the tail or staggered in diagonal rows so that no rings are evident. Hairs grow backward in a fixed pattern around each scale, either from beneath the rear edges of scales or from between them. A magnifying lens is needed to see the pattern. All rodent tails have hairs on them; when a tail is called "naked" it means that the hairs are short, thinly scattered, and inconspicuous. "Hairy" tails may have few, but long, hairs that curl outward and make the tail look bristly or hairy but with the scales still partly visible beneath; or many short, flat hairs that cover the scales; or a complete covering of long, dense hair. Hairy tails may have hairs that extend beyond the tip of the flesh in a tuft or "pencil," which may be either barely visible or large and brushlike. Because there is a continuous gradient of amount of hair on the tail, differences in the degree of hairiness are often hard to describe, yet they are one of the most easily seen external differences between some genera and species. Arboreal, fossorial, and semiaquatic species tend to have hairier tails than terrestrial species.

Squirrels (Sciuridae)

Dental formula: I1/1, C0/0, P1 or 2/1, M3/3 = 20–22. Forefeet with four long toes with claws, and thumb a short stump with a small nail; hindfeet with five long toes with claws. The squirrels of Neotropical rainforests are all tree squirrels; they have broad heads, large eyes, short ears, soft fur, long bushy tails, and long legs. They have flexible ankle joints that can rotate to allow them to descend trees head first, with the hindfeet flat against the trunk. Most species have strong chisel-like incisors and large jaw muscles, which enable them to gnaw open the hardest nuts. They all are excellent climbers, but some species do much of their foraging on the ground. Squirrels are generally omnivorous and feed on nuts, soft fruits, insects, fungi, and sometimes leaves, flowers, and bark. Many Neotropical rainforest species, however, feed chiefly on the nuts of palms and a few other trees with large, hard-shelled seeds.

184

Squirrels can often be located by the sounds of gnawing. Those for which nesting has been described either build a round ball of leaves and twigs in a vine tangle or on a branch, or make a nest of leaves in a tree hole. Litter size is about two for those few Neotropical species for which there is information. There have been few studies of Neotropical squirrels; the natural history of most species is poorly known, and the systematics is unclear in several cases. Squirrels of the genus *Sciurus,* especially Central American species, tend to be geographically highly variable in color and color pattern, and many species have a tendency to produce melanistic or black-pigmented populations or individuals. There are about 50 genera and 265 species of squirrels worldwide, with 4 genera and 19 species in the rainforest region.

Northern Amazon Red Squirrel
Sciurus igniventris
Plate 19, map 147

Map 147

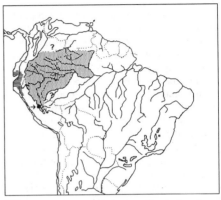

Northern Amazon red squirrel, *Sciurus igniventris*
Guayaquil squirrel, *S. stramineus*
Junín red squirrel, *S. pyrrhinus* (arrow)

Identification. Measurements: HB = 240–295; T = 240–305; HF = 55–75; E = 30–40; WT = 500–900 g.
Back grizzled black and yellowish or orange; hindlegs and forelegs above pure chestnut, rust-red, or orange; or entire animal black. Crown usually blackish; cheeks orange or chestnut; ears large, protruding well above crown, thinly haired, sometimes yellow behind. **Tail usually longer than head and body, very bushy,** exceeds diameter of body, **base black or dark red-brown, orange to yellowish orange distally.** Feet pure red or orange, not mixed with black hairs except in melanistic individuals. **Underparts** thinly haired, sharply demarcated from sides, **contrasting pure pale orange, rust-red, or white,** or black in melanistic individuals. Mammae pairs = 4. A large, long-legged red or black squirrel with a very bushy tail that is often curled up against the back when the squirrel is at rest.
Variation. Animals from north of the Amazon in Venezuela and Brazil are often melanistic, with varying amounts of blackness; melanism is rare in Ecuador and Peru; animals from Ecuador often have a black lateral line separating sides from belly.
Similar species. The northern and southern (*S. spadiceus*) Amazon red squirrels are almost indistinguishable in the field. The northern species is coarsely grizzled yellowish black on the back; the southern appears red, more finely grizzled with black, and has black hairs mixed with red on the feet (this character is visible only with the animal in the hand). Junín red squirrels are uniform chestnut red; all other squirrels in range are small and brown.
Sounds. Alarm call short, low-frequency chatters and chucks; rarely given. Gnaws loudly on palm nuts.

Natural history. Diurnal; arboreal and terrestrial; solitary, but several may feed together at a fruit tree. Feeds chiefly on large palm nuts (*Astrocaryum* spp., *Attaleya* spp.) and other nuts and fruits. These squirrels usually travel in the forest understory or on the ground and can be seen feeding in or near palm trees. They are usually wary and run quickly away, keeping on the ground or low in the undergrowth. They are easiest to locate and approach while they are gnawing on nuts, which they do on an elevated perch. Found in mature and disturbed rainforest.
Geographic range. South America: the Amazon Basin lowlands north of the Amazon from the Rio Negro west through Brazil, Venezuela, Colombia, and Ecuador, and south of the river to N Peru and to the Rio Juruá, Brazil; in S Peru found only in the Andes in upper tropical forest. To 3,300 m elevation.
Status. Common and widespread; sometimes hunted for meat.
Local names. Ardilla (Span); esquilo, quatipuru (Br); huayuashi (Ec).

References. Patton, J. L. 1984. Systematic status of the large squirrels (Subgenus *Urosciurus*) of the western Amazon Basin. *Studies on Neotropical Fauna and Environment* 19:53–72.

Junín Red Squirrel
Sciurus pyrrhinus
Plate 19, map 147
Identification. Measurements: HB = 240–280; T = 208–210; HF = 59, E = 21.
Upperparts entirely uniform dark red, often with small white dots (from scars). **Tail dark chestnut brown at base, orange distally. Underparts sharply contrasting pure rusty orange, white, or orange with white patches medially.** Feet like back.
Similar species. Northern *(S. igniventris)* and southern *(S. spadiceus)* Amazon red squirrels are larger, have tails longer than head and body, and back and top of head grizzled with black.
Natural history. Found in montane forests at 600–2,500 m.
Geographic range. South America: Peru, on the eastern Andean slope in Junín, perhaps to San Martín, but range limits unclear because of confusion with Amazon red squirrels.
Status. Unknown.
Local names. Ardilla.

Southern Amazon Red Squirrel
Sciurus spadiceus
Plate 19, map 148
Identification. Measurements: HB = 240–290; T = 235–300; HF = 59–70; E = 30–37; WT = 600–650 g.
Upperparts dark chestnut red or rusty orange mixed with black on head, neck, and shoulders, becoming pure red on hindquarters; or entire animal black or blackish in melanistic individuals. **Crown often black, cheeks orange;** ears thinly haired, long, protruding well above crown. **Tail very bushy, exceeds diameter of body, black or blackish at base, orange or rusty distally. Feet red with mixture of black hairs. Underparts thinly haired, sharply contrasting pure pale orange, white, or yellowish.** Mammae pairs = 4. **A large, long-legged squirrel,** often holds the tail up over the back.
Variation. Blackish or sometimes pure black individuals are common in some regions of Brazil; they are rare in Peru.

Map 148

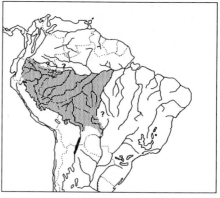

▨ Southern Amazon red squirrel, *Sciurus spadiceus*
▪ South Yungas red squirrel, *S. argentineus*

Similar species. See northern Amazon red squirrel *(S. spadiceus)*.
Sounds. Rarely calls; alarm a snort or sneeze followed by staccato chatter or soft chucks; sneezing calls during mating bouts; gnaws loudly on nuts.
Natural history. Diurnal; terrestrial and arboreal; solitary, but several may feed at the same tree. Feeds chiefly on large hard nuts of palms, especially *Attaleya* spp., and other trees, and other fruits. These squirrels are often seen on the ground and in low undergrowth, or in palm trees; they do not travel in the forest canopy. Loud, rapid gnawing often betrays their location. When alarmed they run away through the undergrowth, sometimes with long leaps between vertical understory trunks. They bury nuts, and several individuals will quickly strip hundreds of nuts from a favored palm. They build a nest of leaves in a tree hollow. During a mating bout, many males will pursue a female. Occupies mature and disturbed evergreen rainforest habitat.
Geographic range. South America: the Amazon Basin of Ecuador, Peru, Bolivia, and Brazil south of the Amazon and west of the Rio Tapajós.
Status. Common and widespread; sometimes hunted for meat.
Local names. Ardilla (Span); masi (Bo); esquilo, quatipuru (Br); huayuashi (Ec).
References. See northern Amazon red squirrel.

Map 149

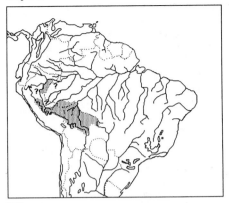

Bolivian squirrel, *Sciurus ignitus*

Bolivian Squirrel
Sciurus ignitus
Plate 20, map 149
Identification. Measurements: HB =
180–195; T = 183–195; HF = 41–52;
E = 20–26; WT = 225–240 g.
Upperparts uniform brown-olivaceous,
finely grizzled. Ears protruding above crown
of head, small buff patch behind ear near
base; eyes surrounded by indistinct pale
ring. Tail quite slender, same color as back,
hairs slightly tipped with yellowish. **Feet
the same color as back.** Chin, throat, and
chest buff to white washed with pale buff;
pale color of chest usually becoming increasingly mixed with gray posteriorly to
**grayish inner thighs and inguinal region
not sharply contrasting with sides.** Underparts well furred. **Mammae pairs = 3.**
Variation. Belly and inner thighs may be
whitish. The small brown squirrels of the
genus *Sciurus* of Peru and Bolivia are confusing and in need of taxonomic revision;
they are often misidentified in museum collections. *S. argentinius* (below) is sometimes
included in this species.
Similar species. Dwarf squirrels (*Microsciurus* spp.) have short ears that do not protrude above the crown; Sanborn's squirrels
(*S. sanborni*) are smaller, have feet paler
than back, sharply contrasting pale underparts, including inner thighs, and bright buff
eye ring; Guianan squirrels are smaller and
slightly darker, probably indistinguishable
from this species in the field.
Sounds. Soft chucks in alarm.

Natural history. Diurnal; arboreal and terrestrial; usually solitary. Feeds on nuts, fruits,
mushrooms, and insects. These squirrels are
most common in dense viny vegetation such
as that found along riversides, where they
use all levels from the ground to the midstory. They are not usually seen high in the
canopy. They make a round nest of green
leaves and twigs hidden about 6–10 m up
in the top of a palm or a dense vine tangle.
Found in mature and disturbed lowland and
montane rainforest.
Geographic range. South America: Amazon Basin of Peru, Bolivia, and W Brazil. To
2,700 m elevation.
Status. Locally common; too small to hunt.
Local names. Ardilla (Span).

Guianan Squirrel
Sciurus aestuans
Plate 20, map 150
Identification. Measurements: HB =
160–202; T = 135–200; HF = 42–50;
E = 16–25; WT = 159–218 g.
Upperparts uniform brown-olivaceous,
finely grizzled. Ears thinly haired, protruding above crown, with or without pale buff
patch behind; eye ring pale buff. **Tail frosted
whitish, yellowish, or reddish and black.
Underparts orange, brightest on chest,
mixed with gray posteriorly, or throat and
inguinal region pale gray or white, or underparts gray. Mammae pairs = 4.**
Variation. There are several forms of this
squirrel that are sometimes recognized as
separate species: *S. a. aestuans,* mainly from
north of the Amazon, has pale ear patches
and a reddish tinge to body and feet; *S. a.
gilvigularis,* mainly from south of the Amazon and Venezuela, has no ear patches; *S. a.
alphonsei,* from coastal Pará south to Pernambuco, has a paler, more yellowish back
and gray underparts; *S. a. ingrami,* from
Bahia to Rio Grande do Sul, is olivaceous
with whitish or whitish and buff underparts.
Similar species. These are the only squirrels
in much of their geographic range. Pygmy
squirrels (*Sciurillus pusillus*) are much
smaller, pale gray, with prominent white
patches behind ears; see Bolivian squirrel
(*Sciurus ignitus*).
Sounds. Alarm calls sharp chucks or chatter,
soft single chips, and high-pitched, twanging
whines.

Map 150

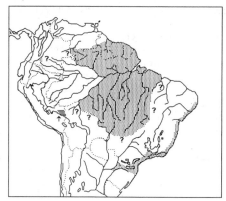

Guianan squirrel, *Sciurus aestuans*

Sanborn's squirrel, *S. sanborni*

Natural history. Diurnal; arboreal and terrestrial; solitary, sometimes pairs. Feeds on
small fruits and nuts and something it scrapes
from tree bark. Guianan squirrels use all levels of the forest but are most often seen in
the low to middle levels. When alarmed,
they run up a tree from the ground with a
chatter and continue to chatter or give sharp
chucks from high in a tree as they hide or
move off through the trees. They are most
common around treefalls and viny vegetation, but also use tall, open forest. They are
active and agile and sometimes travel through
the understory with long leaps from one vertical trunk to another. Found in mature and
disturbed rainforest, secondary forest, and
gardens and plantations.
Geographic range. South America: north
of the Amazon and east of the Rio Negro in
Venezuela, the Guianas, and Brazil; south of
the Amazon west to at least the Rio Madeira,
possibly to Bolivia; and Atlantic coastal rainforest to Rio Grande do Sul.
Status. Widespread and locally common;
not hunted.
Local names. Caixé, serelepe, quatipuru,
caxinguele (Br); eekhoorn (Su); bonboni
(Sar).

Sanborn's Squirrel
Sciurus sanborni
Plate 20, map 150
Identification. Measurements: HB =
152–175; T = 161–184; HF = 44–50;
E = 20–21.
**Upperparts brown-olivaceous. Eye ring
and area around mouth bright pale buff;**
ears thinly haired, **protruding above crown,
bright white or buff patches behind. Feet
yellowish, sometimes distinctly paler than
back.** Underparts including inner thighs
entirely pure white or yellowish orange,
sharply contrasting with sides.
Variation. An animal from the lowlands in
Peru has buff underparts and ear patch; one
from between the Ríos Inambari and Tam-
bopata has those parts white. These squirrels
seem closely related to Guianan squirrels
and may be a subspecies.
Similar species. Bolivian squirrels *(S. ignitus)* are larger, with gray inguinal region.
Natural history. From lowland rainforest.
Geographic range. South America: Peru,
Madre de Dios Department between the
mouth of the Río Manu and the Río Tambo-
pata, 300–580 m elevation. Probably also in
NW Bolivia.
Status. Unknown; apparently rare and with
a small geographic range; but some specimens of *S. "aestuans"* from Bolivia may
belong to this species, or vice versa.

South Yungas Red Squirrel
Sciurus argentinius
Map 148
Identification. Measurements: HB =
199–206; T = 165–190; HF = 52–53;
E = 25–29; WT = 222–261.
**Upperparts blackish suffused with dull
rusty orange; orange tones brightest on
shoulders** and forelimbs; hindquarters more
blackish. Head blackish on crown; **ears
long, stand up above crown, tipped with
fringe of orange hairs; eyes ringed** above
and below **by sharp band of pale orange;**
whiskers black, longest reaching to ear. Feet
grizzled black and orange, the same color
as legs. **Tail bushy,** with long, straight hairs,
colored like back for first one-third, **distal
two-thirds frosted with bright rust-red.**
Underparts well furred, **bright rust-orange,**
sharply demarcated from upperparts, paler
on forequarters, with hair pale gray at base;
darker behind elbows, with hair darker at
base. Females have 1 axial, 1 abdominal, and
2 inguinal = 4 pairs mammae.
Variation. It is said that the ventral color
changes seasonally, from orange in summer
to red-brown *(bayo)* in winter, but we have
not been able to verify this. The specimens
described above were collected in Bolivia
at the end of May. This species has recently
been considered a subspecies of *S. ignitus*,

to which it does not seem closely related. It is highly distinctive, and we give it species status.

Similar species. Southern Amazon red squirrels *(S. spadiceus)* are similar in color, but much larger (twice the weight), brighter red on upperparts, and have pure dark red hindquarters. *S. argentinius* seems to be the only squirrel in its range.

Natural history. Diurnal; arboreal; several may feed together, but probably solitary. Feeds on walnuts *(nogales)*, maize, and probably many other nuts, fruits, plants, and invertebrates. This species is known only from a narrow band of moist montane forest (Bosque Tucumano) on the lower Andean slopes in a region where the lowlands are covered with deciduous dry forest or chaco habitat. Occupies steep, mossy, montane forest with tree ferns in wet ravines and groves of Myrtaceae, walnuts, and feral citrus on the ridges.

Geographic range. South America: from Chuquisaca Department, Bolivia, to Jujuy, Argentina. Known from 1,300 to 2,700 m elevation.

Status. This squirrel is poorly known, but it currently seems recorded only from a tiny geographic range including a narrow elevational band of moist forest, perhaps 10 to 20 km wide and approximately 600 km long. The forest of this area has been severely altered by clear-cutting and burning for pasture, or else intensely grazed in the understory. If this squirrel requires forest, it may be at risk.

Red-tailed Squirrel
Sciurus granatensis
Plate 19, map 151

Identification. Measurements: HB = 200–285; T = 140–280; HF = 40–65; E = 16–36; WT = 212–520 g.

This squirrel has three basic color patterns; there are individuals and populations with intermediate colors between any of the three: (1) **upperparts and tail entirely bright orange-red;** underparts pure white; (2) **upperparts brown-olivaceous; tail** olive at base, heavily frosted **rust-red or orange** distally; underparts pure dark red to orange; (3) **back and head blackish olivaceous, top of head and midback sometimes black; legs and feet red to orange; tail dark at base, orange for most of length, pitch black at tip;** underparts dark red to bright rusty orange.

Map 151

▪ Red-tailed squirrel, *Sciurus granatensis*
▤ Neotropical pygmy squirrel, *S. pusillus*

Generally **a medium-sized squirrel with large ears protruding above crown; a red or brown body; and a bushy red and black or red-tinged tail.** Mammae pairs = 3.

Variation. Color pattern (1) is found in N Colombia east of the Andes, (2) in Venezuela and Central America, and (3) on the west coast of Colombia and Ecuador. There are many intermediate color patterns: squirrels isolated on different sides of a river or mountain range may be different; at higher elevations of the eastern Andes of Colombia, these squirrels are completely olivaceous, with only the faintest reddish tinge to the tail; those on Trinidad are brown-olivaceous, with red-tinged tail; those on Tobago are grizzled brown-yellow-olivaceous, with pale rusty tail with a black tip. Squirrels from coastal Ecuador and Nariño, Colombia, may be speckled with many white spots due to scars.

Similar species. Dwarf squirrels *(Microsciurus* spp.) are smaller, with short ears, and no red on tail; variegated squirrels *(S. variegatoides)* are larger, with black tails heavily frosted white; Deppe's squirrels *(Sciurus deppei)* have blackish tails frosted with white, and grayish bodies.

Sounds. Calls with short chucks; gnaws loudly on hard nuts.

Natural history. Diurnal; arboreal; solitary. Feeds chiefly on large hard nuts of palms and other trees, other fruits, and fungi. Red-tailed squirrels use all levels of the forest, including the ground, but most often the middle levels. They cache nuts by burying them and park fruits to store them briefly in a branch fork or on a vine. They nest in tree holes or leaf nests in vine tangles, but may not always sleep in a nest. During a mating bout several males follow an estrous female. Females defend small territories against other females, but males are nonterritorial and have overlapping home ranges. Found in mature and secondary evergreen and seasonally deciduous forests.

Geographic range. Central and South America: Costa Rica south through Panama; west of the Andes to S Ecuador; in the inter-Andean valleys of N Colombia and NW Venezuela; Trinidad and Tobago. To 2,500 m elevation.

Status. Often common. Widespread and adapts to disturbed vegetation.

Local names. Ardilla chisa, ardilla roja (CR); ardilla común (Co); ardilla colorada (Pn).

References. Nitikman, L. Z. 1985. *Sciurus granatensis.* Mammalian Species, no. 245.

Glanz, W. E., R. W. Thorington, Jr., J. Giacalone-Madden, and L. R. Heaney. 1982. Seasonal food use and demographic trends in *Sciurus granatensis.* In E. G. Leigh, Jr., A. S. Rand, and D. M. Windsor, eds., *The ecology of a tropical forest*, 239–52. Washington, D.C.: Smithsonian Institution Press.

Guayaquil Squirrel
Sciurus stramineus
Plate 19, map 147
Identification. Measurements: HB = 180–320; T = 250–330; HF = 50–65; E = 28–39.

Color variable: (1) Ecuador lowlands: **shoulders coarsely grizzled dirty white and blackish; rump and tail base faint or strong dull orange grizzled black; head and ears black; feet black or white;** tail black frosted with white; underparts dull brown or reddish. (2) Peru, S Ecuador highlands: **upperparts and tail** black heavily frosted with white, **appears pale gray; rump washed with faint or bright orange or buff; neck behind ears pure white or pale yellow; ears and feet pitch black;** underparts gray. Mammae pairs = 4. **A large squirrel with a long, slender gray tail and long black ears.**

Variation. There is much individual variation in color within populations; the body is often spotted with white scars.

Similar species. Red-tailed squirrels *(S. granatensis)* are smaller, and in the area of overlap N of Guayaquil are brownish, with a red-tinged tail.

Natural history. Diurnal; arboreal. Guayaquil squirrels are seen in the trees in both mature and secondary forest and in coffee plantations. Found in humid, dry, and montane forests: in the southern part of its range it occupies humid montane forest along the western Andean slope at 1,400–2,000 m; in the northern part of its range it occurs both in highlands and in humid and dry forests at sea level.

Geographic range. South America: SW Ecuador and NW Peru, around the Gulf of Guayaquil and south along the Andean slope to S Cajamarca. To 2,000 m elevation.

Status. Locally common.

Local names. Ardilla parda, ardilla nucha blanca, ardilla mora, ardilla negra (Pe).

Variegated Squirrel
Sciurus variegatoides
Plate 21, map 152
Identification. Measurements: HB = 220–337; T = 226–325; HF = 45–70; E = 20–35; WT = 428–909 g.

Color highly variable: **upperparts either unpatterned, coarsely grizzled black and gray, yellowish, tawny, or rufous; or patterned, with sides and feet grizzled pale gray or yellowish, center of back from neck to tail a sharply demarcated band of pure dark brown or jet black.** There are pure brown and brownish black populations and individuals. **Ears with prominent pale patch behind,** white in pale animals, tawny to orange in dark animals. **Tail long and bushy,** above **black heavily frosted white,** below with median strip of yellowish to reddish. **Feet and limbs in patterned individuals like back or contrasting red; feet white, buff, blackish, or orange. Underparts pure white, pure rust-red, or red with white patches** on throat and chest and lower belly.

Variation. The 14 named subspecies attest to the great color variation in this species. There is also much variation between individuals within some populations: squirrels from the wetter eastern coasts are generally blackish; those from drier areas are paler gray; those with bright red legs are found in W Costa Rica.

Similar species. Red-tailed squirrels *(S. granatensis)* are smaller, with orange tails; Deppe's squirrels *(S. deppei)* are smaller, with slender tail and gray belly and feet; Yucatán squirrels *(S. yucatanensis)* are gray, with a gray belly; Mexican red-bellied

Map 152

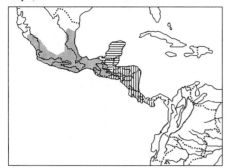

▦ Red-bellied squirrel, *Sciurus aureogaster*
☰ Variegated squirrel, *S. variegatoides*
▥ Yucatán squirrel, *S. yucatanensis*

squirrels *(S. aureogaster)* are more finely grizzled gray, without prominent pale ear patches; the latter two species may not occur together with variegated squirrels.
Sounds. Chucks in alarm.
Natural history. Diurnal; arboreal; solitary. Feeds mainly on soft, juicy fruits, other types of fruits, and flowers. Variegated squirrels are usually seen high in the trees, rarely on the ground. They are most active in the early morning. Found in deciduous forest, open woodland, scrub, and plantations of fruit trees; uncommon in evergreen rainforest.
Geographic range. Central America: Chiapas, Mexico, south to Panama on the east side of the canal. To 600 m elevation.
Status. Locally common; hunted for meat, but adapts to disturbed vegetation.
Local names. Chiza (CR); ardilla negra (Pn).
References. Harris, W. P., Jr. 1937. Revision of *Sciurus variegatoides*, a species of Central American squirrel. *Misc. Pub. Mus. Zool. Univ. Mich.* 38:7–39.

Glanz, W. E. 1984. Food and habitat use by two sympatric *Sciurus* species in central Panama. *J. Mammal.* 65:342–47.

Yucatán Squirrel
Sciurus yucatanensis
Plate 21, map 152
Identification. Measurements: HB = 208–260; T = 220–258; HF = 53–65; E = 17–28.
Upperparts coarsely grizzled or variegated black and gray, olivaceous or tawny. Ears with a pale patch behind. **Tail black frosted with white. Feet often blackish, darker than body, or gray.** Underparts like back,

or pale gray or whitish, or black in melanistic individuals.
Variation. General appearance from blackish brown to gray.
Similar species. See variegated *(S. variegatoides)* and Deppe's squirrels *(S. deppei).*
Natural history. Found in rainforest, lowland broadleaf forest, pine-oak forest, and plantations of coffee and cacao.
Geographic range. Central America: the Yucatán Peninsula, Chiapas, Mexico, Guatemala, and Belize.
Status. Probably hunted for meat.

Deppe's Squirrel
Sciurus deppei
Plate 21, map 153
Identification. Measurements: HB = 163–220; T = 152–190; HF = 46–58; E = 22–26; WT = 190–220 g.
Upperparts finely grizzled brownolivaceous or reddish brown; forelegs from shoulder clear gray. Ears sometimes with small pale patch behind. **Tail slender,** black frosted with white above; brown-olive medially beneath. Underparts pale gray to whitish, sometimes buff. Mammae pairs = 4. **A small brown or gray squirrel.**
Variation. The forelegs in some populations are colored like back.
Similar species. All other squirrels in its range are much larger and have bushy tails.
Sounds. Calls throughout the day with a series of high-pitched, closely spaced notes or birdlike trills.
Natural history. Diurnal; arboreal and terrestrial; solitary; sometimes several call near one another. Feeds on berries, acorns, and fungi. Deppe's squirrels are usually seen on the ground and in lower vegetation levels of dense forest, especially in vine-covered trees. They move quietly and are difficult to observe. They make a leaf nest on a large branch 6 m or more above the ground. Found in dense broadleaf forests and subtropical cloud forests.
Geographic range. Central America: Veracruz, Mexico, south to N Costa Rica. To 2,800 m elevation.
Status. CITES Appendix II (CR). Locally threatened by deforestation.
Local names. Ardilla montañera, ardilla chica (Me).
References. Hall, E. R., and W. W. Dalquest. 1963. The mammals of Veracruz. *Univ. Kans. Pub. Mus. Nat. Hist.* 14:165–362.

In all the plates that illustrate text accounts to the level of genus only (e.g., bats, murid rodents), distinguishing features listed are those useful for identifying the genus, not the species if there is more than one. Likewise, for those examples only, distributions refer to the genus in the rainforest region.

Plate 1

Large Opossums (Didelphidae)
All have prehensile tails; five toes on all feet.

1. Black-shouldered opossum
Caluromysiops irrupta (p. 13)
Black shoulders; tail furred on top to tip.
SA.

2. Bare-tailed woolly opossum
Caluromys philander (p. 12)
Dark stripe down center of face; tail naked except at base. SA.

3. Western woolly opossum
Caluromys lanatus (p. 11)
Dark stripe down center of face; tail furred above for half its length. SA.

4. Central American woolly opossum
Caluromys derbianus (p. 13)
As above; forefeet white; gray patch on midback. C, SA.

5. Bushy-tailed opossum
Glironia venusta (p. 14)
Broad black stripes through eye from crown to nose; tail furred to tip. SA.

6. Common gray four-eyed opossum
Philander opossum (p. 17)
Back uniform gray; cream spots above eyes; tail furred at base for 5–8 cm. C, SA.

7. Anderson's gray four-eyed opossum
Philander andersoni (p. 18)
As above, but midback black with glossy guard hairs. SA.

8. Water opossum
Chironectes minimus (p. 19)
Marbled with broad black bands across back; hindfeet webbed. C, SA.

9. Brown four-eyed opossum
Metachirus nudicaudatus (p. 21)
Back brown; pale yellow spots above eyes; tail naked from near base. C, SA.

10. White-eared opossum
Didelphis albiventris (p. 16)
Like common opossum, except: ears or ear tips white; sharp facial markings. SA.

11. Common opossum
Didelphis marsupialis (p. 14)
Fur in two layers, pale underfur showing through black or gray guard hairs; ears black; white part of tail usually longer than black; cheek below ear not white. C, SA.

12. Virginia opossum
Didelphis virginiana (p. 17)
As above, but cheek below ear white; white part of tail equal to or shorter than black. CA.

Plate 2

Mouse Opossums
(Didelphidae)
All are small; have prehensile tails; no pouch; large ears. (Many species not illustrated.)

1. Brazilian gracile mouse opossum
Gracilinanus microtarsus (p. 33)
Small; fur long and soft; underparts cream.
Southern SA.

2. Delicate slender mouse opossum
Marmosops parvidens (p. 28)
Tiny; legs thin; eye rings indistinct;
underparts white with or without gray
patches. SA.

3. Gray slender mouse opossum
Marmosops incanus (p. 30)
Large; underparts creamy white. Southeast-
ern SA.

4. White-bellied slender mouse opossum
Marmosops noctivagus (p. 26)
Large; underparts white. SA, western
Amazon Basin

5. Red mouse opossum
Marmosa rubra (p. 25)
Medium; upperparts red; underparts
orange. SA, western Amazon Basin.

6. Little rufous mouse opossum
Marmosa lepida (p. 25)
Tiny; red with pinkish underparts; tail very
long. SA, western Amazon Basin.

7. Murine mouse opossum
Marmosa murina (p. 24)
Medium; warm brown with glowing
salmon underparts. SA.

8. Long-furred woolly mouse opossum
Micoureus demerarae (p. 22)
Large; dense woolly fur; tail base densely
furred for 3–4 cm. SA.

9. Robinson's mouse opossum
Marmosa robinsoni (p. 26)
Large; fur short; underparts yellow-orange.
C, SA.

Plate 3

Short-tailed Opossums
(Didelphidae)
All are small; have short, prehensile tails;
short ears; no pouch. (Several species not
illustrated.)

1. Sepia short-tailed opossum
Monodelphis adusta (p. 34)
Tiny; fur extremely short. SA, Andean
region only.

2. Three-striped short-tailed opossum
Monodelphis americana (p. 36)
Three black stripes on back. Eastern SA.

3. Emilia's short-tailed opossum
Monodelphis emiliae (p. 34)
Rump and head forward of ears rufous. SA,
Amazon Basin.

4. Red-legged short-tailed opossum
Monodelphis brevicaudata (p. 36)
Head and rump gray. SA, Amazon Basin
and Guianas.

5. Gray short-tailed opossum
Monodelphis domestica (p. 36)
Large; pale gray. SA, Atlantic Coast and
Paraguay Basin.

Plate 4

Sloths (Bradypodidae, Megalonychidae)

All are large, have long, shaggy hair; hands and feet modified to hooks.

1. **Maned three-toed sloth**
   Bradypus torquatus (p. 44)
   Long black plumes of hair on neck and shoulder. SA, E Brazil.

2. **Pale-throated three-toed sloth** (female)
   Bradypus tridactylus (p. 43)
   Three claws on forefoot; short, stumpy tail; male with orange-and-black spot on back; pale throat. SA.

3. **Brown-throated three-toed sloth** (male)
   Bradypus variegatus (p. 43)
   As above, but throat brown. C, SA.

4. **Southern two-toed sloth**
   Choloepus didactylus (p. 45)
   Two claws on forefoot; no tail; throat the same color as chest. SA.

5. **Hoffmann's two-toed sloth**
   Choloepus hoffmanni (p. 44)
   As above, but throat paler than chest. C, SA.

Anteaters (Myrmecophagidae)

All have long, narrow heads; long tongues; no teeth; greatly enlarged foreclaws.

6. **Southern tamandua**
   Tamandua tetradactyla (p. 39)
   Large, blond with black vest, or pure blond, or black; spotted prehensile tail. SA.

7. **Silky or pygmy anteater**
   Cyclopes didactylus (p. 41)
   Small; dense wooly fur with iridescence; prehensile tail; small eyes and ears. C, SA.

8. **Giant anteater**
   Myrmecophaga tridactyla (p. 38)
   Very large; large bushy tail; black stripe from throat to shoulder. C, SA.

(handwritten margin notes: "PIPELINE ROAD", "CHAGRES FIELD STATION", "CHAGRES FIELD STATION")

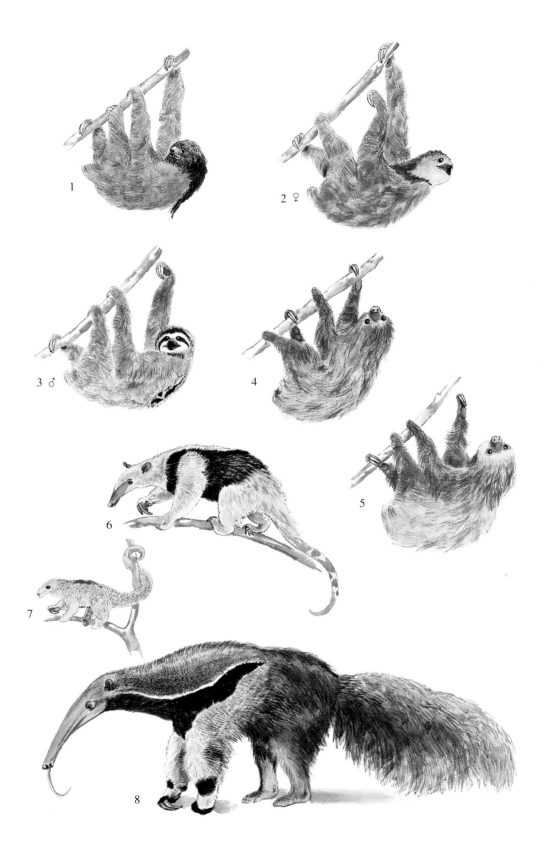

Plate 5

Sheath-tailed Bats
(Emballonuridae)

1. Long-nosed bat
Rhynchonycteris naso (p. 53)
Elongated muzzle; white tufts of hair on
forearm; wavy white lines on back. C, SA.

2. White-lined sac-winged bat
Saccopteryx bilineata (p. 53)
Wavy white lines on back; wing sac close
to forearm bone beyond elbow. C, SA.

3. Ghost bat
Diclidurus albus (p. 57)
White color; short, round ear; glands in
membrane around tail. C, SA.

Bulldog Bats (Noctilionidae)

4. Greater bulldog or fishing bat
Noctilio leporinus (p. 59)
Lip split from below nose; forward-point-
ing nose and ears. C, SA.

Mustached Bat (Mormoopidae)

5. Mustached bat
Pteronotus parnellii (p. 60)
Lips flared forward; tail short, eye small.
C, SA.

Spear-nosed Bats
(Phyllostominae)

6. Fringe-lipped bat
Trachops cirrhosus (p. 68)
Lips and chin studded with long tubercles.
C, SA.

7. Hairy-nosed bat
Mimon crenulatum (p. 66)
Long, narrow noseleaf; large pointed ears;
tail half as long as membrane. C, SA.

Plate 6

Long-tongued bats
(Glossophaginae)

1. Common long-tongued bat
Glossophaga soricina (p. 72)
Lower incisors present, no deep groove on
side of tongue; chin with V bordered by
tiny tubercles, no central wart; fur bicolored, pale at base. C, SA.

Short-tailed Fruit Bats
(Carolliinae)

2. Short-tailed fruit bat
Carollia perspicillata (p. 76)
Short tail present; short, narrow muzzle;
chin with V of tiny warts flanking large
central wart; fur banded. C, SA.

Neotropical Fruit Bats
(Stenodermatinae)

3. Tent-making bat
Uroderma bilobatum (p. 79)
No tail; strong dorsal and facial stripes;
central upper incisors bilobed, with broad
parallel tips, less than twice as long as outer
incisors; tail membrane without hairy
fringe. C, SA.

4. Honduran white bat
Ectophylla alba (p. 83)
No tail; tiny size; white. CA.

5. Large fruit-eating bat
Artibeus jamaicensis (p. 84)
No tail; no dorsal stripe; central upper
incisors bilobed, with broad, parallel tips,
less than twice as long as outer incisors.
C, SA.

Vampire Bats (Desmodontinae)

6. Common vampire bat
Desmodus rotundus (p. 88)
Short M-shaped fold behind nostrils backed
by another fold; large, sharp, triangular
incisors larger than canines; no tail; tail
membrane a complete short band between
legs. C, SA.

1

2

3

4

5

6

Plate 7

Funnel-eared Bats (Natalidae)

1. Funnel-eared bat
Natalus stramineus (p. 90)
Tiny; mouth and ears funnel-shaped; tail
and legs very long; color pale. C, SA.

Sucker-footed Bats
(Thyropteridae)

2. Sucker-footed bats
Thyroptera tricolor (p. 92)
Foot and thumb base with tiny sucker disks;
tail longer than tail membrane. C, SA.

Vespertilionid Bats
(Vespertilionidae)

3. Yellow bat
Lasiurus ega (p. 96)
Ears short and wide; tail membrane thickly
furred above for more than half its length.
N, C, SA.

4. Big brown bat
Eptesicus furinalis (p. 93)
Fur dark at base with pale tips; tail
membrane comes to long point at tip of
tail; no gap between canine and first large
tooth behind it. Genus N, C, SA.

Free-tailed Bats (Molossidae)

5. Brazilian free-tailed bat
Tadarida brasiliensis (p. 100)
Long, naked tail free beyond edge of
membrane; upper lip with vertical wrinkles,
six lower incisors. N, C, SA.

1

2

3

4

5

Plate 8

Marmosets, Tamarins, Goeldi's Monkey
(Callitrichidae, Callimiconidae)

1. Silvery marmoset
Callithrix argentata (p. 106)
Ears naked; white or pale brown; tail black
or brown. SA.

2. Tassel-ear marmoset
Callithrix humeralifera (p. 107)
Ear hidden by long tufts of pale hair; tail
banded. SA.

3. Tufted-ear marmoset
Callithrix jacchus jacchus (p. 108)
Lower back finely banded; tail banded;
white tufts surrounding ear. SA.

4. Golden-headed lion tamarin
Leontopithecus chrysomelas (p. 119)
Black with gold crown and forelimbs. SA,
Bahia, Brazil.

5. Golden lion tamarin
Leontopithecus rosalia (p. 118)
Entirely golden. SA, Rio de Janeiro, Brazil.

6. Goeldi's monkey
Callimico goeldii (p. 120)
Entirely black; head hair in two tiers, a
short, even cap fringed by a long ruff. SA.

Plate 9

Marmosets and Tamarins
(Callitrichidae)

1. Pygmy marmoset
Cebuella pygmaea (p. 105)
Tiny; tawny yellow-gray with fine
striations; tail faintly banded. SA,
western Amazon Basin.

2. Saddleback tamarins (two forms)
Saguinus fuscicollis (p. 110)
Back variegated black and yellowish;
hindquarters reddish; forequarters reddish
(2a, north of Río Marañon) or black
(2b, south of Río Marañon). SA.

3. Golden-mantle tamarin
Saguinus tripartitus (p. 112)
Black head; white muzzle; sharply
contrasting golden shoulders. SA.

4. Black-mantle tamarin
Saguinus nigricollis (p. 113)
Forequarters black or olivaceous;
hindquarters dark red or blackish. SA.

5. Red-chested mustached tamarin
Saguinus labiatus (p. 113)
Sharp white mustache and spot on nape;
underparts bright red. SA.

6. Black-chested mustached tamarin
Saguinus mystax (p. 114)
Sharp white mustache, dark brown
underparts. SA.

7. Golden-handed or midas tamarin
Saguinus midas (p. 112)
Blackish with bright gold hands and feet,
or black hands and feet. SA.

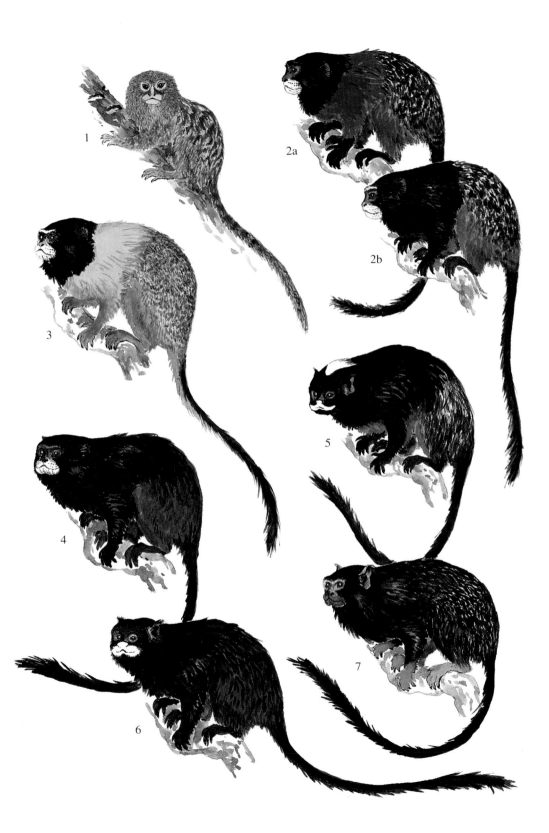

Plate 10

Tamarins (Callitrichidae)

1. Emperor tamarin
Saguinus imperator (p. 115)
Long down-curling white mustaches to
below chin; upperparts pale grayish; tail
orange or orange below. SA.

2. Mottled-face tamarin
Saguinus inustus (p. 117)
Face naked, mottled white and black. SA.

3. Pied bare-face tamarin
Saguinus bicolor bicolor (p. 115)
Face and head to ears naked; tail brown
above, orange below. SA.

4. Silvery-brown bare-face tamarin
Saguinus leucopus (p. 116)
Face naked; ruff dark brown; back silvery
brown; underparts orange. SA.

5. Cotton-top tamarin
Saguinus oedipus (p. 117)
Face naked; crest of long white hair; back
brown; underparts and feet cream. SA.

6. Geoffroy's tamarin
Saguinus geoffroyi (p. 118)
Back variegated black and yellow; face
naked; crown with short white hair; nape
dark red; underparts pale yellow. C, SA.

Plate 11

Small and Medium-Sized Monkeys (Cebidae)

1. Night monkey (red-necked form, Peru)
Aotus sp. (p. 121)
Eyes very large; three black stripes on crown; nonprehensile tail. C, SA.

2. Dusky titi monkey (three forms)
(2a) *Callicebus moloch moloch;* (2b) *C. m. brunneus;* (2c) *C. m. cupreus* (p. 123)
Face small; fur long and dense; tail thickly furred, nonprehensile. SA.

3. Yellow-handed titi monkey
Callicebus torquatus (p. 124)
As (2) above; blackish with yellow collar and hands. SA.

4. Masked titi monkey
Callicebus personatus (p. 125)
As (2) above; wide black fringe encircling face. SA, E Brazil.

5. Common squirrel monkey (two forms)
(5a) *Saimiri sciureus sciureus;* (5b) *S. s. boliviensis* (p. 125)
Small; slender; short hair; white mask around eyes; dark muzzle; nonprehensile tail. SA.

6. Central American squirrel monkey
Saimiri oerstedii (p. 126)
As (5) above. CA.

7. Brown capuchin monkey
Cebus apella (p. 127)
Dark cap, limbs, and tail; dark bar in front of ear; prehensile tail. SA.

8. White-fronted capuchin monkey
Cebus albifrons (p. 129)
Dark cap; pale limbs and tail; pale cheeks to ear; prehensile tail. SA.

9. White-throated capuchin monkey
Cebus capucinus (p. 130)
Black with whitish face, chest, and shoulders; prehensile tail. C, SA.

10. Wedge-capped or weeping capuchin monkey
Cebus olivaceus (p. 128)
Dark cap and line down center of face; no dark bar in front of ear; hands, feet, and tail dark; prehensile tail. SA.

Plate 12

Saki, Bearded Saki, and Uakari Monkeys (Cebidae)

All have nonprehensile tails; long fur; mouth pulled into a frown by large bulging canine teeth under lips; South America only.

1. Guianan saki monkey (male and female)
Pithecia pithecia (p. 131)
Male black with white-rimmed face; female gray-brown. SA.

2. Equatorial saki monkey
Pithecia aequatorialis (p. 132)
Face ringed with white; crown of male with mat of short white hair; underparts orange or brown. SA.

3. Monk saki monkey
Pithecia monachus (p. 132)
Crown almost bald, covered by mop of grizzled hair; chest whitish or buffy. SA.

4. Buffy saki monkey
Pithecia albicans (p. 133)
Ruff, arms, legs, sides, and underparts buffy to reddish; back and tail blackish. SA.

5. Brown bearded saki monkey (male)
Chiropotes satanas (p. 133)
Brown; large tufts over temples; bearded chin; face brown. SA.

6. White-nosed bearded saki monkey
Chiropotes albinasus (p. 134)
Black; triangle covering nose and mouth with red skin clothed with white hairs. SA.

7. Red uakari monkey
Cacajao calvus rubicundus (p. 135)
Tail short; face naked, red. SA.

8. Black uakari monkey
Cacajao melanocephalus (p. 136)
Tail short; face naked, black. SA.

1 ♂ ♀

2

3

4

5 ♂

6 ♀

7

8

Plate 13

Howler Monkeys (Cebidae)
All are large; bearded; with strongly
prehensile tails naked below the tip.

1. **Red howler monkey**
   Alouatta seniculus (p. 136)
   Red to orange. SA.

2. **Red-handed howler monkey**
   Alouatta belzebul (p. 137)
   Entirely black or with hands, feet, tail tip, or
   sometimes forehead or back dull red. SA.

3. **Black howler monkey** (male and
   female)
   Alouatta caraya (p. 138)
   Male black or dark brown; female pale
   yellow or brown. SA.

4. **Brown howler monkey**
   Alouatta fusca (p. 138)
   Brown, red-brown, or yellowish brown. SA.

5. **Mantled howler monkey**
   Alouatta palliata (p. 139)
   Black with pale-frosted fringe on sides or
   cape over back. C, SA.

6. **Mexican black howler monkey**
   Alouatta pigra (p. 140)
   Entirely black. CA.

1 ♂

2 ♂

♀

3 ♂

4 ♂

5 ♂

6 ♂

Plate 14

Spider, Woolly, and Muriqui
Monkeys (Cebidae)
All are large; have strongly prehensile tails
naked under the tip.

1. Black spider monkey
Ateles paniscus (p. 142)
Black; underparts black; face red. SA,
Guiana region only.

2. White-bellied spider monkey
(two forms)
(2a) *Ateles belzebuth chamek;* (2b) *A. b.
belzebuth* (p. 143)
Highly variable; all black, or black or brown
with pale underparts. SA, see range maps.

3. Central American spider monkey
(three forms)
(3a) *Ateles geoffroyi panamensis* (Panama);
(3b) *A. g. geoffroyi* (Nicaragua); (3c) *A. g.
hybridus* (Venezuela) (p. 143)
Highly variable; all spider monkeys in
Central America, or in South America
west of the Andes or in northern Venezuela.
C, SA, see range maps.

4. Muriqui
Brachyteles arachnoides (p. 144)
Entirely pale; ears hairy and protruding
prominently; belly large. SA, E Brazil.

5. Common woolly monkey
Lagothrix lagothricha (p. 140)
Brown, gray, or olivaceous above and
below; robust limbs and tail; round
head. SA.

6. Yellow-tailed woolly monkey
Lagothrix flavicauda (p. 141)
Dark red; yellow under tail tip and on
muzzle. Peruvian Andes only. SA.

Plate 15

Raccoon Family (Procyonidae)

1. Cacomistle
Bassariscus sumichrasti (p. 156)
Tail bushy, brightly banded; white mask
around eyes; dark feet. CA.

2. Olingo
Bassaricyon gabbii (p. 154)
Tail faintly banded, nonprehensile. C, SA.

3. Kinkajou
Potos flavus (p. 155)
Tail prehensile, slightly tapered, not
banded. C, SA.

4. South American coati
Nasua nasua (p. 153)
Muzzle and foreclaws elongated; muzzle
dark; tail banded; tapered. SA.

5. White-nosed coati
Nasua narica (p. 154)
As above, but muzzle white, shoulders
gray. CA.

6. Northern raccoon
Procyon lotor (p. 152)
Tail shorter than head and body, brightly
banded; black mask over eyes continues as
dark bar across throat; forelegs whitish.
N, CA.

7. Crab-eating raccoon
Procyon cancrivorus (p. 151)
As above, but legs and feet dark; mask ends
behind eyes, no brown band on throat.
C, SA.

Dogs (Canidae)

8. Bush dog
Speothos venaticus (p. 148)
Tail and legs short; head and neck pale; fur
long and soft. C, SA.

9. Short-eared dog
Atelocynus microtis (p. 146)
Entirely blackish; tail long enough to touch
ground; ears sharply contrasting pale
brown. SA.

Plate 16

Weasel Family (Mustelidae)

1. Long-tailed weasel
Mustela frenata (p. 158)
Small; dark brown with pale belly; with or
without white facial markings. C, SA.

2. Amazon weasel
Mustela africana (p. 157)
As above, but dark brown stripe down
underparts; no facial markings. SA.

3. Grison
Galictis vittata (p. 159)
Head tricolored; gray crown, white brow
and side of neck, black muzzle and throat.
C, SA.

4. Striped hog-nosed skunk
Conepatus semistriatus (p. 161)
Bicolored black and white; tail bushy,
entirely white distally. C, SA.

5. Tayra
Eira barbara (p. 159)
Head usually pale (may be dark); chest
usually with pale spot. C, SA.

6. Neotropical otter
Lontra longicaudis (p. 161)
Underparts dark or pale, throat not spotted;
tail cylindrical. C, SA.

7. Giant otter
Pteronura brasiliensis (p. 162)
Very large; throat usually spotted brown on
white; tail horizontally flattened. SA.

Plate 17

Cats (Felidae)

1. Jaguarundi
Herpailurus yaguarondi (p. 166)
Small; uniform color, red, brown, gray, or
tawny. C, SA.

2. Oncilla
Leopardus tigrinus (p. 165)
Small; spotted; built like house cat, small
head and feet; hair on neck not reversed.
C, SA.

3. Margay
Leopardus wiedii (p. 164)
Small; spotted; large head and feet; tail
longer than hindleg; neck striped, hair
reversed. C, SA.

4. Ocelot
Leopardus pardalis (p. 164)
As above, but larger; tail shorter than hind-
leg. C, SA.

5. Puma
Puma concolor (p. 167)
Large; uniform color. N, C, SA.

6. Jaguar
Panthera onca (p. 168)
Large; spotted or black; spots, not stripes,
on neck; head and legs large. C, SA.

Plate 18

Deer (Cervidae)

1. Gray (brown) brocket deer (male)
Mazama gouazoubira (p. 178)
Gray-brown; underparts whitish or gray;
small and gracile; males with short, straight,
unbranched antlers; no facial markings;
back slightly humped. C, SA.

2. Red brocket deer (female and young)
Mazama americana (p. 178)
As above, but body and belly red. C, SA.

3. White-tailed deer (male)
Odocoileus virginianus (p. 180)
Gray to red; belly white; back straight;
markings on muzzle and around eyes;
males with branched antlers. C, SA.

Capybara (Hydrochaeridae)

4. Capybara
Hydrochaeris hydrochaeris (p. 223)
Large square muzzle; no tail; webbed feet.
C, SA.

Peccaries (Tayassuidae)

5. White-lipped peccary
Tayassu pecari (p. 176)
Large; narrow, piglike snout; nasal disk
large, protrudes well above muzzle; white
on chin. C, SA.

6. Collared peccary
Tayassu tajacu (p. 175)
As above, but smaller; nasal disk not salient;
faint pale collar on neck; chin not white.
N, C, SA.

Tapirs (Tapiridae)

7. Brazilian tapir (and young)
Tapirus terrestris (p. 173)
Very large; upper lip elongated in proboscis; crown between ears raised in humped
crest. SA.

8. Baird's tapir
Tapirus bairdii (p. 174)
As above, but crown flat. C, SA.

1 ♂

2 ♀

3 ♂

4

5

6

7

7

8

Plate 19

Larger Central and South American Squirrels (Sciuridae)

All have long ears that protrude above crown.

1. Red-tailed squirrel (three forms)
Sciurus granatensis (p. 189)
(1a) parts of N Colombia; (1b) parts of
Venezuela and Central America; (1c) N
coast of Colombia and Ecuador. Medium-
sized; tail reddish. C, SA.

2. Guayaquil squirrel
Sciurus stramineus (p. 190)
Ears black; tail black frosted white. SA.

3. Southern Amazon red squirrel
Sciurus spadiceus (p. 186)
Large; red or rarely black; feet with mixture of black and red hairs (not visible at
distance). SA.

4. Northern Amazon red squirrel
Sciurus igniventris (p. 185)
Large; red or rarely black; feet pure red;
side sometimes with black line. SA.

5. Junín red squirrel
Sciurus pyrrhinus (p. 186)
Medium-sized; back pure red unmixed
with black. SA, Andean Peru.

Plate 20

Small Squirrels (Sciuridae)

1. Sanborn's squirrel
Sciurus sanborni (p. 188)
Large ears; lower belly pure orange or
whitish; feet may be paler than back. SA.

2. Guianan squirrel
Sciurus aestuans (p. 187)
Large ears; lower belly mixed grayish;
females with 4 pairs mammae. SA.

3. Bolivian squirrel
Sciurus ignitus (p. 187)
Large ears; lower belly grayish; feet the
same color as back; females with 3 pairs
mammae. SA.

4. Amazon dwarf squirrel
Microsciurus flaviventer (p. 193)
Small ears; tail shorter than head and body.
SA, Amazon Basin.

5. Western dwarf squirrel
Microsciurus mimulus (p. 193)
As above; midback may have black stripe;
underparts orange. C, SA west of Andes.

6. Central American dwarf squirrel
Microsciurus alfari (p. 194)
As above, but underparts grayish. C, SA to
NW Colombia.

7. Neotropical pygmy squirrel
Sciurillus pusillus (p. 194)
Tiny size; large white patches behind ears;
grayish. SA.

Plate 21

Larger Central American Squirrels (Sciuridae)

Black individuals can occur in most species.

1. Deppe's squirrel
Sciurus deppei (p. 191)
Finely grizzled brownish; forelegs usually gray; underparts gray to white; tail fairly slender. CA.

2. Yucatán squirrel
Sciurus yucatanensis (p. 191)
Coarsely grizzled blackish or brownish; underparts pale gray or white. CA.

3. Red-bellied squirrel (two forms from Mexico)
Sciurus aureogaster (p. 192)
Finely grizzled gray to brown; nape, shoulders, or rump may have orange patches; underparts red to white. CA.

4. Variegated squirrel (three forms)
Sciurus variegatoides (p. 190)
(4a) Nicaragua, El Salvador; (4b) Nicaragua, Costa Rica; (4c) Costa Rica. Highly variable; large; tail very bushy; fur often coarsely variegated; pale patch behind ear; tail black frosted with white above; underparts red to white. CA.

Plate 22

Small Terrestrial Rats
(Muridae, Heteromyidae)
All in the range of about 30–120 g.

1. Grass mouse
Akodon urichi (p. 210)
Tail short; whiskers short. SA.

2. Brazilian shrew mouse
Blarinomys breviceps (p. 207)
Tail very short; eyes and ears tiny. SA.

3. Atlantic forest rat
Delomys dorsalis (p. 205)
Dark midback stripe (not present in all members of genus); whiskers fine, short. SA.

4. Spiny pocket mouse
Heteromys desmarestianus (p. 197)
Spiny fur; externally opening cheek pouches; sharply demarcated white underparts. C, SA.

5. Long-nosed mouse
Oxymycterus inca (p. 211)
Long, narrow muzzle and strongly undershot jaw; long claws on all feet. SA.

6. Macconnell's rice rat
Oryzomys macconnelli (p. 198)
Long fur; orange sides; top of back blackish. SA.

7. Common rice rat
Oryzomys capito (p. 198)
Tawny brown; fur fine and soft; underparts grayish white. SA.

8. Mexican deer mouse
Peromyscus mexicanus (p. 212)
Ears large; dark ring around eye; whiskers long; tail long. CA.

9. Isthmus rat
Isthmomys (p. 213)
Cinnamon or orange; ears large; whiskers long; tail very long. CA.

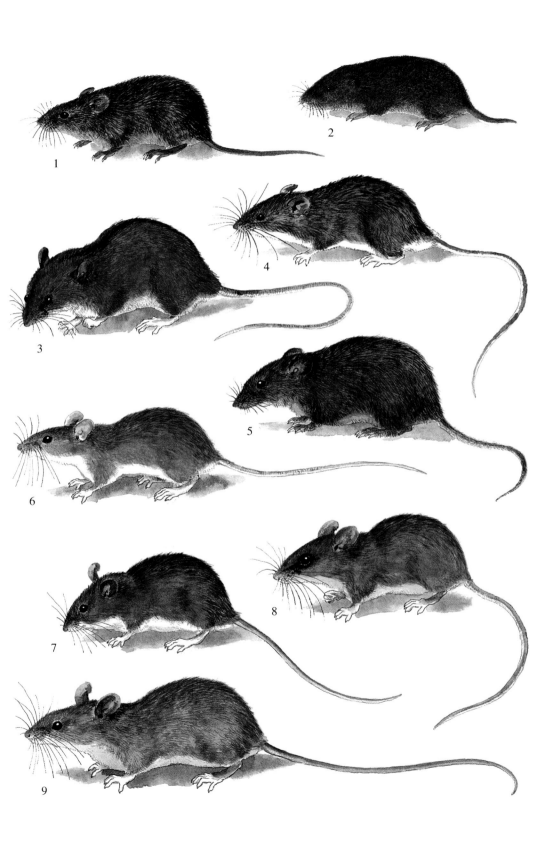

Plate 23

Small or Tiny Mice (Muridae)
All in the range of about 8–30 g.

1. Spiny mouse
Neacomys sp. (p. 201)
Fur spiny; tail as long as head and body;
underparts white or orange. C, SA.

2. Gray spiny mouse
Scolomys melanops (p. 202)
Fur spiny; tail much shorter than head and
body; entirely gray. SA.

3. Pygmy rice rat
Oligoryzomys fulvescens (p. 199)
Tail much longer than head and body; un-
derparts gray; head and neck grayish. SA.

4. Bicolored arboreal rice rat
Oecomys bicolor (p. 200)
Short, broad, pink-soled hindfeet; reddish;
underparts pure white; tail slightly hairy.
C, SA.

5. Harvest mouse
Reithrodontomys sumichrasti (p. 213)
Tail longer than head and body; sides of
head and neck tinged orange; whiskers
long; upper incisors grooved. N, C, SA.

6. House mouse
Mus musculus (p. 215)
Tail about same length as head and body;
whiskers short; underparts not sharply
contrasting with sides.

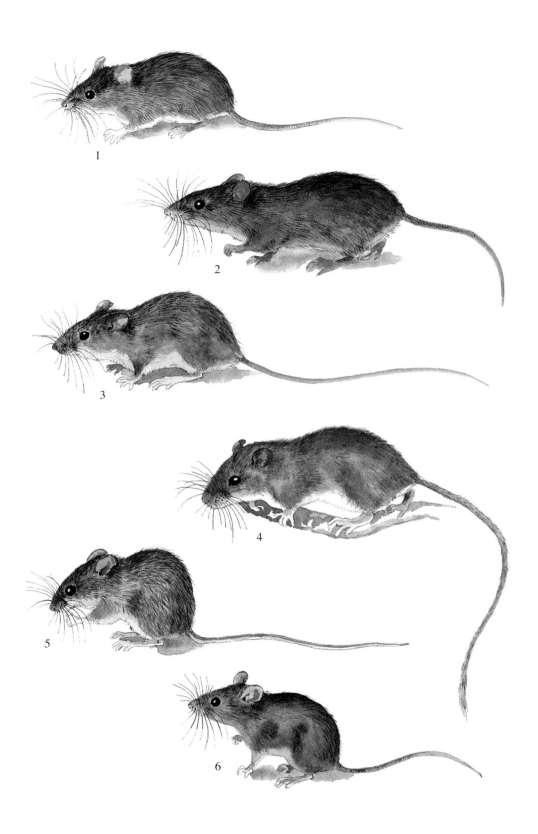

Plate 24

Large Terrestrial and Semiaquatic Rats
(Muridae, Echimyidae)
All in the range of 70–550 g.

1. Water rat
Nectomys squamipes (p. 202)
Hindfoot partly webbed, fringed with hair, heel with scales on sole; ear tips naked; underparts not sharply demarcated from sides; fur soft and dense. SA.

2. Marsh rat
Holochilus brasiliensis (p. 212)
Hindfoot partly webbed, fringed with hair, heel without scales; ear hairy to tip; underparts not sharply contrasting with sides; fur soft and dense. SA.

3. South American water mouse
Neusticomys sp. (p. 207)
Tail completely covered with flat dark brown hair; eyes and ears tiny; hindfoot with fringe of hair but not strongly paddle-shaped; underparts not sharply contrasting with sides. SA.

4. Central American water mouse
Rheomys mexicanus (p. 206)
Tail completely covered with hair; eyes and ears tiny; hindfoot paddle-shaped, fringed with hair; underparts silvery; forefoot with four large pads on palm. CA only.

5. Crab-eating rat
Ichthyomys pittieri (p. 206)
As above, but forefoot with five pads. SA, CA.

6. Black or roof rat
Rattus rattus (p. 214)
Black or tawny; fur coarse and sparse; tail longer than head and body; whiskers long.

7. Norway rat
Rattus norvegicus (p. 215)
Tawny brown; fur coarse and sparse; tail shorter than head and body; whiskers medium.

8. Spiny rat
Proechimys steerei (p. 232)
Fur bristly or spiny, spines usually lie flat; head long and narrow; underparts sharply demarcated, often snow white; tail shorter than head and body. Genus C, SA.

9. Armored rat
Hoplomys gymnurus (p. 233)
As above, but spines on lower back strong, upstanding, geometrically spaced; underparts may be dark. C, SA.

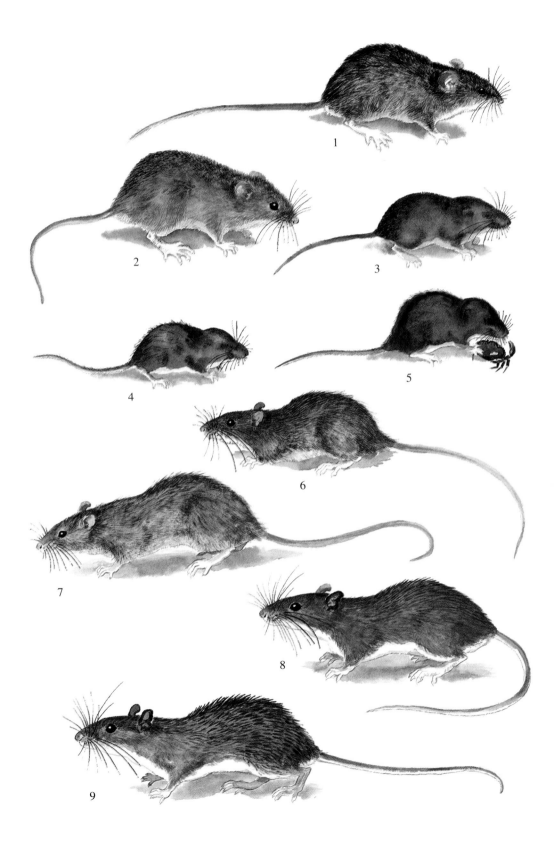

Plate 25

Small and Medium-Sized Arboreal Rats
(Muridae, Echimyidae)
All in the range of 40–300 g; with large
eyes and broad, strong feet.

1. Arboreal rice rat
Oecomys paricola (p. 200)
Whiskers and hair on tail fine; dorsal fur
usually long, with base pale to medium
gray. Genus C, SA.

2. Climbing rat
Rhipidomys nitela (p. 204)
Whiskers and hair on tail coarse and promi-
nent; dorsal hair dark gray at base. SA.

3. Big-eared climbing rat
Ototylomys phyllotis (p. 209)
Large ears; tail naked, shiny, dark. CA.

4. Naked-tailed climbing rat
Tylomys watsoni (p. 208)
Large; tail naked, shiny, white at tip. C, SA.

5. Vesper rat
Nyctomys sumichrasti (p. 209)
Eyes very large; whiskers very long; tail
very hairy. CA.

6. Spiny tree rat
Mesomys hispidus (p. 234)
Heavily spined; ears short; tail sparsely but
prominently haired, with slight tuft. SA.

7. Tuft-tailed spiny tree rat
Lonchothrix emiliae (p. 235)
Heavily spined; ears short; tail scaly
proximally, tip with large tuft of long,
coarse hair. SA.

Plate 26

Larger Spiny Tree Rats
(Echimyidae)

All in range of 220–900 g; have bristly or spiny pelage; short ears; strong, broad feet; South America only.

1. Giant tree rat
Makalata grandis (p. 236)
Large; black and gold; fur bristly, not spiny; tail thickly haired, pitch black.

2. Peruvian tree rat
Makalata rhipidurus (p. 239)
Tail sparsely but prominently haired throughout, or with hairs decreasing gradually down tail above; fur bristly, not spiny; sides of muzzle rusty.

3. Dark tree rat
Echimys saturnus (p. 236)
Large; black and chestnut; tail thickly haired, black with white tip.

4. Speckled tree rat
Echimys semivillosus (p. 237)
Heavily spined; tawny grayish speckled with white tips on spines; tail looks naked.

5. Red-nosed tree rat
Makalata didelphoides (p. 238)
Moderately spined; muzzle to between eyes rusty; tail appears naked but slightly hairy, haired at base for 3–4 cm; whiskers coarse; belly usually grayish.

6. Bare-tailed tree rat
Makalata occasius (p. 237)
Heavily spined; tail conspicuously naked, haired at base for 1–2 cm; underparts pure orange with white patches; muzzle rusty.

7. White-faced tree rat
Echimys chrysurus (p. 235)
Heavily spined; white blaze or spot on face or crown; white or chestnut tail tip.

8. Black-spined Atlantic tree rat
Nelomys nigrispinis (p. 239)
Red-brown streaked with black; tail sparsely haired.

9. Golden Atlantic tree rat
Nelomys blainvillei (p. 241)
Yellow or red-gold; tail thickly haired, with tuft.

10. Pallid Atlantic tree rat
Nelomys lamarum (p. 240)
Heavily spined; drab, yellow ochraceous; tail sparsely haired.

Plate 27

Large Soft-Furred Tree Rats
(Echimyidae)
All in range of 300–700 g; have soft fur and
no spines; small ears; strong, broad feet;
large eyes.

1. Amazon bamboo rat
Dactylomys dactylinus (p. 244)
Large, square muzzle; naked tail; fingers
long, separated by central gap, with nails.
SA.

2. Southern bamboo rat
Kannabateomys amblyonyx (p. 245)
As above, but tail well haired throughout.
SA.

3. Red crested tree rat
Diplomys rufodorsalis (p. 243)
Rust-red; tail completely haired, tip white.
SA.

4. Rufous tree rat
Diplomys labilis (p. 242)
Reddish brown; face grayish with pale spots
at base of whiskers and over eye; tail completely haired. Genus C, SA.

5. Yellow-crowned brush-tailed rat
Isothrix bistriata (p. 241)
Crown with pale yellow patch bordered on
each side by a black stripe; tail fully haired,
dusky and/or orange. SA.

6. Plain brush-tailed rat
Isothrix pagurus (p. 242)
Crown without markings; tail fully haired,
dusky. SA.

7. Painted tree rat
Echimys pictus (p. 238)
Sharply patterned black and white; tail fully
haired, tip white. SA.

Plate 28

Porcupines (Erethizontidae)
All have stout, sharp spines; prehensile tails; feet with long claws; inconspicuous ears.

1. Bicolor-spined porcupine
Coendou bicolor (p. 218)
Hairless; large; black with variable number of white tips on spines; long tail. SA.

2. Brazilian porcupine
Coendou prehensilis (p. 217)
Hairless; large; gray or yellowish. SA.

3. Rothschild's porcupine
Coendou rothschildi (p. 218)
Like (1), but smaller. CA.

4. Mexican hairy porcupine
Coendou mexicanus (p. 222)
Black with pale head; body covered with long hair. CA.

5. Black dwarf porcupine
Coendou koopmani (p. 219)
Small; hairless; blackish. SA.

6. Brown hairy dwarf porcupine
Coendou vestitus (p. 221)
Small; hairy; warm smoky brown; tail short. SA.

7. Black-tailed hairy dwarf porcupine
Coendou melanurus (p. 220)
Small; hairy; black sprinkled with yellow bristles; tail long, black. SA.

8. Frosted hairy dwarf porcupine
Coendou pruinosus (p. 221)
Small; hairy; black frosted whitish; spines very short; tail short. SA.

9. Orange-spined hairy dwarf porcupine
Coendou spinosus (p. 220)
Small; hairy; hair dark at base; some spines orange-tipped; tail usually long. SA.

10. Bahia hairy dwarf porcupine
Coendou insidiosus (p. 220)
Small; hairy; smoky brown to gray; hair usually pale at base; tail usually short. SA.

11. Paraguay hairy dwarf porcupine
Coendou paragayensis (p. 221)
Small to tiny; hairy; neck and shoulders with cape of longer yellow spines; tail long. SA.

12. Bristle-spined porcupine
Chaetomys subspinosus (p. 222)
No hair; nose not bulbous; spines on lower back kinky and flexible like broom straws; short, sharp spines on head and shoulders. SA.

Plate 29

Large Rodents
Paca (Agoutidae)

1. Paca
Agouti paca (p. 224)
Large; red or brown with rows of spots; no visible tail. C, SA.

Pacarana (Dinomyidae)

2. Pacarana
Dinomys branickii (p. 225)
Large; black, with spots from shoulder to rump; prominent tail. SA.

Agoutis and Acouchys (Dasyproctidae)

3. Green acouchy
Myoprocta pratti (p. 231)
Olivaceous; small, thin, white-tipped tail. SA.

4. Red acouchy
Myoprocta acouchy (p. 230)
Red with black rump; tail as above. SA.

5. Brown agouti
Dasyprocta variegata (p. 227)
Uniform brown to orange; tail a short, naked black stub. SA.

6. Black-rumped agouti
Dasyprocta prymnolopha (p. 228)
Small; reddish with black rump; tail as above. SA.

7. Mexican black agouti
Dasyprocta mexicana (p. 230)
Grizzled black; rump black; tail as in (5). CA.

8. Central American agouti (two forms)
Dasyprocta punctata (p. 229)
Either (8a) uniform reddish brown or (8b) blackish with midbody yellowish and rump with long, black, white-tipped hairs; tail as in (5). S, CA.

9. Azara's agouti
Dasyprocta azarae (p. 228)
Tawny gray, midbody washed with orange; tail as in (5). SA.

10. Black agouti
Dasyprocta fuliginosa (p. 227)
Grizzled black; tail as in (5). SA.

11. Red-rumped agouti
Dasyprocta agouti (p. 226)
Brownish with red to orange rump; tail as in (5). SA.

Rabbits (Leporidae)

12. Tapiti or Brazilian rabbit
Sylvilagus brasiliensis (p. 246)
Long oblong ears; short furry tail; reddish nape and feet. C, SA.

CHAGRES
FIELD
STATION

Plate A

Armadillos (Dasypodidae)

All have body covered above with bony armor plates.

1. Yellow armadillo
Euphractus sexcinctus (p. 46)
Many long hairs over armor plates; short, widely set ears; front claws not enlarged. SA.

2. Southern naked-tailed armadillo
Cabassous unicinctus (p. 46)
Large, widely set ears; naked tail; greatly enlarged front claws. SA.

3. Nine-banded long-nosed armadillo
Dasypus novemcinctus (p. 49)
Closely set ears; long, narrow muzzle; 8–10 bands around midbody. N, C, SA.

4. Great long-nosed armadillo
Dasypus kappleri (p. 50)
Closely set ears; long, narrow muzzle; large size; projecting scutes on hind knees; 7–9 bands. SA.

5. Seven-banded long-nosed armadillo
Dasypus septemcinctus (p. 50)
Closely set ears, long, narrow muzzle; small size; 6–7 bands. SA.

6. Giant armadillo
Priodontes maximus (p. 48)
Enormous size; small, widely set ears; greatly enlarged front limbs and claws. SA.

Plate B

Bats (Chiroptera)

Sheath-tailed Bats (Emballonuridae)

All are small, with no noseleaf; tail shorter than tail membrane with tip emerging free above membrane about halfway down its length.

1. Long-nosed bat
Rhynchonycteris naso (p. 53)
Elongated muzzle; white tufts of hair on forearm; wavy white lines on back. C, SA.

2., 2a. White-lined sac-winged bat
Saccopteryx bilineata (p. 53)
Wavy white lines on back; wing sac close to forearm bone beyond elbow. C, SA.

3. Shaggy bat
Centronycteris maximiliani (p. 54)
Long woolly hair; hairy face; no wing sac. C, SA.

4. Doglike sac-winged bat
Peropteryx macrotis (p. 55)
Tuft on crown ends abruptly on naked face; stiff mustache; wing sac reaches forward edge of membrane (propatagium). C, SA.

5. Chestnut sac-winged bat
Cormura brevirostris (p. 56)
Wing sac does not quite reach forward edge of propatagium; wing membrane attaches near base of toe. C, SA.

6. Least sac-winged bat
Balantiopteryx plicata (p. 56)
Tuft on crown ends abruptly on naked face; wing sac in center of membrane. C, SA.

7. Ghost bat
Diclidurus albus (p. 57)
White color; short round ear; glands in membrane around tail. C, SA.

Bulldog Bats (Noctilionidae)

No noseleaf; tail shorter than tail membrane.

8. Greater bulldog or fishing bat
Noctilio leporinus (p. 59)
Lip split from below nose, forward-pointing nose and ears. C, SA.

Leaf-chinned and Mustached Bats (Mormoopidae)

No noseleaf; tail shorter than tail membrane.

9. Mustached bat
Pteronotus gymnonotus (p. 60)
Lips flared forward; eye small. C, SA.

10. Leaf-chinned bat
Mormoops megalophylla (p. 61)
Platelike folds on chin; short rounded ear encircling eye. C, SA.

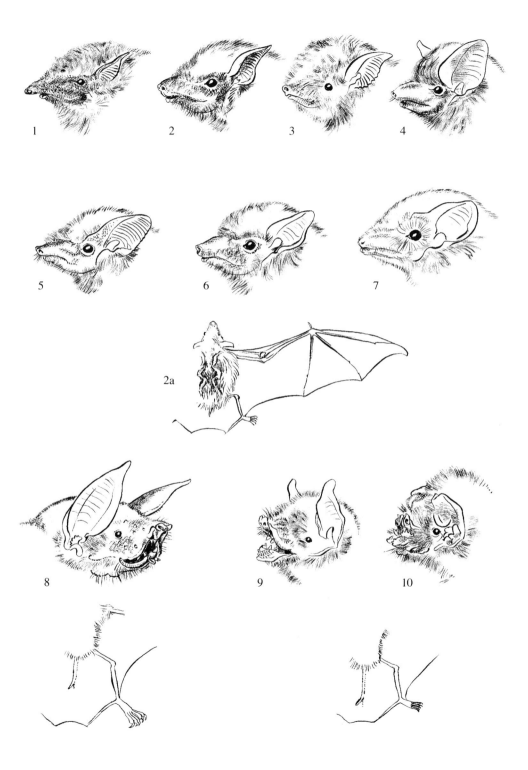

Plate C

Spear-nosed Bats
(Phyllostominae)
All have large noseleaves, narrow muzzles, large ears; tails except where noted.

1. Little big-eared bat
Micronycteris daviesi (p. 62)
Chin with single, smooth pads in V. C, SA.

2. Sword-nosed bat
Lonchorhina aurita (p. 63)
Extremely long, sword-shaped noseleaf; tail to point at edge of membrane. C, SA.

3. Long-legged bat
Macrophyllum macrophyllum (p. 64)
Tail to edge of membrane; membrane square at edge with parallel rows of dots. C, SA.

4. Round-eared bat
Tonatia bidens (p. 65)
Chin with tiny warts in U; large rounded ears. C, SA.

5. Hairy-nosed bat
Mimon crenulatum (p. 66)
Long, narrow noseleaf; large pointed ears; tail half as long as membrane. C, SA.

6. Fringe-lipped bat
Trachops cirrhosus (p. 68)
Lips and chin studded with long tubercles. C, SA.

7. Spear-nosed bat
Phyllostomus hastatus (p. 66)
Short, broad noseleaf and ears; horseshoe below nostrils free from lip. C, SA.

8. Pale-faced spear-nosed bat
Phylloderma stenops (p. 67)
Horseshoe below nostrils continuous with lip at center; skin of face pale and mottled; large size. C, SA.

9. Woolly false vampire bat
Chrotopterus auritus (p. 69)
Noseleaf and horseshoe form continuous hollow cup around nostrils; fur long and woolly; tail tiny. C, SA.

10. False vampire bat
Vampyrum spectrum (p. 69)
Enormous size; noseleaf and horseshoe form hollow cup around nostrils; short fur; midback stripe; no tail. C, SA.

Plate D

Long-tongued Bats
(Lonchophyllinae,
Glossophaginae)
All are small; have elongated, narrow
muzzles; highly extensible long tongues;
short noseleaves and ears; no free horseshoe
at center of lip under nostrils.

1. Common long-tongued bat
Glossophaga soricina (p. 72)
Lower incisors present; tongue with no
groove on side; chin with V bordered with
tiny tubercles; fur bicolored, pale at base.
C, SA.

2. Spear-nosed long-tongued bat
Lonchophylla robusta (p. 71)
Lower incisors present; tongue with deep
groove on side; chin pads slightly rippled
at edge; fur bicolored, pale at base. C, SA.

3. Chestnut long-tongued bat
Lionycteris spurelli (p. 70)
Lower incisors present; fur unicolored;
tongue with groove on side; chin pads
smooth on sides. C, SA.

4. Hairy-legged long-tongued bat
Anoura geoffroyi (p. 73)
No tail or tiny tail; tail membrane reduced
to hairy band down side of leg. C, SA.

5. Dark long-tongued bat
Lichonycteris obscura (p. 74)
No lower incisors; fur tricolored, dark at
base and tip; elbows thinly furred above;
muzzle robust. C, SA.

6. Long-nosed long-tongued bat
Choeroniscus cf. *intermedius* (p. 74)
Muzzle narrow and tubelike; no lower incisors; fur slightly bicolored, pale at base;
first and second thumb joints about equal in
length. C, SA.

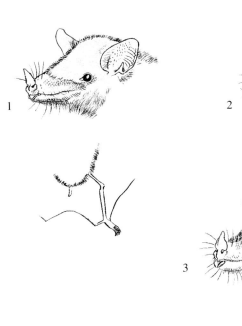

1

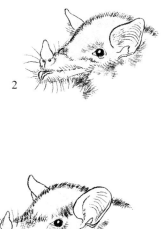

2

3

4

5

6

Plate E

Short-tailed and Little Fruit Bats (Carolliinae)

1. Short-tailed fruit bat
Carollia perspicillata (p. 76)
Short tail present; short narrow muzzle; chin with V of tiny warts flanking large central wart; fur banded. C, SA.

2. Little fruit bat
Rhinophylla fischerae (p. 77)
Muzzle narrow; no tail; fur unicolored; chin with three or four large pads flanking a large central wart. C, SA.

Neotropical Fruit Bats (Stenodermatinae)
All have broad muzzles; no tails; often white stripes.

3. Yellow-shouldered fruit bat
Sturnira lilium (p. 78)
No tail membrane; no stripes; yellow-stained shoulders; fur tricolored, dark at base and tip. C, SA.

4. Tent-making bat
Uroderma bilobatum (p. 79)
Bright facial and midback stripes; central upper incisors bilobed, with broad parallel tips, less than twice as high as outer incisors. C, SA.

5. White-lined fruit bat
Platyrrhinus (tail membrane only) (p. 80)
Facial and midback stripes present; tail membrane with hairy edge; central upper incisors bilobed, more than twice as long as outer incisors, slender and convergent at tips. C, SA.

6. Big-eyed bat
Chiroderma villosum (p. 81)
Fur with dark band at base; palpable depression behind noseleaf; central upper incisors single-pointed, more than twice as long as outer incisors, slender and convergent at tips; facial stripes if present do not extend past middle of ear. C, SA.

7. Large fruit-eating bat
Artibeus jamaicensis (p. 84)
No midback stripe; central upper incisors bilobed, with broad parallel tips, less than twice as high as outer incisors. C, SA.

8. Great stripe-faced bat
Vampyrodes major (p. 81)
Sharp facial and midback stripes interdigitate on head; large size; yellowish ear and noseleaf rims. C, SA.

9. Ipanema or double-lipped bat
Pygoderma bilabiatum (p. 85)
White spot on shoulder; fold from horseshoe under nose to corner of lip. SA.

10. Visored bat (male and female)
Sphaeronycteris toxophyllum (p. 86)
White spot on shoulder; horizontal fold across brow. SA.

11. Wrinkle-faced bat (male)
Centurio senex (p. 87)
Grotesquely wrinkled face; ladderlike pattern in wing. C, SA.

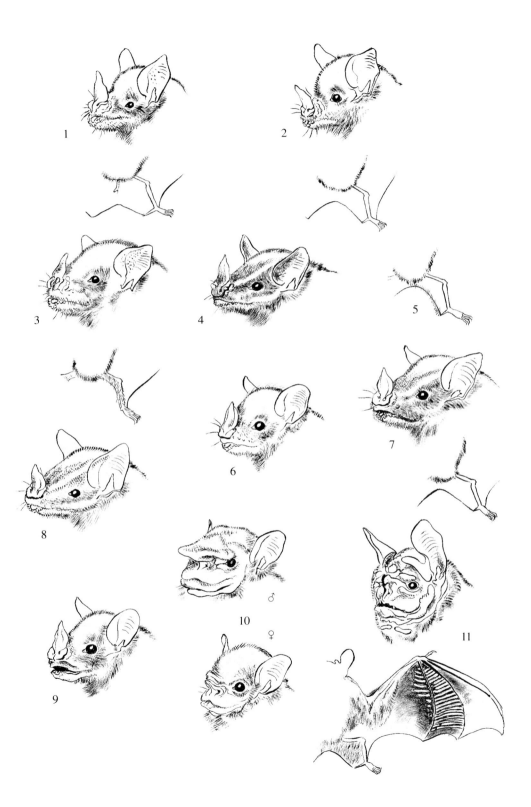

1

2

3

4

5

6

7

8

9

10 ♂

♀

11

Plate F

Vampire Bats (Desmodontinae)

1. Common vampire bat
Desmodus rotundus (p. 88)
Short M-shaped fold behind nostrils backed by another fold; large sharp triangular incisors larger than canines; no tail; tail membrane a complete short band between legs. C, SA.

2. Hairy-legged vampire bat
Diphylla ecaudata (p. 89)
Fold behind nostrils shaped like inverted U; incisors as above; tail membrane almost absent; legs very hairy. C, SA.

Funnel-eared Bats (Natalidae)

3. Funnel-eared bat
Natalus stramineus (p. 90)
Tiny; mouth and ears funnel-shaped; tail and legs very long; color pale. C, SA.

Thumbless Bats (Furipteridae)

4. Thumbless bat
Furipterus horrens (p. 91)
Minuscule (3 g); tail half as long as membrane and entirely within it; thumb a stub with tiny claw. C, SA.

Sucker-footed Bats (Thyropteridae)

5. Sucker-footed bat
Thyroptera tricolor (p. 92)
Foot and thumb base with tiny sucker disks; tail longer than membrane. C, SA.

Vespertilionid Bats (Vespertilionidae)

All have no noseleaf; long tail with membrane coming to a point at tip.

6. Little brown bat
Myotis nigricans (p. 93)
Tiny size (4–7 g); color usually dark; fur unicolored dark or hair tips frosted; calcar shorter than foot; gap between canine and first large tooth behind. C, SA.

7. Big brown bat
Eptesicus furinalis (p. 93)
Larger size (6–16 g); fur dark at base with paler tips imparting rich sheen; calcar longer than foot; no gap between canine and first large tooth behind. C, SA.

8. Big-eared brown bat
Histiotus montanus (p. 95)
Ears enormous. SA.

9. Black-winged little yellow bat
Rhogeessa tumida (p. 94)
Tiny size; fur pale yellow frosted brown; wings blackish. C, SA.

10. Hoary bat
Lasiurus cinereus (p. 96)
Ears short and wide; tail membrane thickly furred above for more than half its length. C, SA.

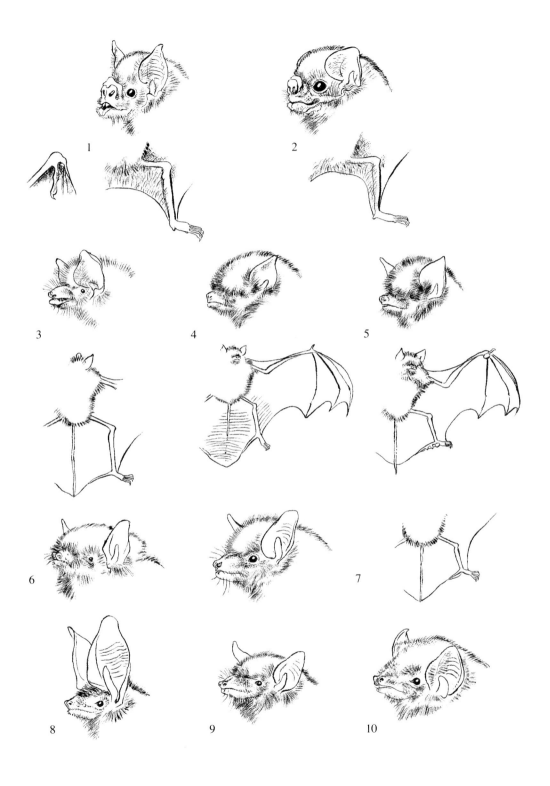

1

2

3

4

5

6

7

8

9

10

Plate G

Free-tailed or Mastiff Bats
(Molossidae)

All have naked tail extending one-third or more of its length beyond membrane; no noseleaf; flattened body; hairy feet.

1. Brazilian free-tailed bat
Tadarida brasiliensis (p. 100)
Upper lip with vertical wrinkles; ears meet but do not join on center of crown; six lower incisors. N, C, SA.

2. Broad-eared free-tailed bat
Nyctinomops laticaudatus (p. 100)
Upper lip with vertical wrinkles; ears join on stalk at center of crown; four lower incisors. C, SA.

3. Flat-headed bat
Neoplatymops matogrossensis (p. 99)
Head and body extremely flattened; forearm skin above sprinkled with tiny bumps. SA.

4. Dog-faced bat
Molossops temminckii (p. 98)
Ears simple, triangular, pointed, widely separated on crown; fur diminishes gradually down face. C, SA.

5. Doglike bat
Cynomops greenhalli (p. 99)
Ears do not meet on crown; fur ends abruptly at naked face; muzzle and chin broad. C, SA.

6. Bonneted bat
Eumops perotis (p. 101)
Large ears flattened forward like hat over brow, join on midcrown; muzzle pointed, slanting sharply back from nose to chin. C, SA.

7. Crested mastiff bat
Promops nasutus (p. 102)
Top of muzzle raised in central ridge; ears just meet with horizontal fold on crown; chin broad and rounded; rear lower edge of ear narrow; four lower incisors; palate inside mouth deeply concave. C, SA.

8. Mastiff bat
Molossus molossus (p. 102)
Top of muzzle raised in ridge; ears just meet with horizontal fold on crown; rear lower edge of ear with broad lateral fold; two lower incisors; palate inside mouth flat. C, SA.

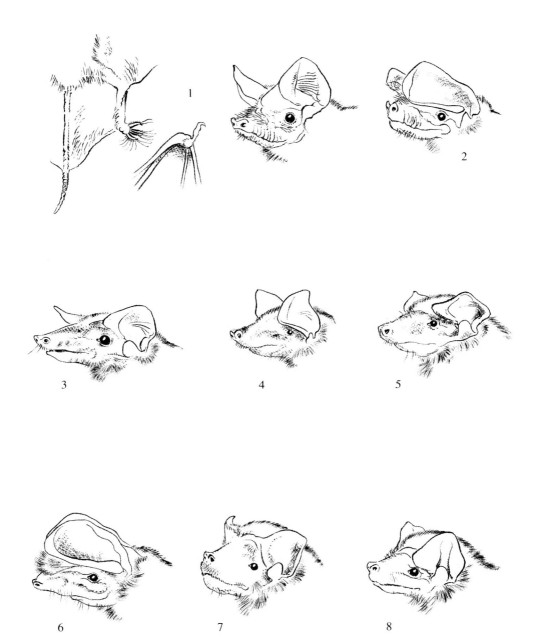

1

2

3

4

5

6

7

8

Map 153

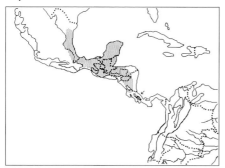

Map 154

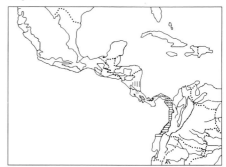

▦ Deppe's squirrel, *Sciurus deppei*
▤ Montane squirrel, *Syntheosciurus brochus*

▦ Central American dwarf squirrel, *Microsciurus alfari*
▥ Richmond's squirrel, *Sciurus richmondi*
▤ Western dwarf squirrel, *M. mimulus*

Richmond's Squirrel
Sciurus richmondi
Map 154
Identification. Measurements: HB =
160–210; T = 130–184; HF = 45–55;
E = 15–25; WT = 235–268 g.
Upperparts entirely uniform brown, tinged
olivaceous; fur short and slightly glossy. Ears
large, no pale patches behind. **Tail shorter
than head and body, fairly slender, brown**
or blackish frosted buff. **Underparts orange**
or orange-washed. Mammae pairs = 3. **A
small squirrel.**
Similar species. Deppe's squirrels *(S. deppei)* have gray or white underparts, usually
gray forelegs.
Natural history. Diurnal; terrestrial and arboreal. This squirrel forages on the ground
or on the main trunks and lower branches
of trees, rarely high in the canopy. Found in
mature and secondary rainforest, gallery forest, and plantations with trees.
Geographic range. Central America: Nicaragua only, the Caribbean coast and drainage.
Status. Unknown, but geographic range
small and much deforested.
References. Jones, J. K., Jr., and H. H. Genoways. 1975. *Sciurus richmondi.* Mammalian
Species, no. 53.

Red-bellied Squirrel
Sciurus aureogaster
Plate 21, map 152
Identification. Measurements: HB =
220–310; T = 206–315; HF = 65–74;
E = 27–35; WT = 432–680 g.
Highly variable: **upperparts finely grizzled
gray, brownish, black,** or intermediate colors; **nape of neck, and/or shoulders, and/**
**or rump sometimes with patch of orange
or chestnut,** or darker black in melanistic
individuals. Tail black frosted with white
above, gray or reddish medially below. **Underparts pure white, deep chestnut red, or
any intermediate color.**
Variation. Animals from the Isthmus of
Tehuantepec northward up the Pacific coast
and to Veracruz have bright chestnut rump
or nape patches. Animals from Veracruz and
Puebla northward are often melanistic.
Similar species. See Deppe's squirrel *(S.
deppei)* and variegated squirrel *(S. variegatoides).*
Sounds. A resonant, harsh, trilling chatter.
Natural history. Diurnal; arboreal; solitary.
Feeds on fruit, seeds including conifer seeds,
conifer leaves, and sometimes corn. Red-
bellied squirrels are usually seen in the trees,
and they rarely come to the ground. They
make leaf nests in trees. Found in a wide
range of habitats from lowland tropical evergreen broadleaf forest to pine-oak woodland
and thorn scrub on the Mexican plateau,
but may be absent from intermediate cloud
forests, where Deppe's squirrel is found.
Geographic range. Central America: Mexico from Nayarit and Nuevo León south to
Chiapas and the highlands of Guatemala. To
about 3,000 m elevation.
Local names. Ardilla, ardilla negra (melanistic individuals), ardilla pinta, ardilla gris
(Me).
References. Musser, G. G. 1968. A systematic study of the Mexican and Guatemalan
gray squirrel, *Sciurus aureogaster* F. Cuvier
(Rodentia: Sciuridae). Misc. Pub. Mus. Zool.
Univ. Mich., no. 137.

Map 155

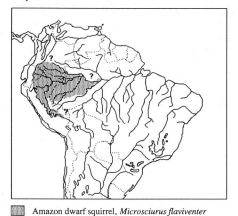

Amazon dwarf squirrel, *Microsciurus flaviventer*

Amazon Dwarf Squirrel
Microsciurus flaviventer
Plate 20, map 155
Identification. Measurements: HB =
120–160; T = 96–150; HF = 35–43;
E = 10–17; WT = 60–128 g.
Upperparts brown, finely grizzled reddish
to olivaceous. **Ears short, not protruding
above crown of head,** pale yellow behind
with short pale hairs not visible above rim
of ear. **Tail slightly shorter than head and
body,** relatively slender, tapered toward tip,
slightly frosted with dirty white. Underparts
thickly haired, deep to pale orange, brightest
on chest, not sharply demarcated from sides.
Mammae pairs = 3.
Variation. Some animals from Ecuador,
Colombia, and Brazil have grayish underparts washed with orange. Several species
may be represented among the nine or so
named forms currently grouped in this taxon.
Similar species. Pygmy squirrels *(Sciurillus
pusillus)* are much smaller and gray, with
prominent white ear patches; Guianan *(Sciurus aestuans),* Sanborn's *(S. sanborni),* and
Bolivian *(S. ignitus)* squirrels have long ears
that protrude well above crown.
Sounds. Soft chucks in alarm; soft, low-
frequency, descending, birdlike trill.
Natural history. Diurnal; arboreal; solitary
and in pairs. Amazon dwarf squirrels use all
levels of the forest from the ground to the
canopy. They can often be seen foraging,
apparently for arthropods, by searching actively over large trunks, vines, and treefalls.
They also feed on a substance scraped from
the bark of trees such as *Inga* spp., as do

other small squirrels. One nest was a ball of
leaves lined with fibers at 3.5 m in the top of
a small palm. Found only in evergreen rain-
forest.
Geographic range. South America: Ama-
zon Basin of Colombia, Ecuador, Peru, and
Brazil west of the Rios Negro and Madeira.
To 2,000 m elevation.
Status. Locally common to rare; not hunted.
References. Allen, J. A. 1914. Review of
the genus *Microsciurus. Bull. Am. Mus. Nat.
Hist.* 33:145–65.

Western Dwarf Squirrel
Microsciurus mimulus
Plate 20, map 154
Identification. Measurements: HB =
128–150; T = 92–150; HF = 33–42;
E = 12–17; WT = 120 g.
**Upperparts dark brown finely grizzled
reddish to olivaceous, with or without distinct broad black stripe down midback;**
top of head the same as back or blackish or
black; **ears short, not protruding above
crown,** with or without pale patch on back.
Tail distinctly shorter than head and body,
in black-striped forms black toward tip,
frosted with gray. Underparts thickly haired,
dull to bright, dark rust-orange brightest on
chest, not sharply demarcated from sides.
Variation. Animals from Nariño, Colombia,
south through Ecuador have a black stripe
down midback; those from El Valle, Colombia, north to Darién, Panama, have uniform
brown backs; those from eastern Panama
are brown-olivaceous and have white behind
the ear tips. Specimens from lowlands often
speckled with small white spots (probably
scars from botflies). *Microsciurus similis*
(not described in this book) inhabits the
higher elevations of Cauca and Nariño in
the Andes of Colombia.
Similar species. These are the only lowland
dwarf squirrels in the southern half of their
range. In Panama, where they occur with
Central American dwarf squirrels *(M. alfari),*
they apparently live at higher elevations, but
some overlap occurs. The species are almost
indistinguishable in the field: the back of
Central American squirrels is more finely
ticked, and the underparts are grayish or
yellowish.
Natural history. Diurnal; arboreal and terrestrial; solitary and in pairs. Forages on
tree trunks and around brush piles, probably
searching for arthropods. Found in evergreen

forest. In Panama only at elevations above 800 m; in Colombia and Ecuador it is found in the lowlands.
Geographic range. Central and South America: Chiriquí, Panama, south to NW Ecuador, and N Colombia. To 1,600 m elevation.
Status. Not hunted, but parts of range extensively deforested.

Central American Dwarf Squirrel
Microsciurus alfari
Plate 20, map 154
Identification. Measurements: HB = 123–159; T = 100–125; HF = 32–42; E = 12–18; WT = 54–100 g.
Upperparts dark brown, very finely grizzled. Ears short, not protruding above crown, with or without pale white or buff patch behind. **Tail shorter than head and body, tapered toward tip,** like back or more reddish. **Underparts well furred, dull gray-tinged, or gray frosted yellowish, not sharply demarcated from sides.** Feet like back or slightly more reddish.
Variation. Animals from Darién, Panama, have white ear patches and yellow-gray underparts; those from Chiriquí, Panama, have no ear patches and gray underparts.
Similar species. See western dwarf squirrel *(M. mimulus).*
Natural history. Diurnal; arboreal and terrestrial. Central American dwarf squirrels use all levels of the forest. Found in cloud forest and dense rainforest.
Geographic range. Central America: S Nicaragua and Costa Rica south to W Colombia. To 1,700 m elevation.
Status. Uncommon.
Local names. Chiza (CR); ardilla voladora (Pn).

Neotropical Pygmy Squirrel
Sciurillus pusillus
Plate 20, map 151
Identification. Measurements: HB = 89–115; T = 89–113; HF = 24–29; E = 10–15; WT = 33–45 g.
Upperparts of body and tail **pale gray** with faint yellowish tinge, but may look brown in poor light in field. Head slightly reddish; **ears short, do not protrude above crown, with prominent white patches behind, white hairs protruding over rim.** Tail slightly longer than head and body. Underparts gray or buff-gray, not sharply demar-

cated from sides, only slightly paler than back. **A tiny squirrel with slender legs.**
Variation. Animals from French Guiana have more conspicuously reddish heads than other populations.
Similar species. These are the smallest New World squirrels and the only pale gray squirrels in the Amazon Basin. Amazon dwarf squirrels *(M. flaviventer)* are larger and brown; all other brown squirrels in the region have ears that protrude well above crown and sharply contrasting paler underparts.
Sounds. A soft to amazingly loud cricket-like chirp, "tsick"; more rarely a soft chatter or trill.
Natural history. Diurnal; arboreal; solitary and females with young; up to at least four may feed on the same tree, or a group of males may chase a female during a mating bout. Feeds chiefly on a substance scraped from the inner surface of tree bark. These tiny squirrels forage from near the ground to so high in the canopy that they disappear from sight. They pull chips of outer bark from certain large trees (often *Inga* spp.), leaving the trunk densely pockmarked, and are most easily seen by waiting near such a tree (but some birds likewise chip bark). They move rapidly, running up and down trunks and hopping and flitting from branch to branch in the canopy like small birds. They always travel from tree to tree high in the canopy. On flat trunks they have a highly characteristic habit of making almost instantaneous 180-degree switches of direction. As they forage on a trunk, the tail is carried stiffly out behind and given little jerks from time to time, but is not brought up over the back. Young foraging near their mothers frequently play by dashing back and forth around the trunk. Guianan squirrels *(S. aestuans)* and dwarf squirrels *(Microsciurus* spp.) sometimes forage on the bark of the same trees used by pygmy squirrels. Their tiny size makes these squirrels difficult to spot, but they are not wary of humans and are easy to observe. Found with a patchy distribution in mature lowland evergreen rainforest only.
Geographic range. South America: known only from three widely separated areas of the Amazon Basin of Peru (near Iquitos) and Colombia; Suriname and French Guiana and adjacent Brazil, and Brazil on the lower Rio Tapajóz.

Status. Locally common, but the distribution is extremely patchy even within its known geographic areas and in individual forests, suggesting specialization on some as yet unknown feature of the habitat, such as particular species of trees.
Local names. Coatípuru-zihno (Br); n'gue, ngua (Sar).
References. Olalla, A. M. 1935. El genero *Sciurillus* representado en la Amazonia y algunas observaciones sobre el mismo. *Rev. Museu Paulista* 19:425–40.
 Anthony, H. E., and G. H. H. Tate. 1935. Notes on South American Mammalia. No. I. *Sciurillus. Am. Mus. Novitates,* no. 780.

Montane Squirrel
Syntheosciurus brochus
Map 153
Identification. HB = 153–185; T = 120–160; HF = 41–48; E = 8–21.
Upperparts uniform finely grizzled brown-olivaceous. Ears tiny, well furred, and inconspicuous, not protruding above crown. Incisors grooved. **Tail thickly furred, shorter than head and body,** frosted with olivaceous-buff above, blackish below. Underparts grayish orange. **Fur thick and soft, giving squirrel a roly-poly look.**
Similar species. Central American dwarf squirrels *(Microsciurus alfari)* are smaller, dark brown, slender-tailed, and with a grayish white belly in area of overlap.
Natural history. Crepuscular and diurnal; terrestrial; in pairs or family groups. This rarely seen squirrel spends most of its time on the ground and in the underbrush, where it hides and is difficult to see. One nest was in a tree hollow about 7 m up. Found in wet montane palm and oak forests.
Geographic range. Central America: Costa Rica and Panama in the Cordillera de Talamanca; scattered localities in a strip less than 250 km long. At 1,920–2,300 m elevation.
Status. Apparently rare or extremely difficult to see; geographic range small and narrow.
Local names. Ardilla (Span).
References. Enders, R. K. 1980. Observations on *Syntheosciurus:* Taxonomy and behavior. *J. Mammal.* 61:725–27.

Pocket Gophers (Geomyidae)

Dental formula: I1/1, C0/0, P1/1, M3/3 = 20. Pocket gophers are large, sausage-shaped, rat-like rodents with short, naked tails, small eyes and ears, and enlarged claws. They have large external cheek pouches, and the skin around the mouth closes behind the incisors, so that the long, broad incisors are always exposed and visible. They are fossorial and spend most of their lives underground, where they make deep, complex burrow systems to live in as well as shallow subsurface foraging tunnels. They feed on underground plant parts or pull aboveground plants down into their tunnels from below. The incisors are used in digging, and the arrangement of the lips closing behind them keeps dirt out of the mouth. Because they live secret, underground lives, their social behavior is not well known, but their population genetics has been extensively studied in North America. They live in small populations that are easily isolated and are prone to strong genetic differentiation, so there are many subspecies. There are 5 genera and 35 species, all in the New World, and 1 genus with 6–7 species in the rainforest region.

Middle American Pocket Gophers
Orthogeomys spp.
Figure 13, map 156
Identification. Measurements: HB = 172–268; T = 81–135; HF = 34–53; E = 5–10; WT = 247–900 g.
Upperparts uniform dark sepia **brown** to beige or chestnut, **with or without white patch on crown or white band around hips, hairs unicolored to base,** overhairs coarse, underfur woolly. **Ears tiny, lost in fur; eyes small. Large cheek pouches open externally; incisors large, broad, with a single groove on face; always exposed as skin of mouth closes behind them.** Body long and cylindrical, legs short; all **claws enlarged, middle claw of forefoot longer than outer claws.** Tail short, robust, cylindrical, naked, blunt at tip. Underparts slightly paler than back, hairs unicolored to roots.

Map 156

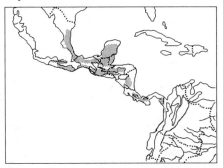

Middle American pocket gophers, *Orthogeomys* spp.

Species. Six or seven of the 11 species can occur in rainforest. *O. cavator, O. darienensis,* and *O. thaeleri,* from SE Panama and NW Colombia, are large, dark animals, blackish or reddish, with sparse fur on underparts. *O. cherrei,* from Costa Rica, have a striking white blaze on crown. *O. underwoodi,* from Costa Rica, are small, with a broad white band all the way around body at hips. *O. hispidus,* from Mexico to Honduras, are chestnut brown, with short, stiff fur, and a few long white overhairs on rump.

Similar species. No other rodents in range have large exposed incisors and greatly enlarged claws. In southern South America, where pocket gophers are absent, their ecological place is taken by the remarkably convergent hystricognath genus *Ctenomys* (tuco-tuco).

Natural history. Probably diurnal and nocturnal; fossorial; solitary. Pocket gophers are herbivores that feed on underground parts of plants, such as tubers and stems, and on aboveground parts pulled down from below. *O. hispidus* sometimes feeds on the roots of understory palms. The presence of pocket gophers is shown by their shallow foraging tunnels, which appear as low humps snaking between damaged plants. Pocket gophers largely occupy open habitats with well-drained soils. Rainforest populations are usually in hilly or montane forests, in agricultural clearings, or in scattered pockets of deep, well-drained soil. They can be serious pests in plantations of bananas or other crops.

Geographic range. Central America: from Taumalipas, Mexico, to Panama and just entering Colombia in N Chocó.

Status. Locally abundant, but populations patchily distributed.

Local names. Tuza, taltuza, salamandra (Span); covatierra, cova-cova (Co); arador, chuchupa, dueño de tierra (Pn); ba (May).

References. Hafner, M. S. 1991. Evolutionary genetics and zoogeography of Middle American pocket gophers, genus *Orthogeomys. J. Mammal.* 72:1–10.

Alberico, M. 1990. A new species of pocket gopher (Rodentia: Geomyidae) from South America and its biogeographic significance. In G. Peters and R. Hutterer, eds., *Vertebrates in the tropics,* 103–11. Bonn: Museum Alexander Koenig.

Figure 13. Middle American pocket gopher, *Orthogeomys hispidus.*

Pocket Mice (Heteromyidae)

Dental formula: I1/1, C0/0, P1/1, M3/3 = 20. This family is characterized by external cheek pouches; small, simple cheek teeth; and a variety of skull features. Many species have large, strong hindlegs and small, weak forelegs and travel by hopping on their hindlegs. All species are small (< 200 g). The family includes kangaroo rats, kangaroo mice, and pocket mice. These rodents are generally seed eaters that use their pockets to transport seeds, which they often store in caches. Most live in arid habitats, a few in humid forest. There are 6 genera and 59 species, all in the New World, with 1 genus and about 3 species in rainforest.

Spiny Pocket Mice
Heteromys spp.
Plate 22, map 157
Identification. Measurements: HB = 108–148; T = 119–175; HF = 27–39; E = 13–22; WT = 49–103 g.
Upperparts and limbs dark brown-black; fur a mixture of **spiny** guard hairs and softer underhairs; spines sometimes partly palé, giving grizzled look; spines not apparent visually from a distance, but are detectable by rubbing fur backward. **Tail bicolored dark above, pale below, usually longer than head and body. Underparts pure white, sharply demarcated from sides. Cheek pouches large, externally opening pockets** extend back to shoulders. All feet white or hindfeet black. Scrotum of adult males greatly elongated behind.
Species. Five of about seven species in the genus occupy different regions of Neotropical rainforest: *H. anomalus, H. australis, H. desmarestianus, H. gaumeri,* and *H. goldmani.* All are similar in appearance. The genus needs revision, and species definitions are likely to change.
Similar species. No other small mice in region have cheek pouches; spiny mice (*Neacomys* spp.) are tiny, with relatively short tails; spiny and armored rats *(Proechimys* spp., *Hoplomys)* are much larger, with tails shorter than head and body; all other mice or rats in range are soft-furred. Another genus of spiny pocket mouse *(Liomys)* occupies dry habitats in Central America; they are paler and have white forelegs.
Natural history. Nocturnal; terrestrial; solitary. Feeds on seeds and berries, including small palm nuts and coffee beans, and on fungi. The cheek pouches are used to transport seeds and other food. Spiny pocket mice are often found near streams and around logs and the bases of trees. They den in burrows and have a home range of 1–2 ha. They are

Map 157

Spiny pocket mice, *Heteromys* spp.

associated with very wet habitats, which occur at sea level in some regions, but mostly in montane forests in Central America. Found mostly in mature forest, but also in coffee plantations and other secondary vegetation.
Geographic range. Central and South America: rainforest species from Oaxaca, Mexico, through Central America to N Ecuador west of the Andes and to N Colombia and Venezuela east of the Andes; Trinidad and Tobago. *H. anomalus* east of the Andes in Colombia and Venezuela; *H. australis* west of the Andes in Colombia and Ecuador; and *H. desmarestianus* in Central America. To about 2,200 m elevation.
Status. Often common.
Local names. Ratón de bolsas, rata espinosa (Span).
References. Rood, J. P., and F. H. Test. 1968. Ecology of the spiny rat, *Heteromys anomalus,* at Rancho Grande, Venezuela. *Am. Midl. Nat.* 79:89–102.
    Fleming, T. H. 1974. The population ecology of two species of Costa Rican heteromyid rodents. *Ecology* 55:493–510.

Murid Rodents (Muridae)

Dental formula: I1/1, C0/0, P0/0, M3/3 = 16, or sometimes fewer molars. Forefeet with four long toes and a short thumb, hindfeet with five toes. This family includes most of the familiar rats and mice. Its members are characterized mainly by skeletal and internal features. They range in size from about 4 to 2,000 g; most are small rats and mice in the 15–300 g range. (There is no taxonomic basis or division between rats and mice; these names are simply traditional words for larger and smaller species.) Murids show a wide array of lifestyles and reproductive traits. A knowledge of the morphological features associated with different ecologies is one of the most useful aids for identifying small rodents. Generally, arboreal species tend to have broad pink-soled feet with sharp curved claws; short legs; long, often robust, hairy tails; smallish ears; large eyes; long, dense whiskers; and somewhat blunt muzzles. Terrestrial grazers tend to have short, fine whiskers; short tails; hispid fur; and short rotund bodies. Mice that tunnel and dig have short tails and ears and long claws. Terrestrial omnivores have long, narrow, black-soled hindfeet; long legs; large ears; well-developed whiskers; long naked tails; and pointed snouts. Semiaquatic species have large paddle-shaped hindfeet with webs between the toes and a fringe of hairs on the side of the foot; slightly to very hairy tails; small ears and eyes; stiff whiskers; and dense, glossy fur. Species that feed on terrestrial invertebrates tend to have long snouts; short round ears; long claws on the forefeet; and short, slightly hairy tails. These are not rigid rules but general tendencies: some species have mixed ecologies and features; other factors such as habitat and climate are also associated with certain morphologies. Most murids make nests of fibers to shelter themselves and their young. Many species have large litters of altricial young that grow quickly and mature early. The life expectancy of individuals of most species seems to be less than a year. This is the largest family of mammals, with about 1,326 species worldwide. These are divided into 17 subfamilies. All of the native genera described in this book belong to the subfamily Sigmodontinae; the introduced house mouse and black and Norway rats belong to the Murinae. There are no published comprehensive keys or detailed descriptions of the species for most genera of rainforest murids: specimens are thus essential for correct identification. We follow the taxonomy of G. G. Musser and M. D. Carleton, "Family Muridae" (in D. E. Wilson and D. M. Reeder, eds., *Mammal species of the world: A geographic and taxonomic reference,* Washington, D.C.: Smithsonian Institution Press, 1993).

Rice Rats
Oryzomys spp.
Plate 22, map 158
Identification. Measurements: about HB = 100–190; T = 79–210; HF = 25–40; E = 15–25; WT = 40–120 g.
Medium-sized to small rats: brown, tawny, reddish, or gray; fur soft, dark gray at base, not usually conspicuously dense. **Ears large, thinly haired;** whiskers long or short; eyes large; eyeshine bright yellow. **Tail usually slightly shorter than head and body,** but may be slightly longer, **naked, slender,** solid or **bicolored pale below, dark above,** not white-tipped. **Hindfeet narrow, usually white above, with black soles.** Underparts grayish white with hairs gray at base. Mammae pairs: 1 pectoral, 1 postaxial, 1 abdominal, 1 inguinal = 8.
Species. A large genus of about 36 species. Up to 4 species may occur in one locality. A few of the most common rainforest species

are described below, but the species cannot be identified with confidence without examining the skull.
O. bolivaris. Dark brown to tawny, long, velvety fur, short ears; distinguished from all others by an exceptionally long whisker above the eye that reaches well back behind the ear. Costa Rica, Panama, the west coast of Ecuador and Colombia.
O. capito (plate 22). Uniform brown, with yellowish sides, darker midback, and gray belly. Young are gray, young adults brown, old adults reddish. The most widely distributed and common species in the Amazon Basin and the Guianas. *O. talamancae* is similar, from Central America, coastal Colombia, Ecuador, and Venezuela. Several other species are also externally similar.
O. macconnelli (plate 22). Back a mixture of tawny and blackish, sides bright cinnamon, fur long and soft; distinctive shape with long hindlegs and feet; long pointed

Map 158

▓ Rice rats, *Oryzomys* spp.

muzzle, long, slender bicolored tail. Found in very dense vegetation such as viny bamboo thickets and large treefalls, apparently always on the ground. Throughout Amazonia and the Guianas.

O. nitidus, O. intermedius. Back uniform reddish, sides often bright orange. *O. nitidus* in the Amazon Basin, *O. intermedius* in coastal forests of SE Brazil.

O. caliginosus (subgenus *Melanomys*). Distinctive and sometimes placed in a separate genus. It is blackish, with reddish tints; underparts about the same as upperparts; it has tail much shorter than head and body and short ears, unlike other species. Honduras south to W Ecuador.

O. buccinatus, O. ratticeps, O. subflavus. This triad of large rats from open, flooded habitats is semiaquatic. They have tawny, yellowish brown fur, a short fringe of white hairs on sides of hindfeet, hairy ears, and tails much longer than head and body (see Water rats, *Nectomys,* for comparisons).

Similar species. Arboreal rice rats (*Oecomys* spp.) have long, dense fur; hair base pale gray; tails unicolored and slightly hairy; short ears, and pink-soled feet. Climbing rats (*Rhipidomys* spp.) have tufted tails and pink-soled feet. Pygmy rice rats (*Oligoryzomys* spp.) are much smaller, with tails longer than head and body. Atlantic forest rats (*Delomys* spp.) sometimes have a black stripe on midback, extreme tip of tail white, and pink soles on toes; may not be distinguishable by external features.

Sounds. No sounds are usually heard in the field.

Natural history. Nocturnal (except *O. caliginosus* diurnal); terrestrial; solitary. Feed on fruit, seeds, and insects. The diet of *O. capito* includes about 30% insects and 67% fruits and seeds. Rice rats are generally most common in areas of dense undergrowth and around fallen logs. They are usually a dominant part of the rainforest small mammal fauna and are easy to see. They run about actively on the ground or on fallen logs; if frightened, they may bounce away with a series of high jumps. After the initial activity of the night, they may climb onto a branch at 1–2 m and rest motionless, head propped on the incisors against the branch. *O. capito* makes a small nest of leaves and twigs in a hollow log or on the ground surface under the litter. Rice rats occupy a wide range of climates and humid habitats. Some rainforest species (e.g., *O. capito*) adapt well to secondary vegetation and plantations; others do not. *O. nitidus* is the dominant rodent of southern dry forests.

Geographic range. North to South America: S United States to N Argentina.

Status. Common to rare.

References. Guillotin, M. 1982. Rythmes d'activité et régimes alimentaires de *Proechimys cuvieri* et *d'Oryzomys capito velutinus* (Rodentia) en forêt Guyanaise. *Rev. Ecol.* 36:337–71.

Musser, G. G., and M. M. Williams. 1985. Systematic studies of Oryzomyine rodents (Muridae): Definitions of *Oryzomys villosus* and *Oryzomys talamancae. Am. Mus. Novitates,* no. 2810.

Pygmy Rice Rats
Oligoryzomys spp.
Plate 23, map 159

Identification. Measurements: HB = 70–110; T = 85–155; HF = 20–30; E = 11–17; WT = 9–40 g.

Upperparts tawny yellow-brown, mixed with black hairs on back producing hispid look; muzzle, **sides of head and neck, and sometimes forequarters tawny gray;** fur short in warm lowlands, long in cold climates. Ears slightly hairy, inner parts thinly covered with pale hairs, especially at base. Whiskers usually short, sometimes reach ear tip but not shoulder; eyeshine bright. **Tail very slender, with one exception much longer than head and body, naked, with a few hairs protruding beyond tip,** gray.

Map 159

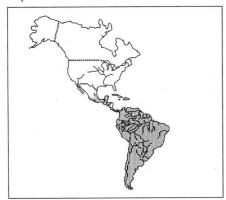

Pygmy rice rats, *Oligoryzomys* spp.

Underparts gray, often washed with buff to orange, with hairs gray to base. Hindfeet narrow. Small to tiny mice with pointed muzzles, flat fur, long tails, and hair inside ears. Mammae pairs: 1 pectoral, 1 postaxial, 1 abdominal, 1 inguinal = 8.

Species. There are about 15 species, all quite similar in external appearance and difficult to distinguish in the field. Only one species usually seems to be found at any one lowland rainforest site, but two or three have been reported from lower montane forests. Species include: *O. fulvescens* (plate 23), in Central America and N South America; *O. microtis,* which has a short tail (equal to head and body length), in Brazil, Bolivia, and Peru; *O. chacoensis,* in Bolivia, Paraguay, and S Mato Grosso, Brazil; *O. destructor* in Andean forests. The taxonomy of this group has not yet been completely defined.

Similar species. Bicolored arboreal rice rats *(Oecomys bicolor)* are reddish brown, with a pure white belly and slightly tufted tail; rice rats *(Oryzomys* spp.) have tails usually shorter than head or body; house mice have a shorter tail and large, rounded ears; spiny mice *(Neacomys, Scolomys)* have spiny fur; pocket mice *(Heteromys* spp.) have spiny fur, cheek pouches, and white bellies. Above 1,000 m in the Andes there are other diminutive rice rats *(Microryzomys* spp.).

Natural history. Nocturnal; terrestrial, but climb well; solitary. Probably feed on seeds, fruit, and insects. These little mice prefer dense and brushy habitats; they are rare in mature terre firme rainforest but do occur there. They are most numerous in open areas such as river edges, gardens, secondary brush, rice fields, and plantations. Pygmy rice rats commonly occupy houses and camps in rainforest clearings. They can be agricultural pests in rice fields and storage barns. *O. fulvescens* builds a nest in the rice plants, 50–80 cm above the ground. Found in a wide range of habitats and climates from rainforest to grassland and from sea level to the high Andes.

Geographic range. Central and South America: Mexico to Tierra del Fuego. To over 4,000 m elevation.

Status. Locally common.

References. Carleton, M. D., and G. G. Musser. 1995. Systematic studies of oryzomyine rodents (Muridae: Sigmodontinae): Definition and distribution of *Oligoryzomys vegatus* (Bangs, 1902). *Proc. Biol. Soc. Wash.* 108:338–69.

Veiga-Borgeaud, T. 1982. Données écologiques sur *Oryzomys nigripes* (Desmarest, 1819) (Rongeurs, Cricétidés) dans le foyer naturel de peste de Barracão dos Mendes (Etat de Rio de Janeiro, Brésil). *Mammalia* 46:335–59.

Arboreal Rice Rats

Oecomys spp.

Plate 23, 25, map 160

Identification. Measurements: HB = 95–175; T = 90–190; HF = 20–32; E = 11–20; WT = 21–88 g.

Upperparts uniform warm reddish to brown, sometimes with gray tones, hair base pale, dove gray; fur soft, dense, dull, often long. Ears medium length, brown, usually without pale hairs inside; **muzzle short,** quite blunt; **whiskers dense but not long or coarse, standing up in a fan above nose;** eyes large and bulging; eyeshine bright. **Tail usually about as long as head and body, brown, lightly haired, with a few hairs extending beyond tip in an inconspicuous tuft.** Feet short, broad, with strongly curved claws; dusky or white above, often with a dark patch across hindfeet above toes; soles pink, with fleshy foot pads. Underparts pure white, pale gray, or gray with orange wash; area around mouth sometimes white. Mammae pairs: 1 pectoral, 1 postaxial, 1 abdominal, 1 inguinal = 8.

Map 160

Arboreal rice rats, *Oecomys* spp.

Map 161

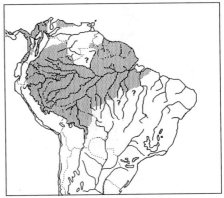

Spiny mice, *Neacomys* spp.

Species. There are about 13 species, up to 4
of which can be found together in lowland
Amazon Basin localities. Bicolored arboreal
rice rats *(O. bicolor)* (plate 23) are distinctive and widespread: small (20–35 g), reddish, with at least part of hair on underparts
white to the roots. Most other species are
larger (40–80 g), have gray-based fur on
underparts, and are difficult to distinguish
in the field. The taxonomy is undergoing
revision.
Similar species. Small species of climbing
rats *(Rhipidomys)* are similar and difficult
to distinguish externally; they are usually
grayer, with a longer, more robust tail with
a more pronounced tuft, and long, coarse
whiskers that reach to shoulder. Females
have 6 mammae.
Natural history. Nocturnal; arboreal; solitary. Feed on fruit and green seeds. These
mice use all levels of the forest, including
occasionally the ground. They are most numerous in dense viny vegetation. At night
they can be seen running rapidly along
pencil-thin vines and slanting branches or
sitting motionless in leafy thickets above
the ground. They nest in tree holes, in dense
masses of vines or epiphytes, and among
palm leaves. They often invade houses in the
forest and seem particularly fond of thatched
roofs. Found in mature and secondary rainforest, dry forest, savanna, scrub, and gardens
and plantations.
Geographic range. Central and South
America: Costa Rica south to Argentina.
To at least 2,200 m elevation.
Status. Often common.

Spiny Mice
Neacomys spp.
Plate 23, map 161
Identification. Measurements: HB =
71–100; T = 70–111; HF = 20–24;
E = 13–19; WT = 15–32 g.
**Upperparts brown or red-brown finely
streaked with black, or head and midback
blackish streaked with orange; sides of
head, neck, shoulders, and lower sides
orange,** often bright; **fur spiny,** with narrow
spines mixed with hairs; spines not visually
apparent, but detectable by rubbing fur backward. Whiskers fine and long, reaching beyond ear. Tail slender, naked, brownish, as
long as head and body. Forefeet white, hindfeet dusky, or legs and feet gray. Underparts
pure white, pale orange, or white along
midline, bordered by orange laterally. Tiny,
sleek-looking mice.
Species. Recent collections suggest that
there are more than the four recognized
species: *N. guianae* is generally brown
above, with underparts whitish medially,
orange laterally; *N. spinosus* is more blackish above, with gray tones on the head,
and white or pale orange underparts (two
species?); *N. tenuipes* is rich reddish brown
above. The other species, *N. pictus,* is
montane.
Similar species. Pocket mice *(Heteromys*
spp.) have cheek pouches; gray spiny mice
(Scolomys spp.) are gray below, with tail
much shorter than head and body; no other
small mice have spiny fur.

Natural history. Nocturnal; terrestrial; solitary. Diet of *N. guianae* is about 60% insects and 40% fruit, seeds, and other plant material. These mice favor areas of dense ground cover and travel on the ground or on logs or vines near the ground. They are restricted to rainforest regions and live in mature rainforest, but can sometimes be found in secondary forest, grassy clearings, or plantations.

Geographic range. Central and South America: E Panama to Ecuador west of the Andes; east of the Andes to the southern Amazon Basin of Brazil and Bolivia. *N. guianae* north of the Amazon and east of the Rio Negro; *N. spinosus* from the rest of the Amazon Basin; and *N. tenuipes* from Panama, Colombia, Ecuador, and Venezuela west of the Andes. To 1,875 m elevation.

Status. Rare in some regions (i.e., Panama), common in others.

Gray Spiny Mice

Scolomys spp.

Plate 23, map 162

Identification. Measurements: HB = 84–105; T = 55–76; HF = 18–23; E = 13–17; WT = 20–33 g.

Upperparts uniformly colored, from pale reddish brown to dark gray or blackish finely ticked with brown or fulvous; fur spiny, spines not apparent visually, detectable by rubbing fur backward. Ears small and stiff; whiskers fine, reaching ear. **Tail much shorter than head and body,** naked. Legs and feet entirely gray, toes whitish; thumb with a flat nail; tip of heel hairy below. **Underparts clear medium gray, paler than back.** Mammae pairs: 1 postaxial, 1 abdominal, 1 inguinal = 3.

Species. The three species are externally similar: *S. melanops* has strongly proodont incisors, the others are orthodont; *S. juruaense* is pale reddish brown to reddish black; *S. melanops* is dark gray to brownish black.

Similar species. See spiny mice (*Neacomys* spp.).

Natural history. Nocturnal, terrestrial. Found in terra firme forest, both in mature forest *(S. juruaense)* and in disturbed forest with dense undergrowth *(S. melanops, S. ucayalensis)*. *S. melanops* has been found at 1,150 m in wet montane forest.

Geographic range. South America: W Amazon Basin in a small region in Ecuador, N Peru, and Brazil at 1,150 m and 250 m elevation.

Map 162

Gray spiny mice, *Scolomys* spp.

Status. Apparently patchy or rare in a small geographic range. Known from fewer than 50 individuals for all three species combined. Two of the species were first described in 1991 and 1995.

References. Patton, J. L., and M. N. da Silva. 1995. A review of the spiny mouse genus *Scolomys* (Rodentia: Muridae: Sigmodontinae) with the description of a new species from the western Amazon of Brazil. *Proc. Biol. Soc. Wash.* 108:319–37.

Water Rats

Nectomys spp.

Plate 24, map 163

Identification. Measurements: HB = 135–232; T = 152–234; HF = 37–53; E = 17–26; WT = 133–288 g.

Upperparts glossy dark brown, a finely grizzled mixture of black and yellowish; center of back darkest, sometimes blackish, **paling to yellowish sides; fur soft and dense, medium to long. Ears well furred around base, naked toward tip;** whiskers short, few or none extending beyond tip of ear when flattened. **Tail robust,** only slightly longer than head and body, **lightly haired with brown hairs densest midventrally, sometimes forming a slight "keel," and extending slightly beyond tip in a small tuft (but looks naked from a distance). Hindfeet wedge-shaped, with narrow heel and broad palm at base of toes, toes partially webbed; sides of feet with fringe of down-curling silvery white hairs; sole of foot to heel covered with roundish scales.** Underparts gray washed dirty yellow, buff,

Map 163

Water rats, *Nectomys* spp.

or pale orange with silvery sheen, grading gradually from sides with no line of demarcation. Young gray-brown. Mammae pairs: 1 pectoral, 1 postaxial, 1 inguinal, 1 abdominal = 8. Large, plain brown rats.

Species. The three species are similar: *N. squamipes* is larger, *N. parvipes* is smaller (the lower measurements given above are from the type and only specimen). *N. palmipes* is found in Trinidad and the nearby mainland. *N. squamipes* varies from blackish to yellowish or grayish brown; it is likely that these rats represent more than one species.

Similar species. Marsh rats (*Holochilus* spp.) are externally similar; they usually have tails slightly shorter than head and body, short, round ears hairy inside and outside at tips, no scales on sole of heel, very short whiskers, and often orange on sides and underparts. The three semiaquatic *Oryzomys* are very like water rats, but they have tails much longer than head and body, and no webs between toes. Large climbing rats (*Rhipidomys* spp.) have tails much longer than head and body, short, broad feet with no webs or fringe, and long, coarse whiskers. Spiny rats (*Proechimys* spp.) and fishing rats (*Icthyomys* spp.) have white bellies sharply demarcated from sides. Norway rats and black rats (*Rattus* spp.) may be similar in color but have coarse, sparse fur, naked ears and tails, and no webs between toes or fringe on foot.

Natural history. Nocturnal; terrestrial and semiaquatic; solitary. Feed on arthropods, crabs, and other invertebrates, fruit, and

fungi. Water rats are adapted for swimming and are almost always found near water. In captivity they have been shown to catch small aquatic prey by probing with the forefeet. They make nests under logs or roots or in dense vegetation; an *N. squamipes* nest of grass in a tree stump in a swamp had several underwater entrances. These rats are most common in areas of dense tall grass near water, but are also uncommonly found in closed-canopy forest. Found in rainforest, gallery forest, and disturbed forest, flooded grasslands, gardens, plantations, and clearings.

Geographic range. South America: east of the Andes from Colombia and Venezuela south to N Argentina *(N. squamipes)*. *N. parvipes* is known only from French Guiana. To 1,100 m elevation, but rare above 500 m.

Status. Widespread and sometimes common.

Local names. Rata de agua (Span); rato d'agua, rato paca (Br); ratón nativo (Ec).

References. Ernest, K. A. 1986. *Nectomys squamipes.* Mammalian Species, no. 265.

Rio de Janeiro Arboreal Rat

Phaenomys ferrugineus

Map 164

Identification. Measurements: HB = 148; T = 187; HF = 33; E = 17.

Upperparts brilliant rusty orange, slightly darker on crown and midback; fur long, soft, dense, and straight, slaty at base. **Ears short, both sides covered with hair, rust-orange;** whiskers fine and long, longest to shoulder. **Tail robust, much longer than head and body,** lightly covered with fine dusky hairs, dark brown. Feet dark reddish brown, soles pink; **hindfeet with a broad white patch across upper foot and ankle. Underparts from chin cream, sharply demarcated from sides, hairs cream to base.** Mammae pairs: 1 pectoral, 1 postaxial, 1 abdominal, 1 inguinal = 8.

Similar species. The only other bright red mice in range are Brazilian arboreal mice *(Rhagomys rufescens),* which are smaller, with tail equal in length to head and body, and an orange belly.

Natural history. Unknown. The morphology of this rat suggests that it is arboreal.

Geographic range. South America: Brazil, known only from the state of Rio de Janeiro.

Status. Extremely rare; known from only one individual collected in the last century.

Map 164

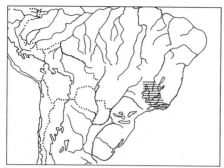

Map 165

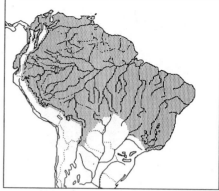

▦ Rio de Janeiro arboreal rat, *Phaenomys ferrugineus*
▤ Brazilian arboreal mouse, *Rhagomys rufescens*

▦ Climbing rats, *Rhipidomys* spp.

Brazilian Arboreal Mouse
Rhagomys rufescens
Map 164
Identification. Measurements: HB = 94;
T = 93; HF = 20; E = 15.
**Upperparts and underparts entirely rich
rust-orange; hairs slate blue at base, rusty
at tips over whole body; fur long and dense.
Ears short, barely projecting above fur,
thickly haired,** red-brown; whiskers fine,
longest to shoulder. Tail equal in length to
head and body, thinly covered with dusky
hairs, slight tuft at tip. Feet pinkish-yellow-
ish, soles pink; first toe of hindfoot with a
rounded nail, not claw. Underparts only
slightly paler than back. Mammae pairs:
1 pectoral, 1 abdominal, 1 inguinal = 6.
Similar species. See Rio de Janeiro arboreal
rat *(Phaenomys ferrugineus)*.
Natural history. Unknown. The morphology
of this species suggests that it is arboreal.
Geographic range. South America: SE
Brazil, known only from Minas Gerais and
Rio de Janeiro.
Status. Rare; known from only a few
individuals.

Climbing Rats
Rhipidomys spp.
Plate 25, map 165
Identification. Measurements: HB =
110–190; T = 130–230; HF = 24–36;
E = 18–23; WT = 35–170 g.
**Upperparts dull yellow-brown, fulvous
brown, or gray-brown; usually slightly
grizzled or streaked with black; fur soft,
dense,** sometimes long; **hairs dark gray at**
base. Eyes large, often narrowly ringed with
dark fur; **whiskers beside nose long, dense,
and coarse, when pressed back reach beyond ear to shoulder. Tail 20–40% longer
than head and body, robust, thinly covered
with coarse, dark hair that forms a small
to prominent tuft at tip. Feet short and
broad, with sharp, curved claws, conspicuously black or dusky across top, usually
with paler toes, soles pink or dark with
pink toes.** Underparts gray, white with
gray base, pure white, or yellowish white.
Mammae pairs: 1 postaxial, 1 abdominal,
1 inguinal = **6.**
Species. There are about 14 species, ranging from medium-sized to large. Ten are
montane, of 1,000–3,000 m cloud forests.
There are four lowland forms:
R. couesi, a large yellowish brown rat with
short fur (near 5 mm on rump), tail tuft
poorly developed, all or some hairs on underparts white to base; from Venezuela to Argentina along the eastern base of the Andes.
R. leucodactylus, a large streaked tawny rat
with long fur (near 10 mm on rump), underparts entirely gray-based, and prominent tail
tuft; from the Guianas and Venezuela to Peru.
R. mastacalis and *R. nitela* (plate 25),
medium-sized (40–80 g) brown to fulvous
rats with white belly and little tail tuft, from
the Guianas, the E Amazon Basin, and SE
Brazil; they are difficult to distinguish
externally.
Similar species. The smaller climbing rats,
such as *R. mastacalis* and *R. nitela,* are similar externally to the larger arboreal rice rats,
such as *Oecomys trinitatus.* Climbing rats

have more robust, thicker tails much longer than head and body, with coarser, longer hairs and terminal tufts; black, thick, coarse whiskers, several of which reach well behind ears if pressed back; and dorsal fur dark gray at base. Members of the closely related genus of Thomas's paramo rats *(Thomasomys)* are found at 1,000–3,500 m in cloud forest or paramo.

Sounds. None usually heard in field.

Natural history. Nocturnal; arboreal; solitary. Climbing rats use all levels of the forest, sometimes including the ground, but they may spend most of their time in the canopy. They seem to favor areas of dense viny vegetation, the tops of palms, tree hollows, or rocks and caves. Found in mature and secondary rainforest, montane and cloud forest, deciduous forest and scrub, and gardens and plantations; also invade rural camps and houses.

Geographic range. Central and South America: barely enters Central America in extreme E Panama; west of the Andes south to NW Peru; east of the Andes widespread in all countries south to N Argentina, but not Paraguay. To 2,400 m elevation.

Status. Widespread and often common.

Atlantic Forest Rats

Delomys spp.
Plate 22, map 166

Identification. Measurements: HB = 102–154; T = 90–150; HF = 25–33; E = 20–23.

Upperparts dark to tawny brown, sides paler; with or without a dark midback stripe (see species below); fur soft, medium length. Whiskers fine, reaching just behind ear tip, no long whisker from behind eye; ears finely haired, appear naked, moderately large. Upper incisors slender and slanted backward. Tail slightly longer to slightly shorter than head and body, naked, tip sometimes white. **Hindfeet long and narrow, with long toes, white above with white hairs extending beyond claws, soles black with pink toe tips;** hindsoles with six fleshy pads. Mammae pairs: 1 pectoral, 1 postaxial, 1 abdominal, 1 inguinal = 8; or lacking pectoral pair = 6.

Species. There are two species:

Delomys dorsalis. Upperparts dark brown, midback with a black stripe from shoulder to rump; sides slightly paler, tawny brown;

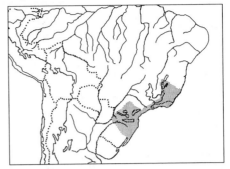

Map 166

Atlantic forest rats, *Delomys* spp.

ears dark gray; eyes sometimes with dark ring or darker area between eye and nose; tail equal to or slightly longer than head and body, gray-brown above, whiter below.

Delomys sublineatus. Upperparts tawny brown, finely streaked with blackish; sides washed with orange, brightest on cheeks and elbows; ears brown; tail slender, shorter than head and body; underparts gray washed with white or orange, with a yellow line separating dorsal and ventral regions.

Similar species. *Oryzomys capito* are similar but have robust incisors and black toe tips; *O. intermedius* are orange-red; climbing rats *(Rhipidomys* spp.) have hairy tails much longer than head and body and broad, dark-banded hindfeet; grass mice *(Akodon* spp.) have short whiskers that do not reach ear and short tails.

Natural history. Nocturnal; terrestrial. These rats can be found in holes around fallen logs and roots and in tunnels under litter. The striped back suggests some diurnal activity. *D. sublineatus* seems to be found in drier or secondary forests, *D. dorsalis* in wet primary forests. Found in Atlantic coastal forests; many localities are lower montane.

Geographic range. South America: E Brazil to NE Argentina; Rio de Janeiro to Paraná *(D. dorsalis),* Espírito Santo, and Rio de Janeiro *(D. sublineatus).* To at least 1,600 m elevation.

Status. May be locally common.

References. Voss, R. S. 1993. A revision of the Brazilian muroid rodent genus *Delomys* with remarks on "Thomasomyine" characters. *Am. Mus. Novitates,* no. 3073.

Map 167

▦ Crab-eating rats, *Ichthyomys* spp.
▤ South American water mice, *Neusticomys* spp.
▥ Brazilian shrew mouse, *Blarinomys breviceps*
▨ Abrawaya's spiny rat, *Abrawayaomys ruschii*

Crab-eating Rats

Ichthyomys spp.
Plate 24, map 167
Identification. Measurements: HB =
110–171; T = 110–155; HF = 28–40;
E = 8–11; WT = 127 g.
Upperparts glossy dark brown, protruding
black overhairs give rump a slightly bristly
look; fur soft, with dense woolly underfur.
**Head with blunt, rounded muzzle; eyes
small; ears small, hairy, inconspicuous,
sometimes white hairs on inside show as
a prominent white spot on side of head.**
Whiskers coarse, stiff, and straight, longest
reaching just beyond ear. **Tail equal to or
shorter than head and body, completely covered with flat hairs, dark brown or sharply
bicolored white below, brown above. Hindfeet wedge-shaped, with narrow heel and
broad palm; outer edges of foot and toes
fringed with down-curling silvery hairs;
toes with partial webs near base; forefeet
with five pads on palm. Underparts silvery gray washed with brown, to pure
white,** sharply demarcated from sides. Mammae pairs: 1 postaxial, 1 abdominal, 1 inguinal = 6.
Species. Three of the four species occur in
rainforest below 1,000 m: *I. tweedii, I. stolzmanni,* and *I. pittieri;* the first two are also
found at higher elevations.
Similar species. Central American water
mice (*Rheomys* spp.) are similar; they have
forefeet with four pads on palm. South Amer-

ican water mice (*Neusticomys* spp.) are
smaller, with dark underparts not sharply
contrasting with sides and hindfeet not
wedge-shaped. Marsh and water rats (*Holochilus* spp., *Nectomys* spp.) are usually larger,
with naked, scaly tails.
Natural history. Probably nocturnal; terrestrial and semiaquatic. Feeds on crabs, other
aquatic invertebrates such as insect larvae,
and occasionally small vertebrates. Wild-
caught captives kill and eat crabs with a
stereotyped behavior: first they bite off the
claws, then the other legs, then they bite
open the body at the back of the shell and
eat the contents. The empty dorsal shell and
legs or leg tips remain. The presence of crab-
eating rats in an area may be shown by the
discarded empty shells of small crabs left on
rocks in streams. Crab-eating rats are always
found near fresh (not salt or brackish) water,
and appear to require clear, fast-flowing
streams. Such streams occur in hilly terrain
and generally have sandy or rocky beds, and
they may be small and shallow and shaded
within tall rainforest, cloud forest, or secondary forest.
Geographic range. Central and South
America: *I. stolzmanni* on the eastern slope
of the Andes in Peru and Ecuador; *I. tweedii*
on the western slope of the Andes from E
Panama to Ecuador; *I. pittieri* in the coastal
cordillera of Venezuela.
Status. Unknown; rarely encountered.
References. Voss, R. S., J. L. Silva L., and
J. A. Valdes L. 1982. Feeding behavior and
diets of Neotropical water rats, genus *Ichthyomys* Thomas, 1893. *Z. Säugetierk.* 47:
364–69.
   Voss, R. S. 1988. Systematics and ecology of Ichthyomyine rodents (Muroidea):
Patterns of morphological evolution in a
small adaptive radiation. *Bull. Am. Mus. Nat.
Hist.* 188:259–493.

Central American Water Mice

Rheomys spp.
Plate 24, map 168
Identification. Measurements: HB =
102–140; T = 109–160; HF = 30–41;
E = 8–12.
Upperparts glossy dark brown, overhairs
protrude to give rump a bristly look; **rump
and sides often with a sprinkling of silvery
overhairs; fur soft and dense, underfur
woolly. Muzzle blunt; eyes tiny; ears small,
inconspicuous, covered with hair;** whiskers

Map 168

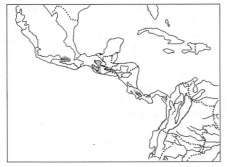

Central American water mice, *Rheomys* spp.

thin and stiff, longest reaching well behind
ear. **Tail equal to or longer than head and
body, completely covered with flat hairs,
dark brown or bicolored, white below, dark
above. Hindfeet wedge-shaped, sides and
toes fringed with silvery hair; forefeet with
four pads on palm. Underparts contrasting silvery gray or white.** Young slate gray.
Species. Two of the five species can occur
in the lowlands. *R. thomasi* is smaller, with
tail < 137 mm; *R. mexicanus* is larger, with
tail > 140 mm; all species are primarily
montane.
Similar species. See crab-eating rats
(*Ichthyomys* spp.)
Natural history. Semiaquatic. Feeds on
aquatic insects and other invertebrates and
occasional small vertebrates. These mice
have mostly been found near clear, tumbling
mountain streams that are shaded by vegetation, but one record is from a polluted
stream in secondary forest. Mainly a genus
of montane and cloud forest species, but also
found where suitable streams occur below
1,000 m.
Geographic range. Central America: Oaxaca, Mexico, south to Darién, Panama. *R.
thomasi* in Chiapas, Mexico, Guatemala, and
El Salvador; *R. mexicanus* in Oaxaca, Mexico. To about 2,000 m elevation.
Status. Unknown; rare or difficult to capture.
References. Hooper, E. T. 1968. Habitats
and food of amphibious mice of the genus
Rheomys. J. Mammal. 49:550–53.

South American Water Mice
Neusticomys spp.
Plate 24, map 167
Identification. Measurements: HB =
100–128; T = 82–111; HF = 24–30;
E = 9–12; WT = 75 g.

**Upperparts glossy blackish brown or buff-
brown; fur soft and smooth, pale gray at
base. Muzzle blunt; eyes small; ears small
but conspicuous, hairy inside and out.**
Whiskers fine and long, several extending
well behind ear. **Tail** equal to or shorter than
head and body, **completely covered with
flat dark brown hairs.** Feet not strongly
wedge-shaped, sides with slight fringe of
brown or silvery hair, blackish or cream
above. **Underparts dark gray or brown
with silvery sheen, not sharply contrasting with sides.**
Species. Three of the four species, (*N. oyapocki, N. peruanus, N. venezuelae*) occur in
the lowlands; the other species is montane.
Similar species. Crab-eating rats (*Ichthyomys* spp.) are similar but have contrasting
pale underparts and strongly wedge-shaped
feet; no other small soft-furred mice have
tails fully covered with smooth hairs that
completely hide the underlying scales, and
small eyes and ears.
Natural history. Semiaquatic. The montane
species feeds on aquatic insects; lowland
species may also eat small crabs. Lowland
water mice have been captured near clear
streams in rainforest.
Geographic range. South America: lowlands east of the Andes and Andes of Colombia and Ecuador; *N. venezuelae* in Venezuela and Guyana; *N. peruanus* in SE Peru;
and *N. oyapocki* in French Guiana. Range
extensions are likely.
Status. Rare or difficult to capture in lowlands; two lowland species are known from
fewer than five individuals each.
References. Musser, G. G., and A. L. Gardner. 1974. A new species of the Ichthyomyine *Daptomys* from Perú. *Am. Mus. Novitates,* no. 2537.
    Voss, R. S. 1988. Systematics and ecol-
ogy of Ichthyomyine rodents (Muroidea):
Patterns of morphological evolution in a
small adaptive radiation. *Bull. Am. Mus. Nat.
Hist.* 188:259–493.

Brazilian Shrew Mouse
Blarinomys breviceps
Plate 22, map 167
Identification. Measurements: HB ≈ 100
mm; T = 30–49; HF = 16–21; E = 8–10.
**Upperparts uniform glossy chocolate
brown. Muzzle blunt; point of nose,
muzzle, and chin white; eyes and ears
tiny, lost in fur.** Forefoot with fifth digit

greatly reduced; forefeet and hindfeet tiny, claws long. **Tail very short, less than half the length of head and body. Underparts gray-brown, hardly differentiated from back. A small cylindrical blunt-nosed mouse with short legs.**
Similar species. All similar mice and short-tailed opossums have conspicuous eyes and ears.
Natural history. Terrestrial. Brazilian shrew mice are probably semi-fossorial; they are said to live in burrows under the leaf litter. Their diet is unknown, but their tooth morphology suggests that they feed on invertebrates. From montane rainforest 650 m and above.
Geographic range. South America: Brazil, coastal mountains from Bahia to Rio de Janeiro and mountains of Minas Gerais.
Status. Apparently rare in a small geographic range. Known from about 40 specimens.
References. Matson, J. O., and J. P. Abravaya. 1977. *Blarinomys breviceps.* Mammalian Species, no. 74.

Abrawaya's Spiny Rat
Abrawayaomys ruschii
Map 167
Identification. Measurements: HB = 128–140; T = 146–150; HF = 29–31; E = 20; WT = 63 g.
Upperparts blackish gray, darkening on head, hairs with ochraceous bands. **Fur densely spiny, with stiff, thin spines on entire upperparts.** Ears thinly covered with dark brown hairs, especially dense on both sides of rims; whiskers short, reach to ear tip when flattened back. **Tail thinly covered with harsh, stiff bristles; brown at base, white at extreme tip, hairs extending in a long (10 mm), thin white tuft beyond bone.** Feet with stiff bristles above, colored like back, toes white, with white hairs extending beyond claws. Underparts gray-based, fur washed with dirty white, paling gradually from sides with no line of demarcation. Young paler, ochraceous gray.
Variation. Only three individuals are known. The head and body and tail measurements of the holotype are problematic, as they do not agree with the others (the length given is likely the total length, not the head and body length, and the tail may be broken). Measurements above do not include the holotype.

Similar species. Echimyid spiny rats are larger: *Proechimys (Trinomys)* have naked ears and bicolored tails; *Euryzygomatomys spinosus* have tail half the length of head and body, and long foreclaws.
Natural history. Unknown. These rats were captured in secondary habitats, but they could occur in nearby forests.
Geographic range. Brazil: known only from a small region of Espírito Santo and Minas Gerais.
Status. Apparently rare in a small geographic range.

Naked-tailed Climbing Rats
Tylomys spp.
Plate 25, map 169
Identification. Measurements: HB = 184–260; T = 191–290; HF = 32–43; E = 20–28; WT = 156–326 g.
Upperparts dull slate gray to warm gray-brown, head often slightly darker than back; **fur soft, dense, and slightly woolly. Ears large, naked, set low, held semi-horizontally, with funnel pointed downward;** muzzle pointed; whiskers black, long, reaching well past ear. **Tail robust, naked, with prominent scales in rings, longer than head and body, dark and sometimes shiny for basal half to two-thirds, long tip sharply contrasting white or yellow. Feet** short and broad, **dark above, usually with sharply contrasting white toes.** Underparts entirely white, or throat, chest, and usually inguinal region white or yellowish and midriff gray. Young slate gray. Large, robust rats.
Species. There are about seven species, all quite similar in size and external appearance. There is individual color variation within a population from gray to brown: adults seem to bleach browner and paler as they age.
Similar species. No other large rats in geographic range have naked, white-tipped tails. The ear posture and white tail tip give these rats a superficial resemblance to some opossums.
Natural history. Nocturnal; arboreal and terrestrial; solitary. Diet described as greenish plant material, possibly lichen or bark. These rats frequent areas of rocks and caves and fallen logs in dense forest, and they are often on the ground as well as in the trees. Found in wet evergreen broadleaf forests.
Geographic range. Central and South America: Mexico to Panama and west of

Map 169

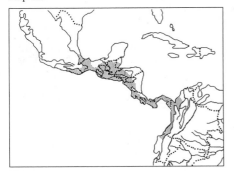

Naked-tailed climbing rats, *Tylomys* spp.

Map 170

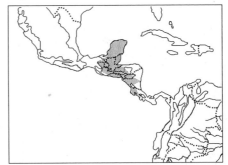

Big-eared climbing rats, *Ototylomys phyllotis*

the Andes in Colombia and N Ecuador. To about 1,100 m elevation.
Status. Apparently uncommon.
References. Hall, E. R., and W. W. Dalquest. 1963. The mammals of Veracruz. *Univ. Kans. Pub. Mus. Nat. Hist.* 14:165–363.

Big-eared Climbing Rat
Ototylomys phyllotis
Plate 25, map 170
Identification. Measurements: HB = 152–170; T = 127–178; HF = 22–29; E = 22–27; WT = 70–130 g.
Upperparts dark to pale gray-brown; fur long and lax, parting easily to give a disheveled look. **Ears very large, naked; eyes large; head and muzzle long, narrow;** whiskers medium to long, to ear or beyond. **Tail** slightly shorter than head and body, robust, **naked, shiny as if varnished, with prominent scales in rings, dark brown or black above, sometimes white below, especially near base.** Forefeet dark above, toes white; hindfeet short and broad, dark above with white toes and edge of foot, or all whitish. Underparts pure white, or white on chin and chest, pale gray posteriorly. A small, slender rat.
Variation. Animals from higher elevations tend to be large, with gray bellies; those from the lowlands are smaller, with white bellies; but there are exceptions.
Similar species. The combination of large ears and robust dark, shiny tail with large scales in rings distinguishes this species from other small rats; naked-tailed climbing rats (*Tylomys* spp.) are larger, with white tail tips.
Natural history. Nocturnal; arboreal and terrestrial; solitary. Feeds on seeds, fruit, and browse. Big-eared climbing rats usually

frequent areas of rocks, caves, or sinkholes, and around fallen trees. They use the lower levels of the forest, generally below 10 m, where they run along vines and branches or the ground. Their nests are at ground level among rocks or under logs, or in low tree holes. This species probably only marginally inhabits rainforest; it is found in a wide variety of habitats from coastal lowland rainforest to montane forest and arid scrub, and in mature forest and secondary vegetation such as roadsides and hedgerows.
Geographic range. Central America: Tabasco and Chiapas, Mexico, south to central Costa Rica. To about 2,000 m elevation.
Status. Locally common.
References. Lawlor, T. E. 1969. A systematic study of the rodent genus *Ototylomys*. *J. Mammal.* 50:28–42.
Lawlor, T. 1982. *Ototylomys phyllotis*. Mammalian Species, no. 181.

Vesper Rat
Nyctomys sumichrasti
Plate 25, map 171
Identification. Measurements: HB = 99–140; T = 102–157; HF = 20–27; E = 15–21; WT = 38–67 g.
Upperparts uniform bright rust-orange or red-brown, sides bright orange; fur short or long, straight and slightly stiff. Ears short, rounded, broad at base; **eyes very large, rimmed with narrow black rings, sometimes faint, that extend forward as a dusky smudge at base of nose and whiskers; whiskers very long, some-**times reaching to midbody. Tail usually slightly longer than head and body, **robust, thickly covered with dull red-brown hairs that curl outward in a sparse brush and**

Map 171

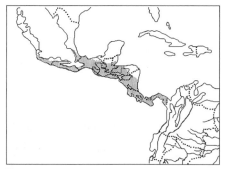

Vesper rat, *Nyctomys sumichrasti*

extend beyond tip in a small tuft. Feet short and broad; forelegs and hindfeet dusky to top of foot; toes white. **Underparts pure white,** including area around mouth and lower cheeks. Young red or gray-brown with fur soft and straight, or distinctly woolly.

Variation. Animals from Panama and Costa Rica are sometimes red-brown, and their young are gray-brown, with woolly fur, until they reach almost adult size; those from Mexico are bright rust, with young the same color as adults.

Similar species. Bicolored arboreal rice rats *(Oecomys bicolor)* and harvest mice *(Reithrodontomys* spp.) are smaller, with slender, smooth tails; rice rats *(Oryzomys* spp.) have naked tails and slender feet. Yucatán vesper rats *(Otonyctomys hatti;* not described in this book) are almost identical externally but brighter red. They are extremely rare, known from fewer than ten specimens from the NE half of the Yucatán Peninsula; their natural habitat is undescribed.

Natural history. Nocturnal; arboreal; solitary. Feeds on fruit, seeds, and flowers. Vesper rats use the middle and upper levels of the forest, above 3 m, and rarely descend to the ground. They nest in tree hollows, and favor those with small entrances 2.5–8 cm in diameter. The nest is a mass of shredded bark within the hollow. When disturbed in her nest by day, a mother vesper rat will flee along horizontal branches, with her young clinging tightly by their mouths to her teats. Found in evergreen lowland and lower montane forests, old secondary and riparian forests, and semi-deciduous forests.

Geographic range. Central America: Veracruz, Mexico, south to central Panama. To 1,600 m elevation.

Status. Usually uncommon or rare, common in a few localities.
References. Genoways, H. H., and J. K. Jones, Jr. 1972. Variation and ecology in a population of the vesper mouse *(Nyctomys sumichrasti).* Occas. Pap. Mus. Texas Tech. Univ., no. 3.

Grass Mice
Akodon spp.
Plate 22, map 172
Identification. Measurements (rainforest species): HB = 90–136; T = 32–98; HF = 17–28; E = 11–21; WT = 19–89 g. **Upperparts uniform dark brown to brown-olivaceous; fur soft, dark gray at base.** Ears rounded, often small; muzzle pointed; **whiskers fine and short, not reaching tip of ear. Tail short, 50–66% of head and body length, robust at base, tapering to thin tip;** black. Feet blackish, narrow; soles black, claws long, with little curve. Underparts dark, about the same color as back, or slightly washed with buff, not strongly differentiated from sides. Body short and rotund. Mammae pairs: 1 pectoral, 1 postaxial, 1 abdominal, 1 inguinal = 8.

Species. The genus is one of the largest in the Neotropics, with over 45 species. Most are montane and live in open grasslands; only a few species are found in rainforest, including *A. urichi, A. cursor, A. dayi, A. nigrita,* and possibly others. The description above applies to species in the rainforest region; some species from south temperate latitudes are chestnut, or washed with pale gray, and may have gray underparts and hairy, bicolored tails. Up to at least four species can occur together.

Similar species. Cerrado mice *(Bolomys* spp.) are externally identical; they cannot reliably be distinguished from these species in the hand. Long-nosed mice *(Oxymycterus* spp.) are usually larger, have very long muzzles with strongly undershot jaws, and hairy tails. In old-field habitats may be confused with cane mice *(Zygodontomys brevicauda),* which have pale gray-frosted underparts. See short-tailed opossums.

Natural history. Terrestrial; mostly diurnal. Probably feed on insects, seeds, and browse. Some grass mice nest in burrows and travel in tunnels under the leaf litter. Found in grassy areas, especially with tall grass, young second growth, gardens, and clearings, but also sometimes in evergreen forest, in coastal

Map 172

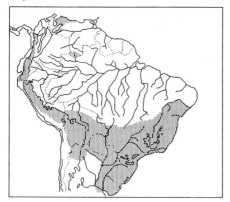

Map 173

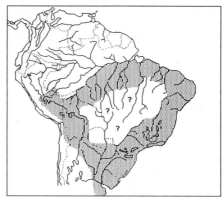

▓ Grass mice, *Akodon* spp. (excluding southern part of range)

▓ Long-nosed mice, *Oxymycterus* spp.

regions of Venezuela and SE Brazil, and in montane habitats.

Geographic range. South America: all Andean countries from N Venezuela to Tierra del Fuego. Absent from the Guiana region. To 5,000 m elevation.

Status. These are often the dominant rodents in cloud forests and grasslands.

References. Myers, P. 1989. A preliminary revision of the *varius* group of *Akodon (A. dayi, dolores, molinae, neocenus, simulator, toba,* and *varius). Advances in Neotropical Mammalogy* 1989:5–54.

Long-nosed Mice

Oxymycterus spp.
Plate 22, map 173

Identification. Measurements: HB = 130–164; T = 70–120; HF = 26–32; E = 17–23; WT = 70–125 g.

Upperparts dark, warm brown on mid-back, grading to olivaceous, or faint to bright fulvous mixed with black hairs on sides of body and head; fur long and sparse, dark gray or blackish at base. **Muzzle extremely long and narrow; chin greatly undershot,** usually pure white; whiskers fine and short, usually not reaching to base of ear. **Tail short, about two-thirds of head and body, thinly covered with short blackish hair.** Feet small and narrow, usually dark brown; soles black, **claws very long, white, with little curve.** Underparts gray washed with yellowish or dull to bright fulvous, not sharply demarcated from sides. **Body short, rotund, and volelike.**

Species. There are at least 12 species; most occur in open habitats in lowland grasslands or above 1,000 m, from cloud forest to puna. *O. inca* and *O. amazonicus* occur in lowland Amazonian rainforest. *O. inca,* in the western Amazon Basin from Acre westward and in lower montane Andean forests, have bright fulvous (near rust-red) underparts and sides; *O. amazonicus,* in Brazil from Mato Grosso eastward, are uniform warm brown above, with orange wash on underparts; several species from the Atlantic forest are olivaceous on sides and larger than Amazonian forms. More than one species can be found together. The genus needs revision.

Similar species. Grass mice (*Akodon* spp.) are similar; they have a shorter muzzle and claws, and are usually smaller. Brazilian shrew mice *(Blarinomys breviceps)* have inconspicuous eyes and ears; other rainforest mice are spiny or have long tails. See short-tailed opossums.

Natural history. Nocturnal and/or diurnal; terrestrial. Feed mainly on insects and other invertebrates and may use their long claws to dig these from the litter. Long-nosed mice are rare in rainforest, where they seem to be found in swampy areas, at least some of which are covered with sedges and have no forest canopy above. Also found in secondary growth and plantations in Amazon Basin rainforest areas.

Geographic range. South America: south of the Amazon from Pará to Peru; south to Argentina and Uruguay. To over 4,000 m elevation.

Status. Some species are locally common.
References. Hershkovitz, P. 1994. The description of a new species of South American hocicudo, or long-nose mouse, genus *Oxymycterus* (Sigmodontinae, Muroidea), with a critical review of the generic content. *Fieldiana Zool.,* n.s., no. 79.

Marsh Rats
Holochilus spp.
Plate 24, map 174
Identification. Measurements: HB = 150–230; T = 140–190; HF = 35–49; E = 15–22; WT = 90–320 g; males reach larger size than females.
Back dull brown, sometimes streaked black and olivaceous, tawny, or orange or sprinkled with gray hairs; **sides of head, neck, forelegs, and body lightly to brightly tinged orange;** fur short and woolly or long and silky. **Ears short and round, hairy inside and out to tips; whiskers short and fine, none usually reaching to ear tip. Tail robust, shorter than head and body,** slightly hairy but without tuft or, usually, ventral keel of hairs. **Hindfeet wedge-shaped, with narrow heel and broad palm; partial webbing between first and second, and third and fourth toes; sides of feet with downcurling fringe of whitish hairs; sole of heel without prominent scales.** Underparts usually gray washed with orange, not sharply demarcated from sides, may also be white. Young like adults, with orange on sides. **Mammae pairs:** 1 pectoral, 1 postaxial, 2 abdominal, 1 inguinal = 10.
Species. There are about four species, two in rainforest regions: *H. sciureus* and *H. brasiliensis.* More species are likely to be described from this group. Some animals from Bolivia have pure white bellies.
Similar species. See water rats (*Nectomys* spp.).
Natural history. Nocturnal and possibly diurnal; terrestrial; solitary. Feed on grass and other green plants. Marsh rats are not found in forest but frequent wet grasslands, marshes, and cultivated areas within the rainforest region. They make oval, two-chambered nests of woven strips about 30 by 20 cm, either on the ground in crevices or under brush piles, or among grass or cane stems up to 2 m above ground. Marsh rats are prodigious breeders, and in some agricultural areas they undergo periodic plague-

Map 174

Marsh rats, *Holochilus* spp.

like outbreaks and become serious agricultural pests in rice, sugarcane, and oil palm plantations.
Geographic range. South America: east of the Andes from Colombia and Venezuela south to N Argentina and Uruguay: *H. sciureus* from the Amazon Basin, and *H. brasiliensis* from SE Brazil south to Argentina. To about 2,000 m elevation.
Status. Widespread and often common.
References. Twigg, G. I. 1965. Studies of *Holochilus sciureus berbicensis,* a cricetine rodent from the coastal region of British Guiana. *Proc. Zool. Soc. Lond.* 145:263–83.

Mexican Deer Mouse
Peromyscus mexicanus
Plate 22, map 175
Identification. Measurements: HB = 116–131; T = 115–135; HF = 25–28; E = 18–22; WT = 29–57 g.
Upperparts brown, tawny in older individuals, slate in younger; **midback dark, paling to cinnamon brown on sides** in tawny individuals; **fur long and soft,** slate gray at base on all of body. **Eye surrounded by dark ring; nose at base of whiskers with darker spot; ears very large,** rounded, gray-brown; **whiskers long, reaching shoulder. Tail slightly longer than head and body, naked, bicolored, gray-brown above, with paler blotches below. Hindfeet narrow and delicate, dark to ankles, then whitish to toes. Underparts pale gray washed with dirty white or yellowish, quite sharply demarcated from sides.** Mammae pairs: 1 postaxial, 1 abdominal, 1 inguinal = 6. Young

Map 175

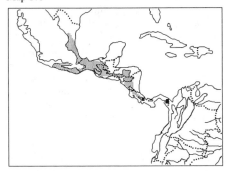

▨ Mexican deer mouse, *Peromyscus mexicanus*
■ Isthmus rats, *Isthmomys* spp.

gray-brown with whitish belly. This is the largest deer mouse in the region.
Variation. Animals from higher elevations are generally darker and more blackish on the back, with longer fur. Those from the lowlands are more tawny. There are about 30 species of deer mice in Mesoamerica; most occupy either high elevations above 1,000 m or habitats drier than rainforest.
Similar species. Rice rats (*Oryzomys* spp.) usually have tail slightly shorter than head and body, smaller ears, no black eye ring, 8 mammae, and often dusky feet or short, fine whiskers that do not reach ear tip.
Natural history. Nocturnal; terrestrial. Feeds on fruit and seeds and makes caches of seeds in crevices. This is chiefly a montane and cloud forest mouse. Found in evergreen forests and thickets, especially near watercourses.
Geographic range. Central America: Mexico, Oaxaca and Veracruz south to the highlands of Costa Rica and Panama. To about 2,400 m elevation.
Status. Uncommon in lowland rainforest; can be very common in humid forests above 1,000 m.
References. Hall, E. R., and W. W. Dalquest. 1963. Mammals of Veracruz. *Univ. Kans. Pub. Mus. Nat. Hist.* 14:165–362.

Isthmus Rats
Isthmomys spp.
Plate 22, map 175
Identification. Measurements: HB = 130–175; T = 174–214; HF = 31–37; E = 24–29.
Upperparts cinnamon brown, sometimes nearly orange, or top of neck and/or shoulders gray-brown; sides paler, bright cinnamon; fur long and soft, dark slate gray at base over whole body. **Ears large, naked;** whiskers long, reaching shoulder. **Tail much longer than head and body, robust, naked, bicolored pale below, dusky above. Feet dusky brown to below wrist or ankle, sometimes to base of toes, then pure white to toe tips, with a sharp straight line across foot where color changes;** hindfeet quite broad and robust. Underparts gray heavily washed with white or yellowish white; **chest often orange.** Young dark brown with velvety fur; can be recognized by their bicolored feet. Large, mouselike rats.
Species. The two species, *I. flavidus* and *I. pirrensis,* are quite similar externally but are separated geographically. The position on the foot of the sharp change from dusky to white varies individually; in a few individuals the line is indistinct.
Similar species. Mexican deer mice *(Peromyscus mexicanus)* have dark eye rings, narrow feet, and shorter tails; vesper rats *(Nyctomys sumichrasti)* have hairy tails, dark eye rings, and snow-white underparts; arboreal rice rats *(Oecomys* spp.) have smaller ears and no sharp line separating colors on feet; rice rats *(Oryzomys* spp.) are smaller and do not have bicolored feet. There are no other large, soft-furred, cinnamon-colored rats where they occur.
Natural history. Terrestrial. Feed on plants and insects. These rats are found under and around fallen logs and tree roots. Occur in evergreen montane forests and cloud forests from about 500 to 1,500 m, most often above 1,000 m.
Geographic range. Central and South America: Panama and Colombia: *I. flavidus* from Chiriquí, Panama, only; *I. pirrensis* around Cerro Pirre, including the Colombian side of the Serrania del Darién.
Status. Each species is uncommon in a small geographic range.

Harvest Mice
Reithrodontomys spp.
Plate 23, map 176
Identification. Measurements: HB = 43–100; T = 66–142; HF = 16–25; E = 12–19; WT = 8–29 g.
Upperparts tawny brown to warm fulvous brown, darkest on midback; sides of head, neck, and body paler, often orange or

Map 176

Harvest mice, *Reithrodontomys* spp.

cinnamon; fur usually long and soft, dark gray at base. **Ears** relatively **large, lightly haired to tips** behind; **whiskers long, several reaching to shoulder;** eyes large; muzzle pointed; **upper incisors grooved.** Tail longer than head and body, lightly haired, solid or bicolored dark above, paler below. Underparts pure white, white with gray base, gray, or gray washed with orange. Hindfeet usually dusky above, with white toes. **Mostly very tiny, delicate mice.** Mammae pairs: 1 postaxial, 1 abdominal, 1 inguinal = 6. *Species.* The are about 20 species; only a few enter the lowland rainforest region. *R. mexicanus,* from Mexico to N Ecuador, is warm fulvous; *R. sumichrasti,* from Mexico to Panama, is tawny brown; *R. paradoxus,* known only from Nicaragua, is ochraceous brown; *R. creper,* from elevations generally above 1,000 m, is blackish brown with red tints and sometimes has a white or spotted tail tip.
Similar species. The following species only partially overlap the geographic range; none have grooved incisors: bicolored arboreal rice rats *(Oecomys bicolor)* are reddish with white bellies, usually weigh more than 20 g, and have short, broad, pink-soled feet; pygmy rice rats *(Oligoryzomys* spp.) have grayish heads and necks; house mice *(Mus musculus)* have tails the same length as head and body.
Sounds. These mice have an extraordinary song of high tonal whistles. They may sing at night from the canopy of fruit trees.
Natural history. Nocturnal; terrestrial and arboreal; solitary. Feed on small seeds, fruit,

and browse. Harvest mice climb well and may be seen running around in the undergrowth above the ground. Some use the forest canopy. On the ground they use networks of runways. Their nests are round balls of grass or fibers, 13–20 cm in diameter, with a small chamber inside. Nests are hidden at ground level under rocks, logs, old boards, and such, or may be above ground, hidden in a small tree or shrub. These mice are characteristic of temperate lowland or montane tropical habitats. In the rainforest region, most are found above 1,000 m, but a few species occasionally live lower. They are most common in overgrown secondary habitats such as old fields, dense roadside weeds, coffee plantations, and orchards, but they may sometimes be found in evergreen forests.
Geographic range. North, Central, and South America: Canada south through Central America to the Andes of Colombia and Ecuador. To 3,100 m elevation.
Status. Common to rare.
References. Jones, J. K., Jr., and H. H. Genoways. 1970. Harvest mice (genus *Reithrodontomys*) of Nicaragua. Occas. Pap. W. Found. Vert. Zool., no. 2.
    Hooper, E. T. 1952. A systematic review of the harvest mice (genus *Reithrodontomys*) of Latin America. Misc. Pub. Mus. Zool. Univ. Mich., no. 77.

Black or Roof Rat
Rattus rattus
Plate 24
Identification. Measurements: HB = 140–200; T = 165–235; HF = 35–40; E = 20–30; WT = 100–240.
There are two colors: (1) **upperparts tawny yellow-brown streaked with black and usually white hairs;** underparts pure white, yellow-white, yellow-brown, or gray; **or** (2) **upperparts black or dark gray streaked with white hairs, especially on rear half of body;** underparts gray. Both forms have **fur long and coarse, but not spiny** or bristly, quite **sparse,** not dense or velvety; a few long guard hairs often protrude to give bristly look; **hairs pale gray at base.** Ears medium, naked; **whiskers long and strong, reaching shoulders. Tail longer than head and body, robust, naked and scaly,** with short, stiff hairs. Feet robust and broad. Young brown or dark gray. **Mammae pairs:** 1 pectoral,

1 postaxial, 2 abdominal, 1 inguinal = 10 (sometimes more).

Variation. The number of black and white hairs mixed in the fur is variable, making individuals more blackish, brownish, or yellowish.

Similar species. Norway rats *(Rattus norvegicus)* are larger (usually more than 200 g), have tails shorter than head and body, short whiskers, and short ears; water rats *(Nectomys* spp.) and marsh rats *(Holochilus* spp.) have soft, dense fur, usually with a gloss or sheen both dorsally and ventrally, ears hairy at base or entirely hairy, short whiskers, paddle-shaped hindfeet with a fringe of hairs along sides; spiny rats *(Proechimys* spp.) have bristly or spiny fur, oblong ears, bicolored tails pale below, and long narrow hindfeet with black soles, and are often glossy chestnut.

Sounds. Gallops and gnaws noisily in the roofs of houses at night.

Natural history. Nocturnal; mostly terrestrial. Feeds on grains, garbage, fruit, carrion, and almost anything remotely edible, such as soap, candles, or leather. This introduced species was probably brought to the New World by the European exploring ships. It is a serious pest worldwide. In the rainforest region it is found only in and around permanent human settlements. It does not appear to invade Neotropical rainforests, although it does occupy such forests in other regions. In some parts of the world, black- and brown-colored forms occupy different habitats. Often found in buildings and granaries. These rats carry disease and often destroy native vertebrate species. They should be eliminated whenever possible.

Geographic range. Worldwide.

Local names. Rata negra, rata doméstica, rata casera (Span); rato-doméstico, caseiro (Br); zwarte rat, ton-alata (Su); rat noir (FG).

Norway Rat
Rattus norvegicus
Plate 24

Identification. Measurements: HB = 196–270; T = 170–235; HF = 41–49; E = 18–24; WT = 200–505 g.

Upperparts tawny yellow-brown streaked with black and/or white hairs; blackest streaks on head and midback; fur texture like black rat. Ears short, naked; whiskers medium length, reaching ear but not shoulder. Tail shorter than head and body, robust and scaly, with short, stiff hairs, sometimes indistinctly bicolored dark above, pale below. Feet large and broad, white above, usually with pink soles. Underparts gray washed with yellow or whitish. Young brown. Mammae pairs: 1 pectoral, 1 postaxial, 2 abdominal, 2 inguinal = 12. Very large, robust, ugly rats.

Similar species. See black rat.

Sounds. Squeals and often fights with other rats.

Natural history. Nocturnal and diurnal; terrestrial. Diet like black rat; also will sometimes bite sleeping humans or livestock. An introduced pest that arrived in the Americas much more recently than the black rat. This rat lives in large colonies and has a complex social organization. It does not invade the Neotropical rainforest but is mainly found in cities and towns, where it reaches high numbers in sewers, canals, waterfronts, fish markets, and other wet and trashy places. It often lives in complex burrow systems in gardens and parks and likes drainage pipes. It can often be seen by day, dashing along a gutter and diving down a drain or scuttling from one hiding place to another. Extremely aggressive, a carrier of disease, and a danger to native fauna, it should be eliminated when possible. Much less common than the black rat in farms, country towns, and plantations.

Geographic range. Worldwide.

Local names. Rata de alcantarilla, rata doméstica (Span); rato-doméstico, caseiro (Br); bruine rat, rioolrat, ton-alata (Su).

House Mouse
Mus musculus
Plate 23

Identification. Measurements: HB = 66–108; T = 60–92; HF = 14–20; E = 11–17; WT = 10–21 g.

Upperparts uniform tawny to slate brown, paling to tawny or grayish sides; fur soft. Ears large, rounded, naked; whiskers fine, reaching to base of ear. Tail slender, naked, about the same length as head and body. Underparts not sharply demarcated from sides, same color as sides, gray, or gray washed with whitish orange. Hindfeet narrow, dusky or whitish. Mammae pairs: 1 pectoral, 1 postaxial, 2 abdominal, 1 inguinal = 10. Tiny mice.

Similar species. Pygmy rice rats *(Oligoryzomys* spp.) have longer feet, and with one exception they and harvest mice

(Reithrodontomys) have tails much longer than head and body. See those species for other comparisons.

Natural history. Nocturnal; terrestrial but climbs well. Feeds mainly on grains and cereals stored by humans, such as rice and corn, supplemented by other stored foods and insects. These mice are introduced pests. In the rainforest region, house mice are rare and restricted to the vicinity of buildings. Most small mice in houses in isolated camps, settlements, and Indian villages in the forest are native species, such as pygmy rice rats and bicolored arboreal rice rats. House mice build a round nest of fibers. Areas contaminated by house mice and trapped individuals have a strong, musky, "mousy" odor stronger than the odor of native species. In agricultural areas they may be found in grain fields and hedgerows as well as buildings. They cause much damage to stored grain.

Geographic range. Worldwide.

Local names. Ratón casero (Span); cachita (Br); huismuis, moismoisi (Su).

Hystricognath Rodents (Suborder Hystricognathi)

Dental formula: I1/1, C0/0, P1/1, M3/3 = 20. The hystricognaths include most of the world's largest rodents and are all relatively large (120 g to 60 kg). Their distinctive features include an outward-flaring structure of the jaw to which the masseter muscles are attached. Externally, they usually have relatively large heads and large eyes, long, stout whiskers, and relatively small ears; many have bristly or spiny fur (many others do not). Most have a sheet of muscle tightly attached under the skin of the back that raises the spines, bristles, or fur. Most have some lateral mammae, on the sides of the body rather than the belly. They have long gestation periods and few young in a litter (usually one to three), which are born with eyes and ears open, fully furred, and able to run around. Many species are grazers or browsers. This suborder of rodents underwent a large evolutionary radiation in South America while it was isolated from other continents. Most living hystricognaths are found in Central and South America; only the porcupines have spread naturally to North America. A few relict genera live in the Old World. The species that occur in the rainforest region can be divided into three main groups: the porcupines, the large cavylike rodents, and the spiny rats. There are 13 families, about 40–45 genera, and 200 or more species of hystricognaths in the New World. It is not yet clear how many species there are in many genera.

Porcupines (Erethizontidae)

All the rainforest porcupines are alike in body shape and many physical features. Their upperparts are covered with thick, sharp, stiff, dangerous spines that are circular in cross section and have barbed tips. The spines detach easily when touched (porcupines cannot "throw" their spines) and embed themselves in the skin of an enemy; the barbs then cause them to work inward into the flesh. The feet all have four large toes with long, strong, curved claws; the hindfoot also has a reduced hallux. The feet are highly modified, with a broad naked pad that flares sideward and covers the area where the first digit would be; the toes and claws flex in a tight pincerlike grip over this pad. The tail is muscular and prehensile, thick at the base and tapering to a thin, flexible tip that is naked on its upper surface and has stiff bristles on its underside. The tail is wrapped "backward" in a spiral twist around branches. The tail is not lashed in defense like that of North American porcupines. The nose and muzzle are greatly swollen and bulbous, naked and pinkish. The head is relatively small and round, with small eyes and inconspicuous ears buried in the spines on the head. The body is thick and heavy, with short legs. Some small dwarf species have long, soft, woolly hair between or over the spines; others do not. Porcupines give birth to a single young. The young of all species are apparently hairy as juveniles and subadults; young of large hairless species are therefore easily confused in the field with hairy dwarf porcupines. In this book species that have no soft fur are described first, followed by hairy species.

The porcupines need taxonomic revision; in particular, no two authorities treat the hairy dwarf porcupines (often placed in their own genus, *Sphiggurus*) in the same way. The accounts in this book are descriptions of specimens from known areas, including many type specimens. The animals described occur in the regions noted, but the scientific names and species delimitations given here may change when the systematics is better understood. In particular, all hairy porcupines from Paraguay and from Bahia to Rio Grande do Sul, Brazil, may intergrade and be a single, extremely variable species. There has been no in-depth field study of any species, and for most there is no published natural history information at all. Many of the dwarf species are known from only a few specimens collected long ago. All Neotropical porcupines except two are described in this book, including two high-elevation species. There are several other forms *(C. richardsoni, C. sanctaemartae, C. centralis)* that may be valid species, and several specimens in collections seem to be undescribed forms. The family is restricted to the New World and includes 4 genera and about 15 species.

Bristle-spined porcupines *(Chaetomys subspinosus)* are morphologically distinct animals with no close living relatives. They are now generally thought to belong with the true porcupines (Erethizontidae), although they also share some features with the spiny rats (Echimyidae). There is one species.

Brazilian Porcupine
Coendou prehensilis
Plate 28, map 177

Map 177

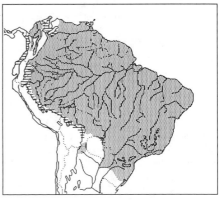

Brazilian porcupine, *Coendou prehensilis*
Bicolor-spined porcupine, *C. bicolor*

Identification. Measurements: HB = 444–560; T = 330–578; HF = 80–95; E = 20–29; WT = 3.2–5.3 kg.
Upperparts covered with strong barbed spines, many spines tricolored with white or pale yellow base and tip, black or dark brown center, colors of roughly equal width; general color black or dark brown heavily overlaid with white or yellowish, from a distance animals look gray or yellow; no soft fur between spines; spines on head, legs, and tail short; spines on back long and thick. Head round, face usually white; nose and lips large, soft, bulbous, pink; ears small, inconspicuous, lost in spines of head; whiskers stout and long, reaching shoulder; eyes small, black, eyeshine faint dull red. Tail shorter or longer than head and body, robust, prehensile, whitish at base, with short spines except distal one-third of top surface naked; tail often curled around branches in an up-curling coil or a twisted, longitudinal spiral. Feet with broad, expanded pads opposing four long, strong, inward-flexing claws; pale gray-brown above. Underparts covered with short, soft spines, pale gray-brown or whitish. Young hairy, with red-brown or brown, long, soft fur partially covering spines. Strong, peculiar odor. A large, whitish porcupine.
Variation. Light parts of spines vary from white to bright yellow, dark parts from brown to black. Animals from the western Amazon Basin in Peru may be dark, with small white tips on spines.
Similar species. Bicolor-spined porcupines *(C. bicolor)* are blackish overall; tail always

longer than head and body; underparts spiny. Black dwarf porcupines *(C. koopmani)* are smaller and blackish; an undescribed whitish dwarf porcupine from Ecuador is much smaller and lacks an inflated nose. Other dwarf porcupines have soft fur covering most of spines as adults. Bristle-spined porcupines *(Chaetomys subspinosus)* have short, soft, brown spines without white tips.
Sounds. No calls are usually heard, but porcupines rattle the leaves when they travel through the trees and drop a patter of fruit fragments when they eat. Social calls include whines and a long, moaning cry of low amplitude.
Natural history. Nocturnal; arboreal; solitary. Feeds on seeds from immature fruits, green or ripe fruits, palm fruits, and occasionally bark and probably leaves. They appear to favor old second growth or

exceptionally viny areas of forest, where they use the middle and upper levels of the vegetation, but they can also be seen in the high canopy of tall, open, mature forest. They do not jump and must descend to the ground to cross a gap between trees. They are usually slow-moving, often immobile and difficult to spot, but they nonetheless are surprisingly agile and can climb swiftly. Their tightly gripping feet and tail allow them to negotiate amazingly thin vines and branches. By day they den in hollow trees (which smell strongly of their odor) or crouch, head between the forelegs, in a shaded place in the branches of a tree. Found in mature and secondary rainforest, deciduous forest, gallery forest, and gardens and plantations.

Geographic range. South America: east of the Andes in all countries of the Amazon Basin, to Paraguay and N Argentina, and Trinidad. To 1,500 m elevation.

Status. Rare in many areas but common, dominant mammals in others. Widespread, sometimes hunted for meat.

Local names. Puerco-espín (Span); ouriço, porco-espinho (Br); erizo (Co); boomstekelvarken, djiendjamaka (Su); cuandu (FG).

References. Montgomery, G. G., and Y. D. Lubin. 1978. Movements of *Coendou prehensilis* in the Venezuelan llanos. *J. Mammal.* 59:887–88.

Charles-Dominique, P., M. Atramentowicz, M. Charles-Dominique, H. Gérard, A. Hladik, C. M. Hladik, and M. F. Prévost. 1981. Les mamifères frugivores arboricoles nocturnes d'une forêt guyanaise: Interrelations plantes-animaux. *Rev. Ecol.* 35:341–435.

Bicolor-spined Porcupine
Coendou bicolor
Plate 28, map 177
Identification. Measurements: HB = 378–493; T = 460–540; HF = 92–100; E = 27–30; WT = 3.4–4.7 kg.
Generally like Brazilian porcupine except: **upperparts blackish; shoulders and back black or dark brown or speckled with some yellow or white tips of spines;** face, sides, tail, and legs usually speckled with pale yellow or white, but may be entirely blackish or dark brown; **large spines on back relatively narrow, often bicolored white for basal half, black or brown for distal half,** pale tip, if present, much shorter than black

section of spine. **Tail longer than head and body.** Feet and underparts pale gray-brown, mixed with white on feet; underparts covered with stiff spines.

Variation. The number of pale-tipped spines varies from none to many. If *C. rothschildi* is a valid species, the *C. "bicolor" richardsoni* west of the Andes may be that species.

Similar species. See Brazilian porcupine (*C. prehensilis*). There is another, distinctive, black porcupine, *C. quichua*, in the Andes of Ecuador. It is the same color as the bicolor-spined porcupine but is smaller, with a short tail (about 180–200 mm measured on skins), conspicuously slender spines, and short whiskers about reaching ear; at close range, sparse, soft black hair can be seen to be sprinkled throughout the dorsal spines (map 179).

Natural history. Nocturnal; arboreal. Adult pairs have been found denning together in tree holes. Found in mature and disturbed lowland and montane rainforest.

Geographic range. South America: east of the Andes in the Andean foothills of Colombia, Ecuador, Peru, and Bolivia. To 2,500 m elevation.

Status. Can be locally common; widespread.

Local names. Puerco-espín (Span); casha cuchillo, erizo, pululo (Pe); puchan (Ec).

Rothschild's Porcupine
Coendou rothschildi
Plate 28, map 178
Identification. Measurements: HB = 332–438; T = 260–413; HF = 61–78; E = 23–26; WT = 2–3 kg.
Similar to bicolor-spined porcupine but somewhat smaller: **upperparts black uniformly speckled with small white tips on some spines; entirely spiny, with no fur covering spines; spines mostly bicolored white at base and black for distal half, some spines with a white tip much shorter than black part. Tail 70–90% of head and body length,** black or speckled with white on sides, black below. Underparts gray-brown. Feet dark brown. Young with spines covered by soft hair.
Variation. Degree of speckling with white varies from almost none to considerable. Sometimes considered a subspecies of *C. bicolor,* but also resembles *C. quichua.* See comments for *C. bicolor.*

Map 178

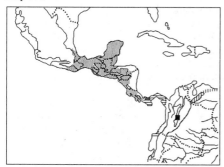

Map 179

▦ Mexican hairy porcupine, *Coendou mexicanus*
▤ Rothschild's porcupine, *C. rothschildi*
▥ Frosted hairy dwarf porcupine, *C. pruinosus*
■ Brown hairy dwarf porcupine, *C. vestitus*

▦ Black-tailed hairy dwarf porcupine, *Coendou melanurus*
▤ Black dwarf porcupine, *C. koopmani*
▥ Paraguay hairy dwarf porcupine, *C. paragayensis*
▥ Bahia hairy dwarf porcupine, *C. insidiosus*
◩ Orange-spined hairy dwarf porcupine, *C. spinosus*
■ Quichua hairy dwarf porcupine, *C. quichua*

Similar species. These are the only heavily spined animals in Central America with no fur covering spines in adults. Mexican hairy porcupines *(C. mexicanus)* have soft, blackish hair, pale spines showing on head.
Natural history. Nocturnal; arboreal. Feeds on fruit and perhaps leaves. Found in lowland evergreen forests, secondary forests, and cultivated areas.
Geographic range. Central America: Panama (but see comments for *C. bicolor*).
Local names. Puercoespín, gato de espinas (Pn).

Black Dwarf Porcupine
Coendou koopmani
Plate 28, map 179
Identification. Measurements: HB = 319–364; T = 280–370; HF = 59–75; WT = 950 g.
Upperparts entirely spiny; no fur covering spines; black overall with yellow or white patches where base of spines visible, especially on head and forequarters; spines may have small white or pale red-brown tips, giving finely speckled appearance; if spine tips are red-brown, these are concentrated around the shoulder and midback; spines generally white or yellow at base, black for two-thirds of distal part, pale tip if present is short. Spines on rump above base of tail short and upstanding, black, with sharp tips worn off (area is black). Ears with naked area behind. Tail 80–90% of head and body length, prehensile, black. Feet black. Underparts dark brown or gray-brown; covered with soft hairlike bristles. Young with rust-red hair

between spines. **A very small, blackish porcupine.**
Variation. Amount of pale tipping on spines varies. A specimen of dwarf hairless porcupine from Ecuador is quite similar, but has long, fine white tips on spines, so that the animal is heavily speckled with white or yellow overall; its nose is not swollen, and it has bright, red-orange eyeshine. It may be an undescribed species, or closely related to the black dwarf, quichua, or black-tailed hairy dwarf porcupines.
Similar species. Brazilian porcupines *(C. prehensilis)* are much larger and have thicker, heavily white-tipped spines and look gray overall; giant tree rats *(Makalata grandis)* have no spines, nonprehensile tails, nonbulbous noses, and gold-black backs.
Natural history. Unrecorded. The short, worn spines at the tail base suggest that this animal rests with its rump wedged against a surface (this feature is not present in all individuals and is also found in several other species). Known only from lowland evergreen rainforest areas.
Status. Unknown; apparently much rarer than Brazilian porcupine, with which it occurs. Known from fewer than 50 specimens.
Geographic range. South America: Brazil south of the Amazon from the Rio Madeira east to Belém.

Local names. Cuandu, ouriço.
References. Handley, C. O., Jr., and R. H. Pine. 1992. A new species of prehensile-tailed porcupine, genus *Coendou* Lacépède, from Brazil. *Mammalia* 56:237–44.

Black-tailed Hairy Dwarf Porcupine
Coendou melanurus
Plate 28, map 179
Identification. Measurements: HB = 280–380; T = 220–363; HF = 52–75 (su); E = 7–24; WT = 1.5–2.4 kg.
Upperparts clothed with soft black hairs overlaid with scattered long, thin, tricolored bristles with long yellow-white tips; fur pale at base, thickest over shoulders, rump with thick yellow spines emerging from fur, or fur may completely cover all dorsal spines. Head finely grizzled, whiskers fine. Eyeshine fairly bright orange. Tail long, about 80% to longer than head and body, pitch black except above base, where colored like back. Legs and feet grizzled dark gray-brown or blackish. Underparts pale gray-brown frosted with whitish.
Variation. This species is sometimes incorrectly called *C. insidiosus.*
Similar species. These seem to be the only hairy dwarf porcupines in their range.
Natural history. Found in lowland rainforest habitat. Probably uses the high forest canopy, where it is difficult to detect.
Geographic range. South America: north of the Amazon in the Amazon Basin of Colombia, probably Venezuela, the Guianas, and Brazil.
Status. Widespread but known from only a few specimens. A recent animal rescue at a flooded dam site found over 100 individuals, implying that the species is common, but extremely cryptic.
Local names. Ouriço cacheiro (Br); listig stekelvarken; djiendjamaka (Su).

Bahia Hairy Dwarf Porcupine
Coendou insidiosus
Plate 28, map 179
Identification. Measurements: HB = 290–350; T = 180–222; HF = 50; E = 16.
Entirely covered with thick, soft, smoky brown fur that completely hides spines except on head, fur unicolored to base or base faintly darker; thick, sharp, hidden spines of back bicolored, basal half whitish, distal half dark brown, or faintly tricolored

with tips faintly orange. Face dark brown with some white showing at bases of spines; whiskers black. Tail short, 50–75% as long as head and body, above haired like body for basal third, dark brown-black for middle third, naked for terminal third. Feet dark brown. Underparts dark to medium brown.
Variation. Pale, almost white or smoky gray animals from NE Brazil probably belong to this species. A gray animal resembling *C. insidious* from Espírito Santo has dark hair base, white-tipped whiskers, a long tail, and faintly tricolored spines (characters of *C. spinosus*); this suggests that the forms may intergrade.
Similar species. Orange-spined hairy dwarf porcupines *(C. spinosus)* have black fur frosted with dirty yellow, and strongly orange-tipped spines.
Natural history. Has been found at over 900 m elevation in Bahia.
Geographic range. South America: Brazil, Bahia and probably Espírito Santo.
Status. Unknown.
Local names. Ouriço cacheiro.

Orange-spined Hairy Dwarf Porcupine
Coendou spinosus
Plate 28, map 179
Identification. Measurements: HB = 300–538; T = 200–378; HF = 46–70; E = 11–25.
Upperparts except head entirely covered with soft black fur with long pale yellow or dull pale orange tips, looks washed overall with those colors, or with little or no hair; spines on head and shoulders tricolored with yellow base, black center, orange tip, sometimes with a fourth black band at tip; spines on rump bicolored, pale at base with short black tip, spines long, to 3.5–6 cm; orange spines visible on head. Whiskers long, reaching behind ear, bicolored dark at base, pale at tip. Tail long or short, 41–100% as long as head and body, rusty orange below and on sides for basal half, naked part of top tip long (10 cm). Feet colored like back. Underparts thickly or thinly furred, with no spines, dirty yellow-orange with sharply demarcated contrasting black or dark brown hair base.
Variation. Extremely variable (or possibly more than one species is involved): animals

from Parana, Brazil *(C. v. roberti),* can be almost hairless or densely haired; they have short black whiskers. An animal from Santa Catarina is dark brown, with only faint yellow tipping of hairs, and long spines (6 cm) on midback. The name *C. villosus* is a synonym.

Similar species. See Bahia hairy dwarf porcupine *(C. insidiosus).* Paraguay dwarf porcupines *(C. paragayensis)* have a cape of long, yellow, exposed spines over head and shoulders.

Natural history. From Atlantic coastal forests.

Geographic range. South America: E Brazil, Rio de Janeiro and probably Minas Gerais, to Rio Grande do Sul. To 1,150 m elevation.

Status. Apparently uncommon.

Local names. Ouriço cacheiro.

Paraguay Hairy Dwarf Porcupine
Coendou paragayensis
Plate 28, map 179

Identification. Measurements: HB = 288–340; T = 228–260; HF = 53–65.
Upperparts blond; strong, pale yellow spines showing on whole back, top of head to shoulders with a "cape" of long (5–6 cm on shoulder) spines, thinly overlaid with blond-tipped soft hairs; rump dark brown-black, with short, black, upstanding spines hidden in dark, yellow-tipped fur. Tail long, 73–90% of head and body length, yellow at sides and below base, dark brown at tip. Feet brown grizzled with pale yellow. Underparts without spines, hair gray-yellow with contrasting dark brown base. A small to tiny dwarf porcupine.

Similar species. See orange-spined hairy dwarf porcupine *(C. spinosus).*

Geographic range. South America: eastern Paraguay.

Brown Hairy Dwarf Porcupine
Coendou vestitus
Plate 28, map 178

Identification. Measurements: HB = 290; T = 130.
Upperparts thickly covered with soft, warm smoky brown hair covering spines except on head and face, where white-tipped spines emerge through hair. Long, thin bristles emerging through fur are bicolored pale for basal half, brown for

distal half, which blends almost invisibly with dorsal fur; spines on back long, to 3.7 cm, pale yellow or white for most of length, with short, dark brown tips; spines on face thick, sharp, and robust. Whiskers fine. Tail short, about 50% of head and body length. Feet colored like back. Underparts thickly haired, with no spines, warm brown, slightly paler than back.

Similar species. See frosted hairy dwarf porcupine *(C. pruinosus).* Andean porcupines *(Echinoprocta rufescens),* found only at high elevations, are largely hairless, with red-brown spines and a slender tail.

Natural history. Has been collected at about 2,600 m elevation, but is said to be found in the warm lowlands.

Geographic range. South America: Colombia, Cudinamarca, the upper Magdalena Valley west of Bogotá (San Juan de Rio Seco), and possibly the lower eastern Andean slopes. Poorly known.

Status. Apparently extremely rare, known from only a few individuals; the region where the specimens described above were collected has been almost entirely deforested.

Frosted Hairy Dwarf Porcupine
Coendou pruinosus
Plate 28, map 178

Identification. Measurements: HB = 330–380; T = 190; HF = 50–60.
Upperparts thickly haired with soft, gray-brown, almost blackish fur that completely hides spines of body; fur sprinkled throughout with prominent, long, thin, white-tipped bristles; fur frosted with whitish, especially on sides. Head gray-brown, finely ticked with white of thin white tips of hairs and bristles; whiskers stiff, black. Spines on back very short, less than 2.5 cm, most spines on back bicolored, pale yellow for most of length, with black tips; spines on head and a few on shoulder tricolored pale-dark-pale. Tail short, 50% of head and body length, naked area on top of tip short (≈ 5 cm); base pale yellow or whitish, grizzled with fine white spines, tip and undersurface black. Feet brown, grizzled whitish. Underparts thickly furred, with no spines; fur dark brown at base, heavily frosted with silver, looks whitish with dark brown showing through parts in fur.

Similar species. Brown hairy dwarf porcupines *(C. vestitus)* are warm brown, without white frosting on hairs above or below, and longer spines; black-tailed hairy dwarf porcupines *(C. melanurus)* have a long tail and long spines.

Natural history. Four frosted porcupines were found together in a nest. They nest in hollows in rotten tree trunks and perhaps in the branches of trees. Found at 2,500 – 2,600 m elevation.

Geographic range. South America: Colombia and Venezuela, known only from the Andes near Mérida.

Status. Apparently rare, known from only a few specimens.

Mexican Hairy Porcupine
Coendou mexicanus
Plate 28, map 178

Identification. Measurements: HB = 350 – 460; T = 200 – 360; HF = 63 – 85; E = 19 – 25; WT = 1.4 – 2.6 kg.

Upperparts from neck to tail base thickly covered with long, soft, pitch black or dark brown fur that covers spines; appears black; scattered parts in fur show yellow spines; spines beneath fur of back long (3 – 3.5 cm), bicolored, yellow with short black tip. Head and neck with bright yellow spines showing through thin hair, or thickly haired, hair covering spines; ears buried in fur but marked by tuft of pale whitish brown fur behind ear; whiskers robust and long, reaching to behind ear or to shoulder; eyeshine dull red. Tail 50 – 80% of length of head and body, covered with stiff black bristles below, above like back for basal third, black bristles for middle third, naked for distal third. Underparts with fur and no spines, brown or brown mixed with whitish. Strong, peculiar odor. A small porcupine that looks black with a pale head.

Variation. Animals from Chiriquí, Panama, are very small, dwarf porcupines and have been considered a separate species *(C. laenatus);* some animals, especially from the Yucatán, have sparse fur and many yellow spines on body exposed.

Similar species. Rothschild's porcupines *(C. rothschildi)* have no soft fur (adults) and are black sprinkled with white.

Natural history. Nocturnal; arboreal; adult pairs have been found together in dens. Feeds on ripe and green fruits and seeds and probably browse when fruit is scarce. By day

dens in hollow trees or sits on a tree branch among dense vines or foliage. Most common in montane forest; uncommon in the lowlands. Found in rainforest and deciduous forest.

Geographic range. Central America: Veracruz, Mexico, south to W Panama. To 3,000 m elevation.

Status. Unknown, unlikely to be threatened. Hunted for meat.

Local names. Puerco espín (Span); kixpach och, citam (May).

Bristle-spined Porcupine
Chaetomys subspinosus
Plate 28, map 180

Identification. Measurements: HB = 380 – 450; T = 260 – 275; HF = 67 – 70; E = 9; WT = 1.3 kg.

Upperparts pale to dark brown; head and shoulders densely and evenly covered with short (1.5 cm), sharp, upstanding, kinky spines; back behind shoulder to rump, legs, and tail base thickly covered with long (to 5 cm), stiff, slightly wavy, dry bristles like thin broom straws, tricolored pale yellow at base, dark brown, then pale brownish yellow at tip; no sharp spines on lower back. Head round, ears buried in spines of head, marked by a small tuft of soft brown fur, the only soft fur on entire body; whiskers medium length, reaching ear; muzzle almost naked, brown, nose not greatly swollen and bulbous. Tail prehensile, curling dorsally, 60 – 70% of head and body length, thick at base tapering to slender tip, above like back for basal fifth, then thinly haired with brown bristles, becoming almost naked near tip, below rusty at base, then dark brown to tip. Feet brown, similar to those of other porcupines. Underparts thickly covered with flat bristles, yellowbrown, slightly paler than back. A thickbodied, cylindrical animal with a long back and short legs.

Variation. Color overall varies from dark brown to pale whitish brown, sometimes with white patches.

Similar species. Other porcupines all have sharp spines on lower back; Bahia hairy dwarf porcupines *(C. insidiosus)* have soft, smoky fur covering spines of back.

Sounds. A raucous puffing call.

Natural history. Nocturnal; arboreal. Feeds

Map 180

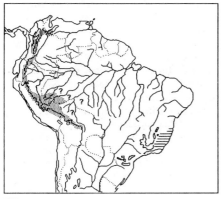

▨ Pacarana, *Dinomys branickii*
▤ Bristle-spined porcupine, *Chaetomys subspinosus*

on fruits, including cacao. Bristle-spined porcupines move slowly and use their tails as do other porcupines. They are excellent climbers that can scale stone walls. Found in mature and secondary Atlantic coastal forests and plantations, especially cacao plantations. *Geographic range.* South America: Brazil, SE Bahia and N Espírito Santo.
Status. US-ESA endangered. Few specimens are known, but apparently healthy populations have recently been discovered.
Local names. Ouriço prêto.
References. Moojen, J. 1952. *Os roedores do Brasil.* Biblioteca Científica Brasileira Ser. A. II.

Large Cavylike Rodents (Agoutidae, Dasyproctidae, Hydrochaeridae, Dinomyidae)

This mixed group of families includes the capybara, agoutis, acouchys, pacas, and pacaranas. All but the last are long-legged, ungulatelike rodents. All have four toes on the front foot; the weight-bearing toes of the hindfoot are reduced to three in capybaras, acouchys, and agoutis, four in pacaranas, and three, with two much smaller side toes, in pacas. The hair is straight, stiff, and bristlelike. All have large heads with bulging jaw muscles, short ears, short to minuscule tails, and cylindrical and sometimes piglike bodies. All have the typical rodent posture of sitting on the haunches; all but the capybara can hold their food in the forepaws. They have precocial young in litters of one or two, except for the capybara, which has litters of up to eight. Most of the genera include only one or two species and do not present taxonomic problems. Agoutis (*Dasyprocta* spp.), however, are highly variable geographically and have not received any modern taxonomic revision. They are here divided into the most generally accepted set of species, but there appear to be intermediate forms where some species meet geographically, and there are many skins that would be difficult to identify without knowing where they came from. There are other species names not used in this book. It is likely that the species limits given here will change when the genus is revised.

Capybara

Hydrochaeris hydrochaeris
Plate 18, map 181
Identification. Measurements: HB = 1,070–1,340; T = 10–20; HF = 220–250; E = 60–70; SH = 500–620; WT = 35–65 kg.
Entirely uniform tan or yellowish brown or rarely dark red-brown; midback sometimes darker than sides; hair coarse and sparse, long on midback. **Head large, rectangular, muzzle square; ears small; eyes small, set high on the head,** giving a supercilious look; eyeshine moderate, eyes look small and yellow or reddish. **Tail rudimen-** **tary, not visible. Feet partly webbed,** four toes on front, three on hindfoot; legs robust. Young like adults but more thickly furred. **Large, stocky, piglike rodents.** Often sit on haunches like a dog. The world's largest rodents.
Variation. Dark red-brown animals (in mixed groups with tan ones) seem to be found more often in rainforest than in open grasslands.
Similar species. Peccaries (*Tayassu* spp.) are black, with a narrow snout and two-toed hooves; pacas (*Agouti paca*) are red with lines of white spots.

Map 181

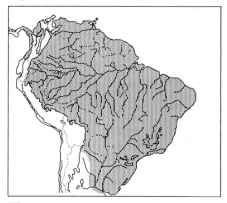

▓ Capybara, *Hydrochaeris hydrochaeris*

Sounds. Sharp yelps in alarm; also call to each other with a variety of small talk (growls, whinnies, twitters) not usually heard from a distance.

Natural history. Naturally diurnal with some feeding at night, but has become nocturnal where intensively hunted; terrestrial; family groups of 2–6, or in open grasslands in herds of up to dozens. Feeds on grass and browse, especially aquatic vegetation. Capybaras are always found near water; in rainforest they are found along rivers and lakes large enough to have open sky above and a margin with aquatic plants or grassy vegetation. The 1–6 young accompany their mother. When alarmed, capybaras usually plunge into the water and swim under the surface to escape. They leave distinctive piles of smooth oval droppings (like giant rabbits) and star-shaped tracks along the water's edge. Found in lowland rainforest, dry forest, scrub, or grassland near water; most numerous in open habitats of seasonally flooded grasslands with permanent water sources (llanos, pantanal), where they are a major prey of jaguar.

Geographic range. Central and South America: Panama to NE Argentina.

Status. Locally common and widespread, but uncommon or rare in Panama and in populated areas of Amazonia; extensively hunted for meat and easily found with dogs. Populations in rainforest are small and narrowly restricted to open watersides. Because rivers are the main travel routes of hunters,

capybaras are no longer seen on river trips in populated areas. Large populations still exist in remote forests and some wet grassland regions (llanos, pantanal). In some places they are not hunted because the meat is said to cause disease. Capybaras are regionally "farmed" for meat and leather, with management of wild herds for sustainable yield; their herbivorous diet, large size, high reproductive rate, and herding behavior make them ideal for this purpose.

Local names. Capibara (Span); carpincho (Ar, Pa); capihuara (Bo, Ec); capivara, cupido (Br); chigüire (Co, Ve); ponche, lancho (Co, Pa); ronsoco, ivéto (Pe); cabiaï, kapoewa (Su, FG).

References. Ojasti, J. 1973. *Estudio biologico del chigüire o capibara.* Caracas: Republica de Venezuela fondo Nacional de Investigaciones Agropecuarias.

Paca
Agouti paca
Plate 29, map 182

Identification. Measurements: HB = 616–775; T = 12–32; HF = 107–130; E = 43–56; SH = 27–32; WT = 5–13 kg.

Upperparts chestnut red or dark brown, sides from neck to rump with three or four lines of large white spots sometimes coalescing into stripes. Hair coarse, sparse, and flat. Head large, cheeks swollen, ears short, brown; whiskers long; eyes large, widely spaced; eyeshine bright yellow to orange. Tail a tiny stump hidden beneath rump hair. Lower cheeks, throat, chest, and underparts white. Forefeet with four toes, hindfeet with three large toes and two tiny toes that usually do not touch the ground. Body heavy, piglike, with large rump. Young like adults.

Similar species. Pacaranas *(Dinomys branickii)* are black, with a conspicuous tail and spots only on the rear of the body; baby tapirs *(Tapirus* spp.) have spotted legs and head; baby deer *(Mazama, Odocoileus* spp.) have thin necks and legs, and distinct tails. Another paca species *(Agouti taczanowskii)* occurs at higher elevations in the Andes from Venezuela to Peru; it has spots on the top of the back and long, dense hair.

Sounds. Pacas walk heavily and noisily in dry leaves. In alarm they usually do not call; rarely they give a hoarse bark or make a deep

Map 182

Paca, *Agouti paca*

rumbling sound by grinding their teeth. The head is modified with expanded cheekbones and pockets in the cheeks that probably act to resonate this sound, which seems to be a social signal.

Natural history. Nocturnal; terrestrial; solitary or rarely in pairs. Feeds on fallen fruits, browse, and some tubers. Pacas are most common near water, from large rivers to small seeps, and in swampy areas and dense thickets, but they are also sometimes seen in open forest far from water. They live in monogamous pairs that share small territories, but they forage alone. By day the male and female usually den separately, in large burrows with a main entrance and several hidden entrances that are plugged with leaves. The burrows are usually in permanently dry banks near water and may be marked by mounds of dirt. At night between bouts of foraging, pacas sometimes lie down to rest in the open. Found in mature, disturbed, and secondary rainforest, montane forest and gallery forest, and gardens and plantations.

Geographic range. Central and South America: SE Mexico to S Brazil and N Paraguay. To 2,000–3,000 m elevation.

Status. CITES Appendix III (Ho). Pacas are the most prized Neotropical game animals for their tender, veal-like meat; they are easily hunted by day with dogs or at night with headlights, and are now scarce or locally extinct in overhunted areas. Where little hunted they are common and easily seen. Because of their huge geographic range and liking for inaccessible swamps, thickets, and second

growth, lowland pacas are not threatened except locally, especially in Central America. Efforts should be made to manage this species for sustainable hunting yield from intact forests.

Local names. Paca (Ar, Br, Su); gibnut (Be); jochi pintado, sari, paca (Bo); guagua, lapa (Co); guanta (Ec); pak (FG); tepezcuintle, haleb (Cent Am); majaz, picuru (Pe); conejo pintada (Pn); water haas (Su); hei (Sar); acutipá (Gua).

References. Marcus, M. 1983. Population density, home range, and foraging ecology of paca *(Agouti paca)* on Barro Colorado Island. Unpublished report, Smithsonian Tropical Research Institute.

Collett, S. F. 1981. Population characteristics of *Agouti paca* (Rodentia) in Colombia. *Pub. Mich. State Univ. Biol. Ser.* 5:489–601.

Pacarana
Dinomys branickii
Plate 29, map 180
Identification. Measurements: HB = 475–513; T = 145–174; HF = 114–123; WT = 13 kg.

Upperparts and legs black to brown-black sprinkled with white hairs; **sides with two wide white stripes grading to rows of spots from behind shoulder to rump, with two disorganized rows of white spots below these. Head large, grizzled gray-black, muzzle blunt; ears short; eyes small;** whiskers long and stout. **Tail about 25% as long as head and body, thick and stumpy, thickly furred,** black or dark brown. Underparts grizzled brown-black variably mixed with gray. Feet with four toes; claws long. **Thick-bodied, short-legged, thick-necked rodents that move slowly and often adopt a sitting posture.** Young like adults.

Similar species. These are the only black animals with white stripes and rows of spots and a conspicuous tail; pacas *(Agouti paca)* are red or brown with no tail.

Sounds. Sounds noted in captivity include foot stamping, tooth chatter, hiss, grunt, and growl.

Natural history. Unknown. Pacaranas are herbivorous, and they are thought to be nocturnal and to live in burrows. One report suggests that they live in family groups including an adult pair and successive litters of young. In captivity they climb well and like to rest on elevated platforms. Found in lowland and montane rainforest.

Geographic range. South America: the eastern foothills of the Andes from Colombia and Venezuela to Bolivia, and the Amazon lowlands of Peru and W Brazil. To 2,000 m elevation.

Status. Appears to be rare throughout its range; hunted for meat.

Local names. Pacarana, paca-com-rabo (Br); guagua loba (Co); machetero, pacarana, picuru maman (Pe).

References. Tate, G. H. H. 1931. Random observations on habits of South American mammals. *J. Mammal.* 12:248–56.

Red-rumped Agouti
Dasyprocta agouti
Plate 29, map 183

Identification. Measurements: HB = 490–640; T = 13–30; HF = 118–148; E = 40–47; SH = 270–360; WT = 3–5.9 kg.

Head and forequarters finely grizzled olivaceous; rump dark red to brilliant orange, covered by long, straight hairs, which overhang rump in a fringe and are usually paler yellow or orange at base; this color visible when the hairs are erect. Top of head, neck, and midback between shoulders sometimes blackish or with crest of longer, pure blackish hairs. Whiskers stiff, black, reaching to base of ear; chin naked; **ears short and roundish, mostly naked; eyes large. Tail naked, a tiny, inconspicuous stub.** Feet dark brown, forefeet with four strong toes, hindfeet with three long, strong toes and hooflike nails. Underparts grizzled brownish orange or orange with white midline. **A large, ungulatelike rodent with long slender legs, large rump, and strongly humped back.** Runs with a bounding, bouncing gait; walks or trots slightly crouching, head held low. In alarm or excitement may raise long rump hairs in a fan.

Variation. Forequarters vary from dull brown-olivaceous to bright orange-olivaceous; some agoutis from Guyana have black crests on neck and shoulders. Those on the Lesser Antilles, Trinidad, and Tobago are generally smaller than mainland animals and are dark brown; on the Lesser Antilles they have long, chocolate brown hairs forming a cape over the neck and shoulders. There are about 10 names (species and

Map 183

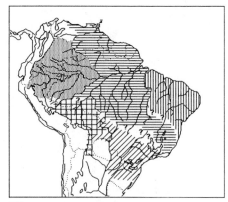

▓ Black agouti, *Dasyprocta fuliginosa*
▤ Red-rumped agouti, *D. agouti*
▥ Black-rumped agouti, *D. prymnolopha*
▨ Azara's agouti, *D. azarae*
▦ Brown agouti, *D. variegata*

subspecies) that have been used for these agoutis. Some recent works use the species name *D. leporina.*

Similar species. Red acouchys *(Myoprocta acouchy)* are smaller, with black rump and a prominent little tail with a white tip. See range maps and descriptions of neighboring agouti species.

Sounds. Alarm often heard: a series of low grunts or loud hoarse squeals or barks; often two individuals call; hindfeet may also be stamped on the ground. Gnaws loudly on hard nuts.

Natural history. Diurnal; terrestrial; often in pairs, sometimes solitary. Red-rumped agoutis feed on fruit and nuts or seeds. They can be found throughout the forest, but most often in areas of good undergrowth cover, around large treefalls, and near streams, rivers, and swampy areas. These agoutis live in monogamous pairs with their offspring on a territory. Pairs travel together; they are wary and are often heard fleeing in alarm but more rarely seen. The best way to observe them is from a blind near a fruit tree, or on roads within forest. They are most active in the early morning and late afternoon and are easiest to approach at dusk, when they either do not see well or are less wary. They may sometimes be active on bright moonlit nights. They bury nuts singly in the ground for retrieval when food is scarce, and this behavior makes them important seed dispersers for many tree species. Found in mature, disturbed, and secondary evergreen, deciduous, and gallery forest, and in gardens and plantations.

Geographic range. South America: the Guianas and Brazil north of the Amazon and east of the Rio Negro, and south of the Amazon east of the Madeira.

Status. Much hunted for meat but usually very common; widespread.

Local names. Cutia (Br); goudhaas, konkoni (Su); picure, acure (Ve).

References. Dubost, G. 1988. The ecology and social life of the red acouchy, *Myoprocta exilis;* comparison with the orange-rumped agouti, *Dasyprocta leporina. J. Zool.* (Lond.) 214:107–23.

Black Agouti
Dasyprocta fuliginosa
Plate 29, map 183
Identification. Measurements: HB = 542–760; T = 20–39; HF = 120–143; E = 36–49; WT = 3.5–6 kg.

Upperparts entirely black, finely grizzled with white (hairs black with tiny white tips); **rump hairs black with inconspicuous white tips on hairs, not usually overhanging rump in fringe;** nape with a slight crest of longer, black hair. Tail black. Feet black. Throat white or strongly grizzled white; underparts dark brown and more or less grizzled with white, midline of belly sometimes white. General shape and posture like red-rumped agouti. Looks black or dark brown from a distance. The largest agouti.

Variation. Some animals from Venezuela have rump hairs heavily frosted with white; some from Venezuela, Brazil, and Peru have hair tips tawny yellow or orange instead of white on foreparts, appearing slightly washed with orange; on close view some animals are dark brown rather than true black.

Similar species. Peccaries (*Tayassu* spp.) are much larger, with thick necks, large heads, and sharply tapering muzzles; acouchys (*Myoprocta* spp.) are smaller and either olivaceous or red with black rump, and have thin, white-tipped tails.

Sounds. In alarm stamps feet and gives a short series of deep grunts or, more rarely, whistlelike calls. Gnaws loudly on nuts.

Natural history. Diurnal; terrestrial; solitary and in pairs. Feeds on fruit and nuts.

Found in mature and disturbed rainforest, deciduous forest, and montane forest.

Geographic range. South America: the western Amazon Basin west of the Rios Negro and Madeira in Brazil, Venezuela, Colombia, Ecuador, and central and N Peru. To above 1,000 m elevation.

Status. Intensively hunted for meat, but persists even close to villages. Geographic range large.

Local names. Cutia (Br, Co); guatín, agutí, picur, conejo negro, ñeque (Co); guatusa (Ec); añuje (Pe); picure, acure (Ve).

Brown Agouti
Dasyprocta variegata
Plate 29, map 183
Identification. Measurements: HB = 445–540; T = 11–38; HF = 94–120; E = 41–45; WT = 3.0–5.2 kg.

Upperparts finely grizzled black and tawny yellow, brown and yellowish, or black and orange; from a distance appears blackish washed with tawny, plain brown, or orange; head often blackish, midline of back usually darker than sides. Feet dark brown except in orange animals, when colored like back. Chin and throat and often midline of venter white; rest of underparts brownish yellow in brown individuals to bright pale orange in orange animals. Shape and posture like red-rumped agouti.

Variation. These agoutis seem to become progressively paler and more orange from north to south and east to west: those from Junín (Peru) are blackish; from the Rio Urubamba, plain brown; from Madre de Dios, yellowish to reddish brown; from Santa Cruz, Bolivia, bright orange, with orange bellies. This group of agoutis is considered by some to belong to the same species as the Central American agouti, by others in a group with the black agouti. It may intergrade with the black agouti to the north and Azara's agouti to the east of its range. The brown and orange forms have also been considered distinct species.

Similar species. See black agouti *(D. fuliginosa).*

Sounds. In alarm rushes away stamping feet and emitting a short series of low grunts or repeated nasal barks, sound between a bleat and a woof. Gnaws loudly on hard nuts, with a slower frequency and deeper resonance than a squirrel.

Natural history. Diurnal; terrestrial; solitary and in pairs. Feeds on fruit and nuts. Brown agoutis favor areas of the forest with dense understory thickets, especially around fallen trees, platanillos in swampy areas, and on watersides. They are wary and are often heard calling in alarm, but they are difficult to see because they will run in alarm many yards from danger. They are most easily approached when they are gnawing on a nut, and are least wary at dawn and dusk, when the light is poor and they seem to venture more out into the open forest or onto forest roads. At night they den under a dense pile of brush, usually a fallen tree with many vines. If the den is disturbed at night (such as by someone walking by), they will rush away calling. Seems most numerous in terra firme forest with many Brazil nut trees or *Attaleya* palms. Found in mature, disturbed, and secondary forest and in gardens and plantations.
Geographic range. South America: east of the Andes from central Peru (Junín) south to Bolivia and N Argentina. To at least 1,500 m elevation.
Status. Common and widespread; hunted for meat.
Local names. Agutí rojizo, akútí (Ar); jochi colorado, jochi calucha (Bo); cutia (Br); añuje (Pe).

Azara's Agouti
Dasyprocta azarae
Plate 29, map 183
Identification. Measurements: HB = 452–575; T = 14–33; HF = 100–120; E = 35–46; WT = 2.4–3.2 kg.
Upperparts gray washed with dull tawny or olivaceous to bright orange, hairs finely banded black, white, tawny yellowish, and orange; **top of head to behind shoulder blackest; midbody broadly washed with orange, flanks and thighs often bright orange, rump hairs black frosted with white or pale orange (looks gray); area around tail orange (viewed from rear).** Sides of head from nose to ear all or partly washed with orange. Legs orange; color ending abruptly at wrist and hindtoes, extremities of feet with blackish "socks." Underparts bright to pale orange, inguinal region often white. Body shape and tail like red-rumped agouti. Young like adults but rump hair short. One of the smallest agoutis.

Variation. This is one of the most distinctive agoutis; the finely banded hair varies from blackish to grayish looking, with orange parts from dull tawny yellow or olivaceous to bright orange. Animals from drier regions (Paraguay, Argentina) tend to be gray or olivaceous.
Similar species. Tapitis(*Sylvilagus brasiliensis*) are smaller, with long ears; green acouchys *(Myoprocta acouchy)* have conspicuous tails.
Natural history. Diurnal; terrestrial. Probably similar to other agoutis. Found in rainforest, cerrado, and chaco; the southernmost agouti.
Geographic range. South America: east of the Andes in a belt from Santa Cruz, Bolivia, east to São Paulo, Brazil, including rainforests of SE Brazil and Mato Grosso, south to N Argentina and Paraguay.
Status. Often common in a large geographic range.
Local names. Agutí amarillento (Ar); jochi (Bo); cutia (Br); acutí (Pa).

Black-rumped Agouti
Dasyprocta prymnolopha
Plate 29, map 183
Identification. Measurements: HB = 450–525; T = 18–30; HF = 95–106; E = 36–43.
Foreparts yellow-orange grizzled with black, becoming dark red-orange from midback aft; rump top covered by contrasting wedge of long, pitch-black hairs, often pale yellow at base, visible when hairs are erect. Crown blackish; neck with crest of longer, black hairs. Feet dark brown. Underparts pale orange sprinkled with white hairs down midline. Young like adults but darker and duller overall. Shape like red-rumped agouti.
Similar species. Red acouchys *(Myoprocta acouchy)* are similar, but smaller, with a thin, white-tipped tail; they may not have any geographic overlap with this species. Red-rumped agoutis *(D. agouti)* have an orange rump.
Natural history. Found in deciduous forest and scrub, including cerrado and caatinga, and probably coastal rainforest habitats.
Geographic range. South America: E Brazil from E Pará south of the Amazon to Bahia. The original locality given for this species was "Guiana"; this was probably erroneous,

since there appear to be no good records from that region.

Local names. Cutia.

Central American Agouti
Dasyprocta punctata
Plate 29, map 184

Identification. Measurements: HB = 480–600; T = 20–55; HF = 120–156; E = 36–47; WT = 3.2–4.2 kg.

There are two basic color patterns: (1) **Upperparts uniformly warm red-brown, yellow-brown, or gray-yellow;** fur banded with fine black striations throughout when viewed closely; **nape and rump hairs not different from rest of upperparts;** chin and inguinal region clear orange or white in gray-yellow animals, chest grizzled like back. (2) **Foreparts brown to blackish, finely grizzled with tawny or olivaceous; crown and nape often blackish; midbody forward of rump with a band of brighter orange-banded hairs; rump hairs long, black, with long yellow to white tips on hair overhanging rump in fringe;** chin and inguinal region whitish; belly brown. Feet of both are black or brown.

Variation. This species is highly variable, with intermediate colors and patterns linking the extremes. Pattern (1) is found on the Pacific slope from Mexico to Panama and Ecuador; gray-yellow animals are from the Yucatán. In pattern (2) the foreparts range from blackish with a paler, more tawny area around midbody just forward of rump, to distinctly orange-olivaceous; some animals from Ecuador and W Colombia have pale yellow frosting on rump, tawny orange foreparts, and necks with grayish sides; color pattern (2) is found along the Atlantic slope from Costa Rica to Panama and Colombia, and in NW Venezuela. Animals from Coiba Island, Panama, have been considered a separate species *(D. coibae);* they are small, pale brown agoutis with color pattern (2).

Similar species. Tapitis *(Sylvilagus* spp.) are smaller, with long ears; capybaras *(Hydrochaeris hydrochaeris)* are much larger and uniform yellow-brown. There are no other medium-sized tailless brown mammals in range.

Sounds. In alarm stamps feet and emits a short series of calls, each consisting of a few low grunts followed by a sharp bark. Gnaws loudly on nuts.

Map 184

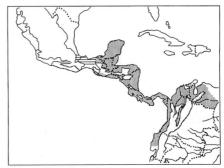

◻ Central American agouti, *Dasyprocta punctata*
▤ Mexican black agouti, *D. mexicana*

Natural history. Diurnal; terrestrial; solitary, or more rarely in pairs. Feeds chiefly on seeds, fruit, and the cotyledons of seedlings, supplemented by fungi, flowers, leaves, and insects. These agoutis favor areas of the forest with dense undergrowth. Although wary where hunted, they are tame and easy to see in some protected reserves with many human visitors. They are most active in the late afternoon just before nightfall. Adults den at night in hollow logs, under fallen brush, or sometimes in burrows. A monogamous pair lives together on a territory, but the couple does not usually travel together. Newborn young live by themselves in a burrow that the mother does not enter: she calls them out for nursing and care. Agoutis bury nuts singly in the ground for retrieval when fruit is scarce, and this activity is important for the dispersal of seeds. Found in mature and secondary lowland and montane rainforest and deciduous forest, and in gardens and plantations.

Geographic range. Central and South America: Chiapas and Campeche, Mexico, southeast through all countries of Central America; NW Venezuela, N and W Colombia, and Ecuador west of the Andes. To at least 1,500 m elevation.

Status. CITES Appendix III (Ho). Intensively hunted for meat and threatened by conversion of forest to pasture, but large populations remain in many areas.

Local names. Indian rabbit (Be); guatusa (CR, Ec); guatín (Co); guaqueque alazán, aguti (Me); ñeque, cuinco (Pn); picure (Ve).

References. Smythe, N. 1978. The natural history of the Central American agouti

(Dasyprocta punctata). Smithsonian Contrib. Zool., no. 257.

Hallwachs, W. 1986. Agoutis *(Dasyprocta punctata):* The inheritors of guapinol *(Hymenaea courbaril:* Leguminosae). In A. Estrada and T. H. Fleming, eds., *Frugivores and seed dispersal,* 285–304. Dordrecht: W. Junk.

Mexican Black Agouti
Dasyprocta mexicana
Plate 29, map 184
Identification. Measurements: HB = 515–560; T = 20–30; HF = 116–127.
Upperparts entirely black or dark brown very finely grizzled with white (hairs have tiny white tips), **except rump hairs long and entirely black,** nape hairs usually long and entirely black, forming a crest. Chin and throat white, rest of underparts dark brown grizzled white or with white stripe down midbelly. Feet entirely black or with pure white toes on hindfeet. Young like adults.
Variation. An animal from Tabasco is grizzled rich brown; possibly a hybrid with the Central American agouti.
Similar species. Peccaries (*Tayassu* spp.) are much larger, with large heads, narrow snouts, and thick necks.
Sounds. Calls in alarm, probably like other agoutis.
Natural history. Diurnal; terrestrial. Found in wet coastal lowland rainforests.
Geographic range. Central America: Mexico, the lowland Atlantic coastal forests of Chiapas, Veracruz, Oaxaca, and Tabasco. To 500 m elevation.
Local names. Cerreti, cuacechi, guaqueque negro.

Red Acouchy
Myoprocta acouchy
Plate 29, map 185
Identification. Measurements: HB = 335–390; T = 51–78; HF = 90–104; E = 25–40; SH = 170–200; WT = 1.05–1.45 kg.
Upperparts dark, chestnut red or orange on sides and legs, grizzled with black and some yellow on crown and neck; **midback and rump glossy black or very dark red; rump hairs long, straight, overhanging rear of body in a straight fringe,** entirely

Map 185

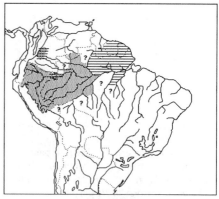

▓ Green acouchy, *Myoprocta pratti*
▥ Red acouchy, *M. acouchy*

dark, not banded. Eyes large, ears quite large, naked, tips high above crown; areas behind ears, around mouth and eyes, and under chin almost naked. Whiskers well developed, black, reaching to behind ear. **Tail short, more slender than a pencil, white below and with a small white tuft at tip, often wagged or flipped or held up, showing the white.** Legs long and thin; forefeet with four toes and a vestigial thumb with a claw, hindfeet with three large, elongated toes with hooflike claws, soles black. **Underparts thinly haired, pale to dark rust-red or orange.** Mammae pairs = 4. A small animal with large hindquarters, humped back, often runs in a slight crouch.
Variation. Some animals are quite grizzled with yellow and appear olivaceous, but they retain the long, dark rump hairs of this species; some hybridization with the green acouchy may have occurred in Colombia in the headwaters of the Río Uaupés (this form has been given the name *M. milleri*). Both red and green acouchys are said to occur in the Serranía de Macarena, Colombia. The red acouchy is sometimes given the name *M. exilis* and the green acouchy the name *M. acouchy.* There seems no good basis for assigning the name *M. acouchy* to the green species.
Similar species. Green acouchys (*M. pratti*) are green-olivaceous, with shorter rump hairs that are distinctly banded with different

colors and are usually the same color as the shoulders or only slightly darker, not black. Red-rumped agoutis *(Dasyprocta agouti)* are much larger, with bright orange rump, and tail a short, thick, naked stump usually not visible. Rabbits *(Sylvilagus* spp.) have long ears, tawny black-streaked backs, and buttonlike tails, and are nocturnal.

Sounds. In alarm stamps hindfeet and emits a squeal or whistle followed by a short series of quickly repeated high-pitched chirps; also a trill-like squawk. Gnaws loudly on nuts.

Natural history. Diurnal; terrestrial; solitary. Feeds on seeds, fruits, and the cotyledons of seedlings. Acouchys bury nuts singly in the forest floor for use when food is scarce. Red acouchys favor zones with dense undergrowth, such as around treefalls. They are nervous and wary and are more often heard calling in alarm than seen, but they generally flee only a few yards and hide in a thicket or behind a log, where they can often be spotted standing motionless. After fleeing in alarm, they will often circle silently and approach a motionless observer from behind. They sometimes run along the tops of fallen logs or sit on them, although they are highly terrestrial and not climbers. They are most active in the early morning and are least wary and most easily seen at dawn or dusk in half daylight. At night red acouchys rest in a nest of leaves within a hollow log or, rarely, in a burrow made by another species; entrances not in use are plugged with leaves. If the den is disturbed at night (e.g., by stepping on the log) they run out, calling in alarm. Found in mature lowland rainforest, not usually in extensive areas of secondary forest.

Geographic range. South America: The Guianas and Brazil north of the Amazon and east of the Rio Negro, and south of the Amazon east of the Tapajós; perhaps an isolated population in Colombia in the Upper Rio Uaupés.

Status. Common and widespread.

Local names. Cutiara (Br); acoechi, maboela (Su).

References. Dubost, G. 1988. The ecology and social life of the red acouchy, *Myoprocta exilis;* comparison with the orange-rumped agouti, *Dasyprocta leporina. J. Zool.* (Lond.) 214:107–23.

Green Acouchy
Myoprocta pratti
Plate 29, map 185

Identification. Measurements: HB = 298–383; T = 40–58; HF = 74–98; E = 31–37; WT = 0.8–1.2 kg.

Upperparts and legs finely grizzled olivaceous, each hair with many narrow, alternating bands of black and yellow; back and sides usually uniform, or rump sometimes darker, almost blackish, but hairs always banded; rump hairs long, but usually not overhanging rump in a fringe; back sometimes washed with reddish tones. Sides of muzzle, cheeks, and patch behind ear rusty orange. Chin and area around eyes almost naked. Underparts pale orange, thinly haired over yellow skin; throat, chest, and midline of belly often white. Small juveniles like adults but with more orange tints. Shape and tail like red acouchy.

Variation. Acouchys from the Venezuelan Amazon may have gray-brown sides and thighs grizzled with white hairs. The name *M. acouchy* is sometimes used for this species.

Similar species. See red acouchy; the agoutis that occur with this species are blackish with black or yellow rumps, red, brown, or yellow-orange.

Sounds. In alarm stamps hindfeet and gives a short series of quickly repeated nasal squawks like a high-pitched party horn. Gnaws loudly on nuts.

Natural history. Diurnal; terrestrial; solitary, or more rarely in pairs. Feeds on seeds and fruit. Green acouchys are most common in mature terra firme forest with dense undergrowth. They nest in hollow logs and sometimes in burrows. Their behavior when encountered is similar to that of red acouchys. Found in mature lowland rainforest habitats.

Geographic range. South America: east of the Andes and west of the Rio Negro, north of the Amazon in southern Venezuela, Brazil, Colombia, Ecuador, and Peru, and south of the Amazon in Peru and Brazil to at least the Rio Madeira.

Status. Often common in a huge geographic range; hunted for meat.

Local names. Cutiara, cutia de rabo (Br); tintin, chacure, coatiara, curi (Co); guatusa pequeña, papali, guatín, tintín (Ec); punchana, añuje menor (Pe).

Spiny Rats and Tree Rats (Echimyidae)

Members of this family are all medium-sized rodents (about 130–900 g) with ratlike shapes. A few have soft fur, but in most it is spiny or bristly. The spines are flat, without barbs, always flexible, and have a longitudinal channel on the upper side; they usually lie flat and look from a distance like normal fur. Most species have tails constructed so that they break off easily if pulled, and as a result many individuals have shortened or entirely missing tails. The ears are short and naked, often with a group of hairs growing from the inside rim, and they have a characteristic irregular rear margin as if they had been damaged. All species have large eyes and long whiskers, including a large group of genal whiskers (from the rear corner of the eye). They have pairs of lateral mammae buried in the body fur along the sides, never on the abdomen, except in the inguinal region. There is a long, narrow skin gland running down the midline of the sternum, especially well developed in males.

The spiny rats (*Proechimys* spp.) are often the most numerous terrestrial mammals in a rainforest; they give birth to about two or three young as often as every two to three months. The tree rats (*Echimys* spp.) are all arboreal; in these animals the nail of the second toe of the hindfoot is flared inward and cup-shaped. They have small litters of one or two young, probably only once or twice a year. Tree rats are difficult to see or capture, and little is known about them, including their precise geographic ranges. Several very beautiful species with no apparent close living relatives seem to occur in relict populations, where they may be rare and perhaps threatened, including the painted tree rat, the red crested tree rat, and perhaps the dark tree rat. All but one species in E Brazil belong to a distinct group, the Atlantic tree rats, *Nelomys* (sometimes placed in the genus *Echimys* or *Phyllomys*).

This is the largest family of caviomorph rodents, with about 15 genera and 60–100 species; all but 4 genera are found in rainforest. The taxonomy of echimyids is still unclear. There are certain to be future changes to the taxonomy, with additional species recognized beyond those described below.

Spiny Rats
Proechimys spp.
Plate 24, map 186
Identification. Measurements: HB = 180–275; T = 124–180; HF = 38–56; E = 20–26; WT = 150–550 g.
Upperparts glossy **chestnut brown to gray-brown; fur stiff, bristly or with narrow flexible spines, spines not visually obvious** except in SE Brazilian species; sides usually grayish or yellowish brown. **Underparts usually snowy white,** occasionally partly gray or buff, **sharply demarcated from sides. Head long and narrow, with pointed snout; ears naked, oblong, dark gray, and medium-sized, protruding above crown;** eyes large, dark brown; eyeshine bright yellow; whiskers long, to shoulder. **Tail naked** or slightly hairy, **bicolored dark gray above, whitish below; breaks off easily and is often shortened or entirely missing.** Large, handsome rats. **Hindfeet usually long and narrow, soles black; feet usually all or partly white above.** Young of most species are gray-brown with white underparts, but young *P. brevicauda* are similar to adults.
Species. Over 70 forms have been described, but there are probably no more than 20–30

species; the taxonomy is still under revision. Color of back varies from dark blackish brown to cinnamon, or grayish or yellowish in species from drier areas. Museum specimens lose their gloss and are often faded to a cinnamon color unlike that of living animals. *P. semispinosus* is the dominant species in Central America and N South America west of the Andes. Other common species include: *P. steerei, P. brevicauda, P. simonsi* (western Amazonia), *P. cuvieri* and *P. guyannensis* (Guianas and eastern Amazon Basin), and *P. guairae* (N Venezuela). There is one heavily spined blackish species *(P. hoplomyoides).* Species in the Atlantic forest of Brazil belong to the subgenus *Trinomys;* these have prominent spines on the lower back, are often grayish brown, and some have hairy or tufted tails. In Amazonia, up to four species can occur in the same locality. Skulls are necessary for identification, and only experts can distinguish between them.
Similar species. Spiny rats are the only large, chestnut rats with bristly fur, ears that protrude above crown, and sharply demarcated snowy underparts. Armored rats *(Hoplomys gymnurus)* are similar, with strong, upstanding spines in a geometric pattern on

Map 186

▨ Spiny rats, *Proechimys (Proechimys)*
▤ Spiny rats, *P. (Trinomys)*

lower back; other large terrestrial rats (water rats [*Nectomys* spp.], marsh rats [*Holochilus* spp.], crab-eating rats [*Ichthyomys* spp.], black and Norway rats [*Rattus* spp.]) have gray-brown backs, solid-colored tails, or grayish bellies, and soft, rather than stiff, fur; tree rats (*Echimys* spp.) have short ears that do not protrude above crown.

Sounds. When alarmed sometimes calls with a musical twittering (especially females with young); almost always flees by suddenly bounding away with a series of about five rapid taps of the hindfeet on the ground (the only common small nocturnal mammal to do this).

Natural history. Nocturnal; terrestrial; solitary. Feeds on seeds, fruit, and fungi (especially mycorrhizal fungi), and a few leaves and insects. Spiny rats are found throughout dryland forest but are most common in dense viny undergrowth and around fallen trees and complex tree roots. They do not climb except onto low fallen logs. They forage slowly, walking around in a small area, and are often seen sitting motionless near logs or roots or under shrubs. Piles of palm nut shells show where they take nuts to eat under roots, logs, or brush. During the day they shelter under fallen brush, in hollow logs, or in holes in the ground. From an early age (one week), the young follow the mother as she forages. Spiny rats are usually the most numerous terrestrial mammals in any rainforest habitat where they occur, and they are easy to see at night. They are the major prey of ocelots, bushmasters, and probably many other predators, and they are prolific breeders. They do not nest in houses, nor do they

usually enter them, but they may live near them. Found in many forest types throughout the entire Neotropical rainforest area, including some dry forests and gallery forests extending into drier habitats.

Geographic range. Central and South America: Honduras to Paraguay. To maximum elevation of 1,800 m, but upper limit is usually much lower.

Status. Almost always common. Spiny rats have excellent meat and, although small, are sometimes trapped for food and may be sold in markets.

Local names. Sauiá, rato-de-espinho (Br); ratón de espina (Co); rata, sacha cuí (Pe); stekelrat, maka-alata (Su); mocangué (Pn).

References. Patton, J. L. 1987. Species groups of spiny rats, genus *Proechimys* (Rodentia: Echimyidae). *Fieldiana Zool.*, n.s., 39:305–45.

Emmons, L. H. 1982. Ecology of *Proechimys* (Rodentia, Echimyidae) in southeastern Peru. *Trop. Ecol.* 23:280–90.

Armored Rat

Hoplomys gymnurus
Plate 24, map 187

Identification. Measurements: HB = 212–300; T = 114–240; HF = 47–61; E = 20–29; WT = 218–790 g; males reach larger size than females.

Upperparts cinnamon brown or almost black with prominent, geometrically spaced, strong, sharp, back-slanting black spines sticking up through fur on back and rump; rump almost black with dense spines; spines on sides may be pale-tipped, or older animals may have a sprinkling of white spines; sides cinnamon, paler than back. **Head long and narrow; ears naked, narrow, standing up above crown;** whiskers long, black, reaching to shoulder or beyond; eyes quite large; eyeshine bright yellow. **Tail shorter than head and body, naked, robust, bicolored dark above, pale below,** especially marked toward tip; often broken off short or entirely missing. **Underparts pure white, sharply demarcated from sides, often with a dark collar across throat, or in black individuals may be dark brown with variable amounts of white splotching.** Feet dusky on outer edge, white on inner side or entirely white or dusky in black animals. Young like adults.

Variation. In some parts of Panama blackish (melanistic) individuals are common,

Map 187

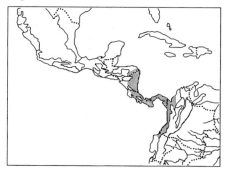

Armored rat, *Hoplomys gymnurus*

Map 188

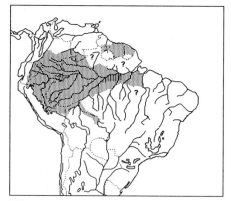

Spiny tree rat, *Mesomys* spp.
Tuft-tailed spiny tree rat, *Lonchothrix emiliae*

with great individual variation in the degree of darkening of the body, underparts, and feet. According to both molecular and morphological evidence this species probably belongs in the genus *Proechimys*.

Similar species. Spiny rats (*Proechimys* spp.) are similar but do not have conspicuous, upstanding spines except in E Brazil; see spiny rat.

Sounds. Makes whining and squealing calls when disturbed.

Natural history. Nocturnal; terrestrial. Feeds on fruit. These rats are the most numerous mammal in the pluvial rainforests of western Colombia, the wettest forest habitat in the New World. They frequent fallen logs, brush piles, rocks, and streamsides in forest and seem to have an affinity for water. Found in mature and disturbed rainforest.

Geographic range. Central and South America: Honduras south along the Atlantic slope to Costa Rica and Panama and west of the Andes on the Pacific slope of Colombia and Ecuador. To about 800 m elevation in Panama, lower in Colombia.

Status. Locally common.

Local names. Rata éspinosa (Span).

References. Tesh, R. B. 1970. Notes on the reproduction, growth, and development of echimyid rodents in Panama. *J. Mammal.* 51:199–202.

Spiny Tree Rat

Mesomys hispidus
Plate 25, map 188

Identification. Measurements: HB = 155–210; T = 140–220; HF = 27–37; E = 12–16; WT = 130–220 g.

Upperparts uniform pale to medium brown; midback often heavily streaked with black; pelage of conspicuous short, wide, flat, flexible spines from shoulder to rump, each spine brown with a pale tip, spines on head and neck are softer and narrower than spines on back. **Muzzle blunt; ears short and rounded,** thinly haired, brown, with a small tuft of longer hairs at base; eyes large, eyeshine moderately bright yellow; whiskers fine, long, reaching shoulder. **Tail robust, brown, thinly covered with long red-brown hairs that do not hide spines and form a sparse tuft at tip; tail often partly or completely missing. Underparts uniform pinkish orange, sharply demarcated from sides.** Feet short and broad, claws sharp, strongly curved; soles pink. Juveniles gray-brown. **A small, robust rat with a long back and short legs.**

Variation. Upperparts vary from pale tan to dark brown with black and reddish tips on spines (some animals from the western Amazon Basin). Underparts of an animal from Colombia are creamy white. A small race or species (*M. stimulax*) is found south of the Amazon and east of the Rio Tapajóz in Brazil, and one with narrow spines (*M. leniceps*) occurs at higher elevations in Peru. It is not yet clear how many species are in this genus, but there are probably at least four or five.

Similar species. Bare-tailed tree rats (*Makalata occasius*) are larger, with a smooth, naked tail; red-nosed (*M. didelphoides*) and

Peruvian *(M. rhipidurus)* tree rats are larger, with inconspicuous spines; tuft-tailed spiny tree rats *(Lonchothrix emiliae)* are gray-brown, with a large tuft on tail; other small rats have soft fur.

Natural history. Nocturnal; arboreal; solitary. Feeds on fruit, insects, and browse. These small rats are often seen in the low understory near the ground on logs, vines, shrubs, or treefalls, but they also use the high canopy. They favor areas of lush, dense undergrowth and vines, such as old treefall gaps and vine forests. They nest in tree holes and sometimes live in the thatched roofs of houses in the forest. From mature, disturbed, and secondary lowland rainforest.

Geographic range. South America: east of the Andes in the Guianas and most of the Amazon Basin of Colombia, Venezuela, Ecuador, Peru, Bolivia, and Brazil.

Status. Generally uncommon, common in a few localities; widespread.

Tuft-tailed Spiny Tree Rat
Lonchothrix emiliae
Plate 25, map 188
Identification. Measurements: HB = 155–220; T = 150–230; HF = 30–37; E = 13.

Upperparts gray-brown, sides and rump speckled yellowish white; body covered with dense wide, flat, flexible spines mixed with fine buffy hairs; wide spines on sides and rump with pale tips. Ears short, naked, brown, buried in spines of head. Tail robust, scaly, speckled with short, dark, evenly spaced, flat, scalelike hairs, tip with a large tuft of long, coarse, red-brown hairs up to 4.5 cm long, but tail tip often missing. Underparts white, grayish, or buff, often tending to buff on midline; chest often pale gray-brown, throat with white-tipped hairs. Feet dirty white; hindfoot broad and short, with pink soles, forefoot small.

Similar species. Spiny tree rats *(Mesomys* spp.) are brown, with sharply contrasting orange belly, and tail entirely hairy, tip with only a small tuft of fine hairs.

Natural history. Collectors' notes suggest that this is a nocturnal, arboreal species that lives in mature forest and nests in tree holes. Found in lowland rainforest.

Geographic range. South America: Brazil, known only from south of the Amazon on the lower reaches of the Rios Tapajós and Madeira.

Status. Probably locally common in a small geographic range.

White-faced Tree Rat
Echimys chrysurus
Plate 26, map 189
Identification. Measurements: HB = 232–300; T = 293–370; HF = 42–54; E = 11–20; WT = 415–890 g.

Upperparts pale gray-brown; densely covered with wide, flat spines from neck to rump, all other pelage stiff bristles. Head with white blaze down center of face from crown to nose, either completely covering face between eyes or a narrow stripe or small white tuft on crown; crown beside blaze usually dark brown; cheeks and sides of neck sometimes reddish. Ears small, thinly haired, brown; eyes dark brown, eyeshine weak, dull red. Tail fully haired, basal half to two-thirds dark brown, distal half to one-third white, yellow, or chestnut. Chin, throat, inguinal region, and band across chest behind elbows white; rest of underparts gray-brown, not sharply differentiated from sides. Feet large, broad, with strong claws, gray-brown above. Young like adults.

Variation. Some animals from the Guianas are tinged reddish. There is much individual variation within populations in color of pale parts of the body (white, yellow, or on tail, chestnut), length of pale tip on tail, and width of white facial stripe; underparts may lack white.

Similar species. These are the only large, heavily spined gray-brown rats in their range with a white blaze and a fully furred, white-tipped tail.

Sounds. Silent when seen; sometimes drops pieces of fruit.

Natural history. Nocturnal; arboreal; usually solitary. Feeds on fruit and probably leaves. White-faced tree rats use the middle and upper levels of the forest, especially in viny areas, but also in open forest. These rats can run quickly through the branches, but they usually move slowly and are difficult to spot. When disturbed they creep quietly into thick vegetation and sit motionless for many minutes, then sneak away silently. They make nests of leaves in canopy tree holes. Found in mature rainforest.

Map 189

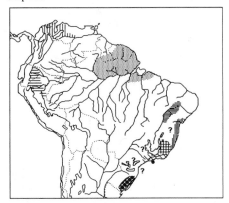

	White-faced tree rat, *Echimys chrysurus*
	Dark tree rat, *E. saturnus*
	Speckled tree rat, *E. semivillosus*
	Golden Atlantic tree rat, *Nelomys blainvillei*
	Pallid Atlantic tree rat, *N. lamarum*
	Black-spined Atlantic tree rat, *N. nigrispinis*
	Rusty-sided Atlantic tree rat, *Nelomys* sp.
	Drab Atlantic tree rat, *N. dasythrix*
[●]	Giant black-spined Atlantic tree rat, *N. thomasi*

Geographic range. South America: the Guianas and Amazon Basin of Brazil east of the Rios Negro and Xingu.
Status. Unknown, usually uncommon; widespread.
Local names. Sauiá, conoco, bandeira (Br); maka-alata (Su).
References. Miles, M. A., A. A. De Souza, and M. M. Póvoa. 1981. Mammal tracking and nest location in Brazilian forest with an improved spool-and-line device. *J. Zool.* (Lond.) 195:331–47.

Dark Tree Rat
Echimys saturnus
Plate 26, map 189
Identification. Measurements: HB = 240–335; T = 280–295; HF = 48–51; E = 17–18.
Upperparts glossy brown; head, midback, and tail base deep glossy black; sides chestnut; pelage stiff bristles, center of back with wide, flat spines (not extending to rump). **Ears short, buried in hair of head.** Underparts white, white-spotted, or whitish with buff tinge; chin dark. **Tail densely haired, black at base with variable amount of white on or near tip.** Feet broad and strong, claws strong.

Variation. Some animals are entirely dark brown and black.
Similar species. The only large blackish rat in its range with a fully furred tail. Black (melanistic) squirrels have prominent ears and bushy tails; Peruvian tree rats *(Makalata rhipidurus)* are dull brown, with an almost naked tail (the furry-tailed variety does not occur in same area).
Natural history. Found in premontane and lowland rainforests.
Geographic range. South America: Ecuador and N Peru, in the eastern Andean foothills and adjacent lowlands. To at least 1,000 m elevation.
Status. Rare, known from fewer than 10 individuals, from a small geographic region.

Giant Tree Rat
Makalata grandis
Plate 26, map 190
Identification. Measurements: HB = 260–330; T = 260–320; HF = 54–62; E = 18–22; WT = 582–586 (and probably much heavier).
Upperparts golden black, a coarse mixture of black, yellow, and orange-tipped hairs, sometimes rusty; **midback blackest, paling to brownish yellow on sides; pelage of dry, stiff, straight, coarse bristles; shines with gold iridescence at some angles to light.** Head with blackish face, crown, and sides; ears short, thinly haired, dark gray; whiskers long, coarse, black. **Tail fully haired with smooth, straight, flat hairs; same color as back for first one-sixth, abruptly changing to pitch black for distal five-sixths.** Underparts dirty whitish yellow, sometimes tinged with buff, often gray or pale brown on throat; inguinal region and ventral tail base usually rusty. Feet grizzled black and yellow, large, broad, and strong, toes long, claws thick, sharp and curved; hindlegs thick at ankles. A large, squirrel-like rodent. Young like adults.
Variation. Rare individuals have a white or rusty patch on the nape, a 1–2 cm white tip on the tail, or more blackish or rusty backs.
Similar species. Brush-tailed tree rats *(Isothrix* spp.) are olivaceous, with soft fur; black (melanistic) squirrels have longer ears, bushy tails, and are diurnal.
Natural history. Nocturnal; arboreal. Nests in tree holes, often near water. Found only in lowland rainforest, probably restricted to floodplain forests.

Map 190

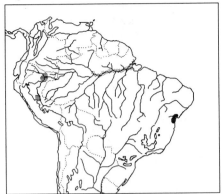

Map 191

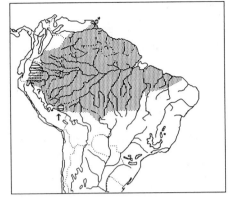

▦ Giant tree rat, *Makalata grandis*
● Painted tree rat, *Echimys pictus*
▤ Peruvian tree rat, *M. rhipidurus*

▦ Red-nosed tree rat, *Makalata didelphoides*
▤ Bare-tailed tree rat, *M. occasius*

Geographic range. South America: Brazil along the Amazon on both banks from the Rio Negro to the Isla Caviana at the mouth, and the lower Rio Negro; known only from near the main river.
Status. May be locally common in a small geographic range.
Local names. Toró preto.

Bare-tailed Tree Rat
Makalata occasius
Plate 26, map 191
Identification. Measurements: HB = 218; T = 225; HF = 37; E = 40.
Upperparts dull buff-brown streaked with dark brown; body from shoulder to rump on both back and sides thickly covered with wide, flat, sharp, flexible spines, which lie flat and may not be apparent from a distance; spines on back brown or black, many with rusty tips, especially on rump; tips of spines not drawn out into hair-like processes; spines on sides gray-brown; fine, soft, sparse orange hair between spines on body. Head clothed with stiff, thin bristles; ears short, naked; tip of muzzle and cheeks reddish; **whiskers fine, thin, but long,** reaching shoulder. **Tail robust, long, equal to or longer than head and body length; body hair ending abruptly 1–2 cm from base, rest of tail conspicuously naked, scaly. Underparts clear orange, often with white chest and inguinal region; undersurface of tail base and rear of thighs not rust-red.** Feet brownish with whitish toes.

Similar species. Red-nosed *(M. didelphoides)* and Peruvian *(M. rhipidurus)* tree rats have coarse, thick whiskers, tail always hairier than the present species and usually shorter than head and body, base of tail thickly haired for about 3 cm, and underparts usually gray-brown, but these species may not be distinguishable in the field; spiny tree rats *(Mesomys* spp.) are smaller, with long reddish brown hairs on tail.
Natural history. Unrecorded. A rat apparently of this species was photographed as it sat motionless on a branch at 5 m height in whitewater floodplain forest at night. From lowland and perhaps montane rainforest.
Geographic range. South America: Ecuador and Peru east of the Andes. The type specimen is said to have come from the western slope of the Andes in N Ecuador at 1,300 m elevation, but this seems erroneous since all others known are from the lowlands of the eastern side.
Status. Rare; known from fewer than 10 individuals.

Speckled Tree Rat
Echimys semivillosus
Plate 26, map 189
Identification. Measurements: HB = 200–268; T = 210–261; HF = 36–43; E = 17–22; WT = 194–407 g.
Head, and often neck and shoulders, gray streaked with black; midback gray washed with fulvous; **rump or rump and midback well speckled with white-tipped spines;** sides like back or gray; **shoulder to rump**

heavily spined with flexible, flat spines.
Ears small, often with pale patch behind and
below; whiskers numerous, long, to behind
ear, black. **Tail robust, 80–120% of head
and body length, appears naked,** scaly,
thinly covered with stiff, pale brown hairs.
Feet broad, gray or gray-yellow above. Underparts white or pale orange, sometimes
with gray across abdomen. Young gray with
gray-yellow-brown rump, soft fur, and no
spines.
Variation. Animals from Colombia have a
white muzzle and an almost black crown.
Similar species. See red-nosed tree rat
(*M. didelphiodes*).
Natural history. Probably nocturnal; arboreal. Speckled tree rats den in tree holes.
Found mostly in gallery forest, thorn forest,
dry forest, and trees in llanos, but rarely also
in rainforest.
Geographic range. South America: NE
Colombia and NW Venezuela. To at least
600 m elevation.
Status. Locally common.

Painted Tree Rat
Echimys pictus
Plate 27, map 190
Identification. Measurements: HB =
250–295; T = 273–325; HF = 43–47;
E = 16.
Upperparts with striking black and white
markings; entirely dirty white except for a
sharply defined glossy black saddle on top
of back from shoulders to tail and across
upper arms, and a wide black stripe on
neck and crown forming points on nape
and between eyes; fur coarse but not bristly,
dense, of uneven lengths, brown at base
with white or black tips, underfur long
and wavy. Ears short, naked, not protruding
above crown; whiskers very fine, longest to
shoulder. **Tail much longer than head and
body, thickly furred, black with tip shiny,
silky white above, golden yellow below.
Feet extremely broad, toes long,** whitish
or gold above, soles black to base of toes.
Variation. This species was originally described as brown and white; this may have
been due to fading of the fur, but possibly
there are brown-and-white animals. This rat
does not seem closely related to any other
species; it has variously been placed in *Echimys,* to which it seems closest, *Nelomys,* and
Isothrix.

Similar species. Hairy dwarf porcupines
(*Coendou* spp.) have spines, prehensile tails,
and no striking black markings. They look
like giant hair-balls with huge pink noses.
Natural history. Found in Atlantic forest
and cocoa plantations.
Geographic range. South America: E Brazil, known only from S Bahia.
Status. One of the rarest rodents; known
from only a few individuals from a small region. An apparently healthy population has
recently been discovered.

Red-nosed Tree Rat
Makalata didelphoides
Plate 26, map 191
Identification. Measurements: HB =
200–260; T = 160–330; HF = 37–45;
E = 16–21; WT = 340–405 g.
**Upperparts drab reddish brown to yellow-
ish brown heavily streaked with black, becoming more rusty on rump and tail base;**
midback and rump finely speckled by paler,
reddish brown tips on blackish bristles; **dorsal fur a mixture of thin bristles and flat,
flexible, sharp spines,** sometimes with thin
rusty hairs between spines. **Muzzle faint to
bright rust-red, including top to between
eyes;** whiskers coarse, reaching shoulder,
brown; ears small, thinly haired; eyes dark
brown, eyeshine faint dull red. **Tail usually
shorter than head and body, thickly haired
at base for 3–4 cm, rest almost naked,
scaly, thinly haired with inconspicuous
brown hairs that do not extend beyond
tip in a tiny tuft.** Underparts pale orange,
usually mixed with gray or brown at midbody, or entirely gray-brown, sometimes
with white areas; rear of thighs and haired
tail base below faint to bright rust. Feet short
and broad, pale reddish or yellowish. Young
like adults but without spines.
Variation. Animals from eastern Amazonia
and the Guianas are small and more reddish,
with brighter rusty parts of body (typical
M. didelphoides); those from central Amazonia are more yellowish gray, with indistinct reddish muzzle and thighs; animals
from Colombia, Venezuela, Guyana, and the
central Amazon Basin are large, with heavily black-streaked backs and entirely gray-
brown underparts. These larger forms may
each represent species (*M. macrurus, M.* c.f.
obscurus.). The systematics of this group of
forms is undergoing revision.

Similar species. See Peruvian tree rat (*M. rhipidurus*). Bare-tailed tree rats (*M. occasius*) have fine whiskers, dense spines, and a conspicuously bare tail equal to or longer than head and body; speckled tree rats (*E. semivillosus*) are yellow-gray with white speckles; brush-tailed rats (*Isothrix* spp.) are olivaceous, with soft fur and a bushier tail; spiny tree rats (*Mesomys* spp.) are smaller, more heavily spined, with hairy tails.

Sounds. Silent when encountered; sometimes drops fruit fragments.

Natural history. Nocturnal; arboreal; solitary. Feeds on fruit and unripe seeds. Red-nosed tree rats use the canopy and middle layers of the forest but sometimes descend close to the ground. They favor dense viny vegetation and have a habit of sitting motionless among thick foliage. When alarmed they will either sit motionless where they are or creep stealthily into a thicket and hide for many minutes before sneaking away upward. They travel quietly, without jumping or disturbing the vegetation. By day they nest in tree holes. Common in some areas, but difficult to see. Found in mature and secondary rainforest and gardens, usually near water and perhaps largely limited to floodplain forests.

Geographic range. South America: east of the Andes from Colombia and Venezuela south to Peru and Bolivia, the Guianas, the Amazon Basin of Brazil, and Trinidad and Tobago. To about 700 m elevation.

Status. Widespread and locally common.

References. Charles-Dominique, P., M. Atramentowicz, M. Charles-Dominique, H. Gérard, A. Hladik, C. M. Hladik, and M. F. Prévost. 1981. Les mamifères frugivores arboricoles nocturnes d'une forêt guyanaise: Inter-relations plantes-animaux. *Rev. Ecol.* 35:341–435.

Emmons, L. H. 1993. On the identity of *Echimys didelphoides* Desmarest, 1817 (Mammalia: Rodentia: Echimyidae). *Proc. Biol. Soc. Wash.* 106:1–4.

Peruvian Tree Rat
Makalata rhipidurus
Plate 26, map 190
Identification. Measurements: HB = 210–247; T = 180–215; HF = 31–45; E = 11–17; WT = 315 g.
Upperparts reddish or yellowish brown, finely or heavily streaked with black hairs;

head and forequarters more grayish or yellowish, rump and tail base more rusty or reddish; **pelage slightly glossy, stiff, narrow bristles and flexible spines with tips drawn out into hairlike processes; without prominent sharp, flat spines on forequarters or sides of body;** rump sometimes lightly sprinkled with a few white speckles from white-tipped spines. **Base of whiskers dull rust-red, but not top of muzzle between eyes;** whiskers coarse, numerous, black, long, reaching shoulder; ears small, almost naked. **Tail robust, 70–95% of head and body length, furred like back for 3.5 cm, rest either covered to tip with short, glossy, dark brown or grayish bristly hairs that do not quite hide scales, or with a short line of such hairs on top surface of base, these diminishing in size for about half the length of tail, rest of tail thinly covered with inconspicuous pale brown hairs that form a tiny tuft beyond tip,** and from a distance tail appears naked. **Underparts entirely clear pale whitish, yellow, or orange or mixed with gray-brown, especially around midbody;** tail base below and rear of thighs dull to bright rust-red. Hindfeet short and broad, reddish above; forefeet small, gray above. Young less red than adults; underparts and hair on tail gray.

Variation. Large, hairy-tailed animals are from the Río Amazonas near Iquitos; small, "naked"-tailed animals are from the Río Ucayali Basin near Pucallpa; they may represent two species.

Similar species. Red-nosed (*M. didelphoides*) and bare-tailed (*M. occasius*) tree rats are probably indistinguishable in the field from the naked-tailed form. The hairy-tailed forms are the only large red-brown rats with a hairy tail.

Natural history. Known only from lowland rainforest.

Geographic range. South America: Peru, central and northern Amazon Basin.

Status. Apparently uncommon.

Black-spined Atlantic Tree Rat
Nelomys nigrispinis
Plate 26, map 189
Identification. Measurements: HB = 215; T = 240; HF = 40; E = 18.
Upperparts warm reddish brown evenly and lightly streaked with black; sides brighter fulvous; **pelage a mixture of stiff**

bristles and flexible spines with tips drawn out in hairlike processes, **spines not visible, even at close range looks like normal coarse fur. Ears short,** naked; eye sometimes surrounded by dusky ring; whiskers medium to long, reaching ear or shoulder. **Tail longer than head and body, moderately robust, covered with smooth, shiny, pale brown hair that forms a smooth tuft at tip, tail looks paler than back.** Feet and hands dusky brown, toes and outside of hindfoot silvery white. Underparts pale orange, hairs white at base; chin, chest, and inguinal region may be white. Young like adults.
Variation. Another species or subspecies *(N. medius),* from Santa Catarina, Brazil, is larger, but externally identical.
Similar species. Drab Atlantic tree rats *(N. dasythrix)* are mousy gray-brown with dirty yellow or gray underparts; pallid Atlantic tree rats *(N. lamarum)* are dirty yellow not streaked with black, with a thinly haired tail; rusty-sided Atlantic tree rats *(Nelomys* sp.) have heavily spined backs, bright rusty sides, gray belly, and tail shorter than head and body; southern bamboo rats *(Kannabateomys amblyonyx)* are larger, with square muzzles; other plain large rats in region have large ears and naked tails.
Natural history. Found in mature and secondary Atlantic coastal rainforest.
Geographic range. South America: SE Brazil, São Paulo and Rio de Janeiro. To at least 850 m elevation.

Giant Black-spined Atlantic Tree Rat
Nelomys thomasi
Map 189
Identification. Measurements: HB = 270–287; T = 270–340; HF ≈ 42–48; E = 17–20.
Upperparts rich reddish brown streaked with black; sides cinnamon; fur a mixture of bristles and flexible spines, spines not apparent. Ears short, with a few black hairs growing from inner rim. **Tail robust, thickly covered with dark brown hair, tail sometimes broken off or missing.** Underparts yellow-orange grading gradually from sides. **Very large.**
Variation. Sometimes considered to be a subspecies of the black-spined tree rat.
Similar species. These are the only giant tree rats in their range. Norway rats *(Rattus norvegicus)* have naked tails and large ears.

Natural history. Said to frequent coconut palms and to be considered an agricultural pest that eats mandioca.
Geographic range. South America: Brazil, São Paulo, Ilha São Sebastião only.
Status. Unknown; geographic range minuscule, but there is currently considerable forest remaining on the island, and the species apparently adapts to disturbed or secondary vegetation.
Local names. Cururuá, cururuá-sem-rabo.

Pallid Atlantic Tree Rat
Nelomys lamarum
Plate 26, map 189
Identification. Measurements: HB = 195–230; T = 210–233; HF = 32–35; E = 13–16.
Upperparts uniform drab, pale, grayish yellow-brown (like clay or ocher); sides dirty yellowish; **fur conspicuously spiny, with wide, flat, sharp, flexible spines mixed with hairs, spines visible on rump.** Ears small; whiskers fine, long. **Tail thinly haired with short pale brown hairs that do not hide scales, slight tuft at tip darker brown;** tail of large adults usually 90–96% of head and body length, longer in young. Hands and feet yellowish white. **Underparts pure white, the white sometimes restricted to a narrow line down midline or in splotches.** Medial glandular stripe on midline of chest naked, surrounded by stained fur.
Variation. In the literature and in museums this species is generally confused with the golden tree rat *(N. blainvillei).* The descriptions here are based on the type specimens.
Similar species. See black-spined Atlantic tree rat *(N. nigrispinis).* Golden Atlantic tree rats *(N. blainvillei)* are less spiny, brighter gold, and have a thickly haired tail.
Natural history. From lowland Atlantic coastal forest.
Geographic range. South America: E Brazil, coastal Bahia at 300 m elevation.

Rusty-sided Atlantic Tree Rat
Nelomys sp.
Map 189
Identification. Measurements: HB = 215; T = 205; HF = 38; E = 14; WT = 225 g.
Crown and neck dull orange-brown heavily but finely streaked with black; back and top of rump black coarsely streaked and speckled with rust; thickly covered

with wide, sharp, flexible spines, black tipped with rust, intermixed with pure rusty bristles; sides of body and rump and thighs bright rust; spines on entire body, including underparts, are pale gray at base. Muzzle and sides of head somewhat grayish; whiskers long, reaching shoulder; ears short, blackish. **Tail shorter than head and body, moderately covered with brown hair, scales show through near base, hairs denser in slight tuft at tip.** Feet brown washed with gray-yellow, toes and side of hindfoot white. Underparts gray washed with whitish buff.
Variation. This rat has been called *Phyllomys brasiliensis* in the literature, but its correct name is not clear.
Similar species. See black-spined Atlantic tree rat *(N. nigrispinis)*.
Natural history. Found in Atlantic forest.
Geographic range. South America: E Brazil, Minas Gerais and perhaps coastal forests from S Bahia to Rio de Janeiro.

Golden Atlantic Tree Rat
Nelomys blainvillei
Plate 26, map 189
Identification. Measurements: HB = 192–250; T = 155–220; HF = 33–40; E = 14–20.
Upperparts entirely pale yellow-orange to darker red-gold, head and back darkest, sides paler; pelage stiff bristles and soft, flat, flexible spines, spines not apparent. **Ears conspicuous, stand out from, but not above, crown;** whiskers dense and long, to shoulder. **Tail robust, 80–100% of head and body length, thickly haired, with longer, slightly wavy hairs forming bushy tuft at tip; either dark brown or dusky, darker than back, or pale brown close to color of back, but tip browner;** silver below. Toes white. **Underparts pure yellow-white.**
Variation. This species has generally been confused with the pallid Atlantic tree rat *(N. lamarum).*
Similar species. These are the only gold rats with a thickly haired tail.
Natural history. Arboreal; probably nocturnal. Feeds on fruit. Golden tree rats live in small groups in holes in trees or in palms. Primarily found in caatinga vegetation.
Geographic range. South America: E Brazil, Ceará and Bahia.

Local names. Rabudo vermelho, ratocuandu, rato-de-espinho.

Drab Atlantic Tree Rat
Nelomys dasythrix
Map 189
Identification. Measurements: HB = 180; T = 210; HF = 35; E = 16.
Upperparts entirely dull mousy gray-brown, slightly paler on thighs; some ocher-yellow spine tips on rump; spines inconspicuous, soft. **Tail longer than head and body, well covered with brown hairs that nearly cover scales and form a distinct pencil at tip.** Underparts dirty yellow-white to buff, hairs gray at base, except throat, armpits, and inguinal region may be white.
Similar species. Black-spined Atlantic tree rats *(N. nigrispinis)* are reddish; pallid Atlantic tree rats *(N. lamarum)* have conspicuous, wide spines.
Natural history. Found in Atlantic forest.
Geographic range. South America: E Brazil, São Paulo to Rio Grande do Sul.

Yellow-crowned Brush-tailed Rat
Isothrix bistriata
Plate 27, map 192
Identification. Measurements: HB = 240–275; T = 270–300; HF = 47–50; E = 16–22; WT = 320–570 g.
Upperparts olivaceous gray-brown, rump warmer brown with rusty tints; **fur dense and soft,** with no bristles. **Head with pale yellow medial spot on crown, bordered by a wide black stripe above each eye, stripes broadening behind to nearly coalesce on nape;** muzzle and cheeks pale gray; whiskers long, black, reaching shoulder; ears very short, sometimes outlined behind and below by yellowish; eyes large, dark brown; eyeshine bright yellow. **Tail slightly longer than head and body, thickly and evenly covered with hair that curls outward with a bottle-brush effect, bright rust-orange or yellowish for basal 10–50%, black for distal 50–70%, or entirely orange or dusky.** Feet grayish. Underparts pale yellow or orange with slate gray base of hair. Young like adults, but duller. **Large, squirrel-like rats.**
Variation. There is much individual variation in the amounts of black and orange on the tail; some animals have completely blackish or orange tails. Crown patch is whitish in some populations or individuals. There

Map 192

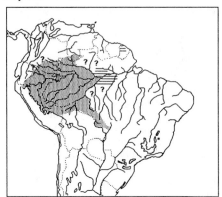

▦ Yellow-crowned brush-tailed rat, *Isothrix bistriata*
▤ Plain brush-tailed rat, *I. pagurus*

seems to be more variation between individuals than between populations.
Similar species. Squirrels *(Sciurus* spp., *Microsciurus flaviventer)* have no facial stripes and bushier, olivaceous tails.
Natural history. Nocturnal; arboreal. Three individuals seen together sat on a branch 6 m high, two of them side by side in contact. They sat motionless for many minutes, behaving like other tree rats. They were in luxuriant floodplain vegetation. By day brush-tailed rats den in tree holes, often in hollow palms, on the borders of rivers. They are said to sit at the entrances of their dens in late afternoon with their heads poking out. Found in lowland evergreen rainforest, perhaps restricted to igapó or other floodplain habitats.
Geographic range. South America: the Amazon Basin of Venezuela, Colombia, Ecuador, Peru, Bolivia, and Brazil west of the Rios Negro and Upper Madeira.
Status. Unknown; appears uncommon everywhere.
Local names. Conocono pequeño (Pe).
References. Patton, J. L., and L. H. Emmons. 1985. A review of the genus *Isothrix. Am. Mus. Novitates,* no. 2817.

Plain Brush-tailed Rat
Isothrix pagurus
Plate 27, map 192
Identification. Measurements: HB = 180–241; T = 170–271; HF = 37–45; E = 16.

Forequarters grizzled yellowish or brownish gray; hindquarters dull rusty brown; fur dense and soft. Head without prominent markings, face reddish; ears short; eyes large, eyeshine bright, yellow; whiskers long, reaching shoulder. Tail much longer than head and body, grades from gray-brown at base to yellow-brown toward tip, completely covered with hairs that curl outward, on undersurface hairs grow laterally so naked tailbone is visible in field; sometimes curls tail in a loose coil. Underparts yellowish or pale orange with hair gray at base. Young like adults except tail gray-brown to dark brown.
Variation. A recent specimen from French Guyana is externally almost identical to *I. pagurus,* but differs chromosomally, and is being described as a new species.
Similar species. Local squirrels *(Sciurus aestuans)* have ears that protrude above crown and bushier tails that are flipped upward against the back; giant tree rats *(Makalata grandis)* have black head and tail; white-faced tree rats *(Echimys chrysurus)* have white facial blaze and white tail tip; no other large rodents in range have hairy tails.
Natural history. Nocturnal; arboreal. Plain brush-tailed tree rats will sit motionless for many minutes when disturbed. They den in tree holes and can be found in terra firme forest far from water, as well as near rivers. Found in lowland evergreen rainforest.
Geographic range. South America: Brazil, Amazon Basin east of the Rios Negro and Madeira; limits poorly known.
Status. Uncommon.
References. See yellow-crowned brush-tailed rat.

Rufous Tree Rats
Diplomys caniceps, D. labilis
Plate 27, map 193
Identification. Measurements: HB = 212–320; T = 178–267; HF = 40–48; E = 12–20; WT = 359–430 g.
Upperparts dull rusty red-brown faintly streaked with dark brown hairs, reddish brightest on rear half of body; forequarters sometimes fulvous; fur soft and dense. Head forward of ears gray streaked with black; face marked with small whitish spot over eye, white streak at base of whiskers, and pale yellow spot behind and

Map 193

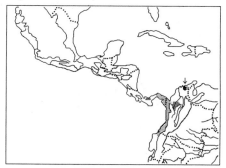

▨ Central American rufous tree rat, *Diplomys labilis*
▤ Colombian rufous tree rat, *D. caniceps*
⊡ Red crested tree rat, *D. rufodorsalis*

below ear; eyes large, eyeshine weak, dull red; ears short, not protruding above crown; whiskers coarse, long, reaching shoulder, nose brown. **Tail 70–105% of head and body length, robust; completely, moderately covered with dark brown hairs that curl outward to give a bristly look; scales of tail just visible beneath hair.** Hindfeet broad; all feet yellow-brown, gray, or brown above. Chin often white; throat whitish or orange; rest of underparts pale to rich orange; midline of lower abdomen, underside and sides of tail base often dark rust-orange. Young fulvous-brown, usually without red tones.
Variation. The two species are externally similar; their skulls appear to differ. There is much individual variation in the amount and brightness of the rusty colors of the back; the entire back may be rusty or only the rear half; or the entire back may be yellow-brown; facial markings may be prominent or indistinct. Some variation may be due to age.
Similar species. These are the only large reddish rats in their range with short ears and a hairy tail.
Sounds. When disturbed makes a high-pitched whining sound.
Natural history. Nocturnal; arboreal; solitary. Feeds on fruit and leaves. By day rufous tree rats den in tree holes, usually in trees near water. If disturbed they will stick their heads out the entrance (and can then be identified by their facial markings). They are agile climbers and aggressive when captured. Found in mature and secondary rain-

forest and deciduous forest, lowland and montane forest, mangroves, and plantations with trees. The information above refers to *D. labilis;* the natural history of *D. caniceps* is unrecorded.
Geographic range. Central and South America: Panama *(D. labilis)* and the Pacific coast of Colombia and N Ecuador (probably *D. labilis);* Colombia, near Medellín, the Cauca Valley, and the Cordillera Occidental of the Andes *(D. caniceps).* To at least 1,500 m elevation.
Status. Locally common; much of probable geographic range has been deforested.
Local names. Ratón maranero (Pa).
References. Tesh, R. B. 1970. Observations on the natural history of *Diplomys darlingi. J. Mammal.* 51:197–99.

Red Crested Tree Rat
Diplomys rufodorsalis
Plate 27, map 193
Identification. Measurements: HB = 190; T = 267; HF = 40; E = 12.
Upperparts bright rust-red; sides and legs paling to yellowish orange with gray underfur showing through parts; fur long and soft. **Head with crest of long fur on crown between ears, bright red like back;** muzzle and sides of head tawny yellow-gray streaked with black; whiskers fine, relatively short, to behind ear; ears short, almost naked, thin tufts of long brown hair sprouting from inner rim. **Tail much longer than head and body, robust, thickly haired with rusty fur for 2 cm at base, abruptly changing to fine brown or black hair that covers scales for three-fifths of length, terminal two-fifths pure white.** Feet brown washed with silver. Throat and chest pale orange, tip of chin white, belly gray washed with orange.
Similar species. Red-tailed squirrels *(Sciurus granatensis)* have much bushier tails and ears that protrude above crown.
Natural history. This beautiful rodent may be one of the rarest Neotropical mammals. It was reported from localities at about 700 m and 2,000 m elevation. It is presumably largely a montane species of humid forests.
Geographic range. South America: Colombia, known only from the Sierra Nevada de Santa Marta. To at least 2,000 m elevation.
Status. Rare, apparently known from fewer than five specimens; geographic range tiny,

but much of it is within Tayrona National
Park and Sierra Nevada Biosphere Reserve.

Amazon Bamboo Rat
Dactylomys dactylinus
Plate 27, map 194
Identification. Measurements: HB =
280–330; T = 345–437; HF = 50–65;
E = 15–22; WT = 600–700 g
Upperparts grizzled yellow-olivaceous
streaked with black, midback with blackish
hairs; sides paler, grayish; **rear of thighs
and sides and bottom of tail base rust-
orange; head and back of neck sometimes
with brown "hood" or nape patch; fur
long and soft, without bristles. Head gray
to beige, often with an indistinct dark
stripe** between eye and ear, and a paler stripe
above it; **ears small;** eyes brown, eyeshine
faint, dull red; whiskers long; **upper lips
enlarged to give blunt, square profile. Tail
robust, longer than head and body, furred
at base, the rest naked, scaly. Hands and
feet monkeylike; hands with four long
fingers, middle fingers separated by a gap,
digits with nails, not claws.** Underparts
white. **Strong, musky odor** that usually
permeates their living areas. Young like
adults. **Very large, stocky rats with square
muzzles.**
Variation. Museum specimens bleach severely with age and show great color variation, with hood or nape patch from almost
white to brown or tan and back sometimes
with rusty central stripe; the true variation is
therefore difficult to determine; but live animals in Ecuador, SE Peru, and E of the Xingu
in Brazil appeared similar (to Emmons), and
there may be relatively little color variation
among living animals. There may be two or
three species among the forms in this taxon.
Similar species. Red-nosed tree rats (*Makalata* spp.) are smaller, with stiff or spiny fur,
usually a reddish muzzle, fingers with claws,
upper lips not enlarged, and warm brown
upperparts; larger tree rats in region all have
hairy tails. Montane bamboo rats *(Dactylomys peruanus)* occur at 1,000–3,000 m elevation on the eastern Andean slopes of SE
Peru and Bolivia; they are olivaceous, with
completely furred tails.
Sounds. Calls from trees and thickets at
night with loud, froglike staccato bursts of
about 3–46 single or double pulses, each
burst about 5–20 seconds in length, given

Map 194

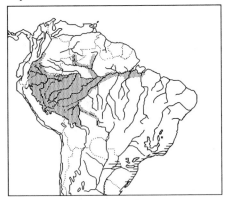

▓ Amazon bamboo rat, *Dactylomys dactylinus*
☰ Southern bamboo rat, *Kannabateomys amblyonyx*

once and not repeated. Often two rats duet,
one giving loud staccato pulses, the other
answering with soft, single pulses. Calls are
most frequent from 7:00 to 7:30 P.M. Loud
calls vary geographically in number and
timing of pulses. When disturbed, the rats
may call with series of faint, single grunts.
Natural history. Nocturnal; arboreal; solitary and in pairs. Feeds on leaves and shoots
of a variety of plants, and on bamboo shoots
and stems. Usually quite slow-moving, bamboo rats often "freeze" for many minutes
when disturbed. Their grasping hands and
feet make them adept at climbing thin vertical vines and slick bamboos, over which
they move extremely quietly. By day they
rest hidden in thick vine tangles or clumps
of vegetation, but they will run out readily
if disturbed. They can be active in late afternoon and are thus sometimes seen by day.
They are difficult to spot at night and are
much more often heard calling than seen;
they can sometimes be found by careful
search of an area where their odor is powerful. Where they feed in bamboo, the stems
are broken with large, jagged gnawed holes.
In most regions they are found only in patches
of bamboo or in dense riverside or lakeside
vegetation, especially canebrakes; but on rich
soils in Ecuador they can be found throughout the high forest canopy, or in disturbed
vegetation. Found in rainforest only.
Geographic range. South America: the
Amazon Basin of Colombia, Ecuador, Peru,
Bolivia, and Brazil.

Status. Locally common where habitat is suitable, but distribution highly patchy, absent over large areas of unfavorable forest.

Local names. Toro (Br, Pe); cocopitzu (Ec); cono cono, pacamama (Pe).

References. Emmons, L. H. 1981. Morphological, ecological, and behavioral adaptations for arboreal browsing in *Dactylomys dactylinus* (Rodentia, Echimyidae). *J. Mammal.* 62:183–89.

Southern Bamboo Rat

Kannabateomys amblyonyx

Plate 27, map 194

Identification. Measurements: HB = 230–347; T = 300–420; HF = 45–57; E = 16–27; WT = 350–570 g.

Upperparts uniform glossy olivaceous to fulvous, finely streaked with black; sides of body and tail sometimes fulvous; fur long, soft, hairs blackish with buff tips. Head large, muzzle square; whiskers long, dense, and coarse, reaching shoulder; cheeks grayish. Tail much longer than head and body, robust, furred to tip, tapering from thick body hair on base to sparse hair distally, bicolored paler below, dark brown above at base paling through dusky to white or buff near tip, tip sometimes with tuft of dark brown hairs. Hands and feet monkeylike; hands with four long fingers, center two separated by a wide gap, nails, not claws, on digits. Underparts orange with variable amount of white on throat and chest; lips around mouth white.

Variation. Animals from Argentina and Paraguay are dull, buff yellowish, with belly and tail almost white.

Similar species. These are the only large arboreal rats in their range with square muzzle and four long fingers with nails. Atlantic tree rats (*Nelomys* spp.) are smaller, with tails shorter or only marginally longer than head and body.

Sounds. Calls infrequently at night, singly or in duets, with repeated raspy pulses spaced at about half-second intervals. Calling can continue for up to several minutes.

Natural history. Nocturnal; arboreal; solitary and in pairs. Feeds on young shoots and leaves of bamboo and other plants. Inhabits thickets of bamboo, especially at watersides, and dense thickets without bamboo in swamps. Said to make large nests. Found in Atlantic coastal forests, inland rainforests, wet gallery forests, and bamboo patches such as planted bamboo hedgerows between fields.

Geographic range. South America: SE Brazil, SE Paraguay, and Misiones, Argentina.

Status. Apparently common in local patches, but such patches may be uncommon.

Local names. Rato-do-taquara (Br); rata de las taquaras, guaiquica (Ar).

References. Olmos, F., M. Galetti, M. Paschoal, and S. L. Mendes. 1993. Habits of the southern bamboo rat, *Kannabateomys amblyonyx* (Rodentia, Echimyidae) in Southeastern Brazil. *Mammalia* 57:325–35.

Crespo, J. A. 1982. Ecología de la comunidad de mamíferos del Parque Nacional Iguazu, Misiones. *Rev. Mus. Argent. Cien. Nat. "Bernardino Rivadavia"* 3:48–162.

Rabbits (Lagomorpha)

Leporidae

Dental formula: I2/1, C0/0, P3/2, M3/3 = 28. Forefeet with five toes, hindfeet with four toes. Members of this order are superficially rodentlike. They can by distinguished from all other mammals by the pair of miniature upper incisor teeth hidden behind the larger front incisors. Rabbits and hares (Leporidae) have long oblong ears, flexible necks, slitlike nostrils that can be opened and closed, large eyes, soles of feet thickly furred, long hindlegs, small button tails, and soft fur and fragile skin. They are small to medium-sized and run with a jumping gait, propelled by simultaneous strong kicks of the hindlegs. They are grazers and browsers especially fond of tender, protein-rich young leaves. They give birth to altricial or slightly precocial young. Members of the genus *Sylvilagus* are generally solitary and make nests on the ground surface. They shelter in thickets of dense vegetation, and their feeding areas are marked with many groups of round pellets. There are about 10 genera and 43 species worldwide, with only one species in Neotropical rainforest.

Tapiti or Brazilian Rabbit
Sylvilagus brasiliensis
Plate 29, map 195
Identification. Measurements: HB = 268–395; T = 10–35; HF = 64–85; E = 40–61; WT = 450–1,200 g.
Upperparts dark, variegated black and tawny, sometimes with a reddish tinge; **nape of neck behind ears pure russet. Ears long** (though relatively short for a rabbit), **oblong, set close together on the top of the head,** brown. **Tail inconspicuous, buttonlike, very short to minuscule,** brown above, somewhat paler below. **Lower legs and feet russet or red-brown.** Underparts white. Small young dark brown with russet nape, back not variegated. Eyeshine bright red to yellow; animal often seen from side with only one red eye visible.
Variation. Body and foot color more or less strongly reddish.
Similar species. These are the only rabbits in most of their range, and the only species in rainforest. In savannas, grasslands, and fields of Venezuela and Colombia and Central America there are cottontails *(S. floridanus),* which are larger and paler, with conspicuous tails with white tuft beneath, longer ears, and whitish feet. Acouchys *(Myoprocta* spp.) have long, thin legs, small ears, and pencil-thin tail and are diurnal.
Sounds. None usually heard in field.
Natural history. Nocturnal; terrestrial; solitary. Feeds on grass and browse. These little rabbits are most commonly seen just after nightfall or before dawn in grassy areas around houses or in gardens and plantations.

Map 195

Brazilian rabbit or tapeti, *Sylvilagus brasiliensis*

When disturbed they hop into a thicket and crouch under thick vegetation. In deep rainforest away from human disturbance they occur in swamps and along river edges. They are not often found among the trees of mature terra firme rainforest far from a habitat edge. They are attracted to salt and human urine (which contains salt). Found in forested habitats from the lowlands to cloud forest, dry inter-Andean valleys, paramo, and chaco.
Geographic range. Central and South America: Atlantic coast of Mexico from Taumalipas south through wetter parts of Central America; in South America to Tucumán, Argentina, and Rio Grande do Sul, Brazil. To 4,500 m in the northern Andes.
Status. Widespread and often common; distribution in the Amazon Basin rainforest is highly patchy.

Local names. Conejo (Span); coelho (Br); tapetí (Bo, Co, Su, Pa).

References. Hershkovitz, P. 1950. Mammals of northern Colombia. Preliminary report no. 6: Rabbits (Leporidae), with notes on the classification and distribution of the South American forms. *Proc. U.S. National Museum* 100:327–75.

Appendix A
Glossary

Adult. A fully developed individual capable of breeding; usually with all its permanent teeth erupted.

Agouti. A grizzled color produced by bands of alternating black and yellow or orange on individual hairs; also the name of an animal.

Allopatric. Of or relating to populations that are geographically separated such that they cannot potentially interbreed.

Altricial. Born in an early stage of development, usually with eyes and ears sealed, body naked, and unable to walk (but usually able to crawl and squeal).

Arboreal. Living and finding food above the level of the ground in the trees or shrubs.

Banded. Having lines or streaks running around the body perpendicular to the axis of the backbone.

Basal. Of or relating to the part of a structure (e.g., a hair or tail) near its base; see also **proximal.**

Biogeography. The geographic distribution of species.

Blackwater. Streams and rivers with clear dark brown or tea-colored water (stained with organic acids; e.g., the Rio Negro). Blackwater originates in areas of poor, leached-out soils, often white sands. Forests seasonally flooded with blackwater are called **igapó** in Brazil.

Broadleaf forest. In Central America, used to distinguish evergreen forests of non-coniferous trees from coniferous forests, which are also evergreen.

Browse; browser. Leaves, shoots, and twigs of plants other than grasses; an animal that eats these.

Caatinga. An arid region of eastern Brazil with highly unpredictable rainfall, containing thorny woodlands and grasslands.

Campo. Literally, field. Small savannas usually surrounded by forest, often seasonally wet, but may be dry. Can also refer to large grasslands.

Cerrado. A dry forest region south of the rainforest, containing a wide diversity of habitats from open grasslands **(campo limpo)** to dense, closed, thickety savanna-woodlands **(campo sujo),** as well as evergreen gallery forests along rivers; where there are rainforest mammal species.

Chaco. A large, flat, semi-desert region of Brazil, Paraguay, and part of Bolivia, containing dry, thorny shrub woodlands with cactus, marshes, evergreen gallery forest, and palm savannas.

Character. Any discrete feature of an organism, especially one that might be used to distinguish it from other similar organisms.

Clearwater. Rivers and streams with clear water with no sediment load or dark color from organic acids (e.g., the Rio Tapajós); see also **whitewater** and **blackwater.**

Cloud forest. High-elevation forest that is almost perpetually wet owing to the rain and dense mists caused by cool temperatures. These forests have short trees, deep mats of moss, and heavy growth of epiphytes on branches and trunks.

Conspecific. Belonging to the same species.

Crepuscular. Most active in reduced daylight at dusk and/or dawn.

Cursorial. Adapted for fast running on the ground. Cursorial animals usually have long, thin legs and walk on their toes.

Deciduous forest. Forest in which most of the trees lose their leaves simultaneously in the dry or cold season.

Digitigrade. Standing so that the body weight is borne only on the toes, and the heel does not touch the ground. Typical of long-legged, cursorial species.

Dimorphic Having individuals in the same population, such as males and females, that fall into two classes that differ in any feature, such as size, color, or shape. Often used of species or populations that are sexually dimorphic (e.g., Guianan saki monkey males are black, females are gray; male deer are antlered, females are not).

Distal. Of or relating to the part of a structure (e.g., a hair or tail) farthest from the center of the body.

Disturbed forest. Mature forest that has been disturbed artificially by people cutting some of the natural vegetation (as by selective logging or small garden patches); or naturally, as by floods, hurricanes, or fires.

Diurnal. Active in the daytime.

Dorsal. Of or relating to the back, usually top, side of the body.

Edaphic. Of or relating to soil type.

Endemic. Found only in a given region.

Evergreen forest. Forest in which the canopy of leaves is intact and green year-round, and only a small proportion of the trees lose their leaves at any one time.

Eyeshine. The reflection from an animal's eyes when light is shined on them at night. It is not visible at extremely close range. Its color varies depending on the angle of the eye to the light: paler, more whitish when the animal is viewed from directly in front, more reddish when seen from the side or rear. Eyes do not themselves shine.

Fauna. The collection of animal species that together inhabit a particular place.

Floodplain. The land on the sides of a river, or the river valley bottom, that is susceptible to being flooded during high water and across which the river meanders through time. Its soils are alluvial.

Form. One or a population of variant individual(s); often used as a way of avoiding a statement of its rank as a species or subspecies (e.g., a small form of raccoon lives on Cozumel Island).

Fossorial. Adapted for digging and living in burrows and tunnels underground. Fossorial animals have short legs and tails, long claws on their forefeet, and small eyes and ears.

Frugivore. An animal that eats mostly fruit.

Fulvous. Reddish-yellow.

Gallery forest. Forest that grows alongside a river or stream and on its floodplain. Regions that are too dry to support a forested habitat overall will often have gallery forests along watercourses between savannas. These constitute corridors where forest animals can live in otherwise inhospitable habitat.

Grazer. An animal that eats mainly grasses, sedges, and forbs.

Grizzled. Having a color formed from a mixture of separate colors (usually black, white, yellow, or red), either by bands of different colors on each hair (most common), or by intermixed hairs of different colors.

Hallux. The first toe of the hindfoot.

Harem. A (polygynous) social organization in which a single male is attached to a group of several females; he usually defends them by driving away other males.

Hispid. Bristly or spiny, or appearing that way because of coarse streaks of black in fur.

Home range. The area of land (or water or canopy) regularly used by an animal, in which it lives and finds all the necessities of life. It may or may not also be a territory.

Igapó. Forest that is flooded by blackwater for a large part of the year.

Insectivore. An animal that eats mostly insects and other invertebrates.

Juvenile. A young animal that is still nursing or has just been weaned.

Llanos. Large grassland regions in Colombia, Venezuela, and Brazil that are partly flooded with standing water during the rainy season; forest grows on hummocks of high ground, and gallery forest along rivers.

Lumping. Jargon for the grouping of two or more formerly separate forms (e.g., species) in a single species. See also **splitting.**

Mammae. Nipples or teats.

Mature forest. Forest old enough to be more or less at its climax; its species composition will not change much over time. See also **primary forest;** much primary forest is also mature forest.

Meatus. An opening (e.g., auditory meatus: the hole in the head that leads to the inner ear).

Melanistic. Darker than normal or black. Some species have occasional melanistic individuals (e.g., jaguars, squirrels).

Mesic. Moist or humid; as opposed to xeric, or dry (often incorrectly used as median).

Monogamy. A social system in which a male-female pair forms a long-lasting, exclusive relationship.

Morphology. Structure; an animal's structure or any portion of it (e.g., ear morphology).

Nocturnal. Active in the nighttime.

Olivaceous. A greenish yellow color usually produced by a mixture of black and yellowish or fulvous hairs, or hairs individually banded with those colors.

Opistodont incisors. Incisor teeth that slant backward.

Orthodont incisors. Incisor teeth that point straight down, at close to a right angle to the skull.

Pantanal. A large region of seasonally flooded grassland interspersed with forested high ground in southwestern Brazil and Bolivia.

Parous female. A female that is reproducing or has reproduced.

Plantigrade. Standing with the heels and palms touching the ground and bearing weight.

Platanillos. Large-leaved undergrowth plants, usually bananalike *Heliconia* spp., that may dominate the understory, especially in swampy or river-edge areas.

Polyandry. A social system in which a female mates with more than one male.

Polymorphism. The coexistence in the same population of individuals of different discrete types (e.g., blue-eyed and brown-eyed).

Population. A group of individuals of the same species that live in the same area and interbreed.

Precocial. Born at an advanced stage of development, with eyes and ears open, body covered with hair, and able to walk or run shortly after birth.

Prehensile tail. A muscular tail that can wrap around objects and grasp them tightly, supporting the weight of the body. Such a tail usually has a naked gripping surface near its tip. An adaptation of arboreal animals.

Primary forest. Forest that has not been cut over or disturbed by man; may also be mature forest.

Proodont incisors. Incisor teeth that point forward.

Proximal. Of or relating to the part of a structure (e.g., a hair or tail) closest to the center of the body.

Savanna. Natural grassland, often containing trees, especially palms, and fire-resistant species.

Scansorial. Living both on the ground and in the trees, as do many opossums and rodents.

Scat. Feces; often used for carnivore droppings.

Secondary forest. Young forest that has grown back over areas formerly cut by man or stripped by natural causes. The plant species composition of such forest will change as it matures. It is often dense and scrubby.

Splitting. Jargon for the separation of two or more taxa formerly grouped together (e.g., changing two subspecies to full species).

Subadult. An adolescent between the juvenile and adult stages, with permanent teeth not completely erupted and usually not yet of breeding age.

Sympatric species. Species whose geographic and habitat ranges overlap such that individuals of each can come into contact during the breeding season.

Synonyms. Two or more taxonomic names that apply to the same organism. The earliest published takes precedence.

Syntopic. Of or relating to animals that live in the same habitat, on a fine scale, such that their individual home ranges overlap.

Systematics. The study of the evolutionary relationships of organisms, usually entailing a process of organizing taxa to show their relationships.

Striped. Having lines or streaks running lengthwise parallel to the axis of the backbone.

Taxon. Any standard unit of classification (e.g., species, genus, family, order); the plural is **taxa.**

Taxonomy. The description and classification of organisms.

Terra firme. High ground that is never flooded during the rainy season. Many species of plants and some animals are restricted to this ground.

Terrestrial. Living and finding food on the ground.

Territory. A space (often the whole home range) that is actively defended by its owner to exclude other individuals from using it.

Ticked. Having a fine mixture of two or more colors, usually caused by pale tips on individual hairs; see also **grizzled.**

Type. The individual animal, usually a museum specimen, from which a species was described for the first time, and to which the species' scientific name forever afterward refers. The type locality is the place where the type specimen came from.

Variety. An imprecise term for a distinctive population, often used when it is not clear to what rank it should belong (species, subspecies, etc.).

Várzea. Forest that is flooded by whitewater for a large part of the year.

Ventral. Of or relating to the belly or undersurface.

Whitewater. Streams or rivers that carry a heavy load of sediment and deposit rich alluvial mud along their beds (e.g., the Rio Amazonas). They drain more fertile soils and are completely opaque and usually pale brown. Most originate in the Andes. Forest seasonally flooded with whitewater is called **várzea** in Brazil.

Appendix B

Keys to the Families and Genera of Rainforest Mammals

The following keys are based on external features that can be seen on a live animal, but the animal must be in the hand for close inspection and usually for measurement. In some cases it is necessary to open its mouth and look at the teeth. A metric ruler, pocket hand lens, or small balance is needed in some cases. The characters necessary to identify small mammals accurately often include details of the skull and teeth not usually visible on a whole animal in the hand. Because we often cannot see the truly diagnostic features externally, the keys below may be ambiguous in some cases; they are not intended for rigorous scientific identification, which requires a prepared study specimen, but as a general guide. The keys are designed for use with Neotropical rainforest mammals only; they may or may not work at any level outside the region.

Orders of Mammals

1 Forelimbs in the form of wings with membranes . . . . Bats (Chiroptera)
1' Forelimbs not in the form of wings 2
2 Entirely aquatic, with no hindlimbs but with a horizonal tail paddle or flukes 3
2' At least partly terrestrial, with hindlimbs and no tail paddle or flukes 4
3 With dorsal fin, breathes through blowhole on back of head . . . . . . Dolphins (Cetacea)
3' Without dorsal fin, breathes through nostrils on muzzle . . . . Manatees (Sirenia)
4 Hindfeet with two large, approximately equal-sized, weight-bearing hoofed toes . . . . . .
. . . . Peccaries and deer (Artiodactyla)
4' Hindfeet with three or more weight-bearing toes or claws 5
5 Upper lip elongated into a proboscis, adult weight > 80 kg . . . . . Tapirs (Perissodactyla)
5' Upper lip not a proboscis, weight < 80 kg 6
6 Incisor teeth absent . . . . Anteaters, sloths, armadillos (Xenarthra)
6' Incisor teeth present 7
7 Canine teeth absent 8
7' Canine teeth present 9
8 One large pair of chisel-like incisors with a tiny second pair hidden behind them; incisors
separated by a large gap from row of cheekteeth . . . . Rabbits (Lagomorpha)
8' Same as above but only one pair of large incisors, no second pair hidden behind them
. . . . Rodents (Rodentia)
9 Muzzle short, mouth wide and short, wider than long, nose set on flat face
. . . . Monkeys (Primates)
9' Muzzle long, mouth longer than wide, nose on pointed muzzle 10
10 First toe of hindfoot opposable, set widely apart from four other toes . . . .
. . . . Opossums (Marsupialia)
10' First toe of hindfoot not opposable . . . . Carnivores (Carnivora)

Genera of Opossums (Marsupialia, Didelphidae)

1 Tail much shorter than head and body length . . . .
. . . . Short-tailed opossums, *Monodelphis* (p. 33)
1' Tail equal to or longer than head and body 2
2 Hindfeet with complete webs between all toes, back with four wide black bands from side
to side across top connected by a single stripe down midback . . . .
. . . . Water opossum, *Chironectes* (p. 19)
2' Hindfeet not webbed, back without four black bands from side to side 3

253

3   Top of tail thickly haired for at least nine-tenths of its length 4
3'   Tail tip naked for at least terminal one-third 5
4   Two wide black stripes running from foreleg over front of shoulder and down each side
     of spine, underside of distal third of tail naked, no black stripes through eye . . . .
     . . . . Black-shouldered opossum, *Caluromysiops* (p. 13)
4'   No black stripes over shoulder and down back, underside of tail hairy to near tip, face
     with a prominent black stripe through each eye onto crown . . . .
     . . . . Bushy-tailed opossum, *Glironia* (p. 14)
4"   No black stripes on head or shoulders; tail completely haired above and below to near tip,
     with short hairs on distal two-thirds. Body long and weasel-like, ears short, bare . . . .
     . . . . *Lutreolina* (p. 20)
5   Size large (HB usually > 300 mm), fur with two distinct layers visible from a distance:
     short, dense, woolly yellow or white underfur thinly and incompletely covered by long,
     coarse, stiff, black or white overhairs . . . . Common opossums, *Didelphis* (p. 14)
5'   Size usually smaller (HB < 300 mm), fur uniform, soft and dense, without two prominent
     layers visible from a distance 6
6   Face with a prominent pale yellow or cream spot over each eye 7
6'   Face without pale spots over eyes 8
7   Color gray or black, rarely brownish, fur finely grizzled with white, tail thickly furred for
     first 5–8 cm, tail tip usually pure white separated by a sharp line from dark gray base,
     females have a pouch . . . . Gray four-eyed opossums, *Philander* (p. 17)
7'   Color brown, fur finely grizzled with buff, not white, tail thickly furred at base for less
     than 4 cm; if tail tip pale, it becomes so gradually, without a sharp line; females have
     no pouch . . . . Brown four-eyed opossum, *Metachirus* (p. 21)
8   Face with a dark line extending up middle between eyes to crown, with or without dark
     eye rings, tail may be thickly furred on top for half its length . . . .
     . . . . Woolly opossums, *Caluromys* (p. 11)
8'   Face without a dark line extending between eyes to crown, dark eye rings always present,
     tail never thickly furred for more than one-fourth of its length, size usually quite small,
     ratlike or mouselike 9
9   Second tooth behind upper canine larger to much larger than third tooth, upper canine
     tooth small and usually slender, less than three times as tall as tooth behind it, triads of
     hairs between scales on tail with outer two hairs broad and petiolate, or all hairs slen-
     der and about equally thick 10
9'   Second and third upper teeth behind canine about equal in size, or third tooth slightly larger,
     upper canine tooth large and robust, at least three times taller than tooth behind it, tri-
     ads of hairs between scales on tail with all hairs slender and about equally thick . . . 11
10   Fur not conspicuously long, triads of hairs between scales on tail with outer two hairs
     broad and petiolate (except *M. invictus*); color often gray-brown or gray with white or
     gray belly . . . . Slender mouse opossums, *Marmosops* (p. 26)
10' Fur long and lax, triads of hairs between scales on tail with all hairs slender and about
     equally thick, color often warm brown to reddish, belly orange, size always tiny (HB =
     72–116), South America only . . . . Gracile mouse opossums, *Gracilinanus* (p. 31)
11   Size large (HB > 150 mm), fur long, dense, and woolly, tail tip may be white or spotted, tail
     base thickly furred for 2–4 cm . . . . Woolly mouse opossums (part), *Micoureus* (p. 22)
11' Fur short and velvety or size much smaller than HB = 150 mm, tail thickly haired at base
     for less than 2 cm, color usually warm brown to reddish with pink, orange, yellow, or
     white belly 12
12   Tail dusky, without spots or pure white tip, underparts orange with at least centerline of
     midbody without gray-based hair, eye rings prominent . . . .
     . . . . Mouse opossums, *Marmosa* (p. 24)
12' If tail dusky, then underparts with hairs at midbody entirely gray-based; if tail with pure
     white distally or spotted with white, underparts at midbody with or without entirely
     gray-based hairs . . . . Woolly mouse opossums, *Micoureus* (p. 22)

Families and Genera of Anteaters, Sloths, and Armadillos (Xenarthra)

1 Head and body covered with armor plates, usually little or no hair . . . .
. . . . Armadillos (Dasypodidae) . . . 6
1' No armor plates on body, head and body covered with hair 2
2 Muzzle elongated, head conical, limbs short and thick, no teeth . . . .
. . . . Anteaters (Myrmecophagidae) . . . 3
2' Muzzle short, head round, limbs long and thin, teeth present . . . .
. . . . Sloths (Bradypodidae, Megalonychidae) . . . 5
3 Size very large (HB > 1,000 mm), tail bushy with hanging plume . . . .
. . . . Giant anteater, *Myrmecophaga* (p. 38)
3' Size smaller (HB < 700 mm), tail slender, not bushy 4
4 Size large (HB > 500 mm), fur stiff, glossy bristles . . . .
. . . . Tamanduas, *Tamandua* (p. 39)
4' Size small (HB < 300 mm), fur soft and woolly . . . . . Pygmy anteater, *Cyclopes* (p. 41)
5 Forefeet with three claws, short stumpy tail, face flat, without large caninelike teeth . . . .
. . . . Three-toed sloths (Bradypodidae), *Bradypus* (p. 43)
5' Forefeet with two claws, no visible tail, muzzle protruding, with large, caninelike teeth
. . . . Two-toed sloths, (Megalonychidae), *Choloepus* (p. 44)
6 Ears arising close together on top of head, bases nearly touch . . . .
. . . . Long-nosed armadillos, *Dasypus* (p. 49)
6' Ears set far apart on sides of head, wide armored crown between 7
7 Upperparts sprinkled with long, conspicuous, stiff, whitish hairs, foreclaws not greatly
enlarged . . . . Yellow armadillo, *Euphractus* (p. 46)
7' Upperparts without conspicuous hairs, foreclaws greatly enlarged 8
8 Size enormous (HB > 700 mm), hindlegs disproportionately large and thick . . . .
. . . . Giant armadillo, *Priodontes* (p. 48)
8' Size smaller (HB < 500 mm), hindlegs not enlarged . . . .
. . . . Naked-tailed armadillos, *Cabassous* (p. 46)

Families of Bats (Chiroptera)

1 With a freestanding, spear-shaped noseleaf behind nostrils, or if without, with naked,
grotesquely wrinkled face and ladderlike pattern in wing . . . .
. . . . Leaf-nosed bats (Phyllostomidae)
1' Without spear-shaped noseleaf or wrinkled face and ladder pattern in wing 2
2 With a pair of low, horseshoe- or M-shaped folds behind nostrils, one behind the other,
and central incisor teeth larger than canine teeth . . . . . Vampire bats (Phyllostomidae)
2' Without any folds behind nostrils, incisors smaller than canines 3
3 With long, robust tail projecting free beyond edge of tail membrane for at least one-third
of its length . . . . Free-tailed or mastiff bats, (Molossidae)
3' With tail either completely or almost completely (more than three-fourths of its length)
within tail membrane, or shorter than tail membrane and with any free tip protruding
above the middle of the membrane 4
4 With sucker disks on base of thumb and on ankle . . Sucker-footed bats (Thyropteridae)
4' Without sucker disks on thumb or ankle 5
5 Chin with a large concave plate on each side, or lips meaty and flared out forward, and
wings sometimes meeting at midline of back so that back looks naked . . . .
. . . . Leaf-chinned and mustached bats (Mormoopidae)
5' Lips and chin not as above, back not naked 6
6 Upper lip split near nose, forming drooping, bulldoglike folds, size large (FA > 60), fur
extremely short, one pale stripe down midback . . . . Bulldog bats (Noctilionidae)
6' Upper lip not split and drooping, size smaller (FA < 58), fur not extremely short, back
not striped or with two stripes 7

7   Thumb a rudimentary stub encased in membrane, emerging claw tiny, tail half as long as tail membrane, entirely enclosed within it .... Thumbless bats (Furipteridae)
7'   Thumb with well-developed digit and claw outside membrane, tail either as long as tail membrane or, if shorter, with tip protruding free above it 8
8   Tail much shorter than legs or tail membrane, tip sticks up free above from middle of membrane, glandular sac sometimes present in propatagium in front of elbow, or beside tail .... Sheath-tailed bats (Emballonuridae)
8'   Tail longer than legs, reaches edge of tail membrane, wing sac never present in propatagium 9
9   Two phalanges on third finger, legs very long, ears funnel-shaped, wings and body always pale .... Funnel-eared bats (Natalidae)
9'   Three phalanges on third finger, legs not greatly elongated, ears not funnel-shaped, wings and body black to pale .... Vespertilionid bats (Vespertilionidae)

Genera of Sheath-tailed Bats (Emballonuridae)

1   Color white, grayish white, or whitish buff .... Ghost bats, *Diclidurus* (p. 37)
1'   Not whitish 2
2   Nose elongated in a proboscis, forearm with spaced tufts of white hair ....
     .... Long-nosed bat, *Rhynchonycteris* (p. 53)
2'   Nose not in a proboscis, forearm without white tufts 3
3   Wing membrane attaches high on foot near base of toe 4
3'   Wing membrane attaches at ankle 5
4   Fur long and shaggy, covers face and chin, ear tips pointed, no wing sac, color yellowish to orangish .... Shaggy bat, *Centronycteris* (p. 54)
4'   Fur not shaggy, face not hairy, ear tips rounded, large wing sac present, blackish to reddish brown .... Chestnut sac-winged bat, *Cormura* (p. 56)
5   Wing sac absent, ears short and rounded .... Smoky bat, *Cyttarops* (p. 58)
5'   Wing sac present or indicated by a fold, ears triangular and slightly pointed 6
6   Wing sac close to bone of forearm beyond elbow, back with a pair of wavy pale lines often present, fur on forehead gradually decreases down face and muzzle ....
     .... White-lined sac-winged bats, *Saccopteryx* (p. 53)
6'   Wing sac in middle or forward edge of membrane, back without white lines, tuft of fur on crown ends abruptly on line between ears 7
7   Wing sac extends from forward edge of propatagium, muzzle short and broad, calcar twice as long as foot .... Doglike sac-winged bats, *Peropteryx* (p. 55)
7'   Wing sac in center of propatagium, muzzle narrow and pointed, calcar only slightly longer than foot .... Least sac-winged bats, *Balantiopteryx* (p. 56)

Genera of Leaf-chinned, Mustached, and Naked-backed Bats (Mormoopidae)

1   Chin with large, platelike folds, ears short, rounded, and joined by a band across forehead
     .... Leaf-chinned bats, *Mormoops* (p. 61)
1'   Chin without plates, ears pointed and widely separated, not joined by band across brow
     .... Mustached and naked-backed bats, *Pteronotus* (p. 60)

Subfamilies of Leaf-nosed Bats (Phyllostomidae)

1   With a pair of low, horseshoe- or M-shaped noseleaves one behind the other behind nostrils, and central incisor teeth larger than canine teeth ....
     .... Vampire bats (Desmodontinae)

1′   With a freestanding, spear-shaped noseleaf behind nostrils, or if without, with naked, grotesquely wrinkled face and ladderlike pattern in wing, incisor teeth smaller than canine
teeth 2

2   Muzzle elongated, narrow, lower jaw often longer than upper, tongue very long, noseleaf
short, spear-shaped, with no free horseshoe below and beside nostrils, size small (FA
≤ 45) . . . . Long-tongued bats (Glossophaginae)

2′   Muzzle and tongue not greatly elongated; if spear-shaped noseleaf present, nostrils at
least partly flanked by free fold or horseshoe, size small to large 3

3   Tail membrane nearly as long to longer than legs, tail usually present, ears usually large,
muzzle narrow . . . . Spear-nosed bats (Phyllostominae)

3′   Tail membrane much shorter than legs or absent, tail usually absent, ears medium, muzzle
usually broad 4

4   Tail always absent, muzzle short and broad, chin tip without large central wart, pale facial
stripes and/or midback stripes often present . . . .
. . . . Neotropical fruit bats (Stenodermatinae)

4′   Tail present or absent, muzzle narrow, chin tip with large central wart, stripes never present, size always small (FA ≤ 45) . . . .
. . . . Little spear-nosed and short-tailed fruit bats (Carolliinae)

Genera of Spear-nosed Bats (Phyllostominae)

1   Tail absent or a short stub hidden in base of membrane, size large (FA > 77; HB > 94),
noseleaf and horseshoe raised to form a deep hollow cup with unbroken rim around
nostrils 2

1′   Tail always present and obvious, size large to small, noseleaf not forming an unbroken
cup around nostrils 3

2   Giant size (FA > 98; HB > 135; WT = 125 g), fur medium length, dark brown to orange,
pale stripe down back . . . . False vampire bat, *Vampyrum* (p. 69)

2′   Large size (FA = 77–87), fur long and woolly, gray-brown, no stripe on back . . . .
. . . . Woolly false vampire bat, *Chrotopterus* (p. 69)

3   Tail longer than legs, reaches edge of tail membrane 4

3′   Tail much shorter than legs and tail membrane 5

4   Noseleaf extremely long, narrow, sword-shaped, as long as ears, ears very large, pointed,
tail membrane comes to a point at tail tip . . . . Sword-nosed bats, *Lonchorhina* (p. 63)

4′   Noseleaf broad, spear-shaped, much shorter than ears, ears large and rounded, tail membrane square at end, does not come to a point at tail tip . . . .
. . . . Long-legged bat, *Macrophyllum* (p. 64)

5   Chin, lips, and muzzle around nose studded with long projecting tubercles . . . .
. . . . Fringe-lipped bat, *Trachops* (p. 68)

5′   Lips and muzzle around nose not studded with projecting tubercles 6

6   Chin tip with single, long, smooth pads angled to form a V, size usually small . . . .
. . . . Little big-eared bats, *Micronycteris* (p. 62)

6′   Chin tip with rows of flat, roundish warts, size usually medium to large 7

7   Noseleaf very long, lance-shaped, much longer than broad, ears very large, pointed . . . .
. . . . Hairy-nosed bats, *Mimon* (p. 66)

7′   Noseleaf spear-shaped, not much longer than broad; ears, if large, rounded; if medium-
sized, pointed 8

8   Ears large, rounded, hair long but not dense, tail membrane longer than legs . . . .
. . . . Round-eared bats, *Tonatia* (p. 65)

8′   Tail membrane equal to or shorter than legs, fur medium to short and velvety, ears
broadly triangular 9

9   Horseshoe around nostrils with free flange completely around base, lips and noseleaf
dark . . . . Spear-nosed bats, *Phyllostomus* (p. 66)

9'   Horseshoe around nostrils with free flange only on sides, continuous with upper lip in center, lips and noseleaf pinkish mottled brown . . . .
. . . . Pale-faced spear-nosed bat, *Phylloderma* (p. 67)

Genera of Long-tongued Bats (Glossophaginae)

1   Tail absent or greatly reduced, tail membrane virtually absent, reduced to a narrow, hairy band much shorter than knees . . . . . . Hairy-legged long-tongued bats, *Anoura* (p. 73)
1'   Tail always present (though sometimes short), tail membrane longer than knees 2
2   Lower incisors present 3
2'   Lower incisors absent 5
3   Tongue with deep groove on side 4
3'   Tongue without deep groove on side, dorsal fur bicolored, noseleaf short, does not reach eye, wing attached at ankle opposite calcar, color pale brown or gray . . . .
. . . . Common long-tongued bats, *Glossophaga* (p. 72)
4   Dorsal fur bicolored, pale at base and dark at tip, muzzle only moderately elongated, noseleaf large, reaches eye when flattened . . . .
. . . . Spear-nosed long-tongued bats, *Lonchophylla* (p. 71)
4'   Fur unicolored chocolate brown or chestnut to base, noseleaf short, wing attached half-way down foot near base of toe beyond origin of calcar . . . .
. . . . Chestnut long-tongued bat, *Lionycteris* (p. 70)
5   Dorsal fur tricolored, dark at base (sometimes faint, like a shadow) and tip, pale in middle 7
5'   Dorsal fur bicolored pale at base and dark at tip 6
6   Distal thumb joint (outside membrane) longer than proximal (inside membrane), long whisker in front of ear, color blackish, South America only . . . .
. . . . Ega long-tongued bat, *Scleronycteris* (p. 74)
6'   Thumb joints about equal in length, no long whisker in front of ear, border of deep chin notch rippled with tubercles on edge throughout length, color dark brown, Central and South America . . . . Long-nosed long-tongued bats, *Choeroniscus* (p. 74)
7   Muzzle robust, deep chin notch extends under chin as a slit without rippled border of tubercles, elbows thinly furred above, underparts paler than back, Central and South America . . . . Dark long-tongued bats, *Lichonycteris* (p. 74)
7'   Muzzle narrow and tubelike, border of deep chin notch with long, smooth pads near lip that extend under chin as deep slit with edges of rippled tubercles, elbows thickly furred above and below, underparts about the same as back, Central America only . . .
. . . . Underwood's long-tongued bat, *Hylonycteris* (p. 75)

Genera of Short-tailed and Little Fruit Bats (Carolliinae)

1   Tail present, tip sticks up freely from middle of tail membrane, Central and South America . . . . Short-tailed fruit bats, *Carollia* (p. 76)
1'   Tail absent, South America only . . . . Little fruit bats, *Rhinophylla* (p. 77)

Genera of Neotropical Fruit Bats (Stenodermatinae)

1   Tail membrane virtually absent, a narrow hairy band down leg, no facial or midback stripes, no white spots where wings join shoulder, fur prominently tricolored dark at base and tip, pale in middle, central upper incisors much longer than outer incisors, broad and buck-toothed . . . . Yellow-shouldered bats, *Sturnira* (p. 78)
1'   With tail membrane or, if almost without, with facial stripes and/or short central incisors 2
2   Pure white spot where wing meets shoulder 3
2'   No white spot on shoulder 6

3   Noseleaf well developed, spear-shaped, no folds on brow 4
3'   Noseleaf not spear-shaped, or noseleaf absent, folds present on brow 5
4   Upper lip with fold from base of noseleaf to corner of mouth forming a "double" lip ...
      .... Ipanema bat, *Pygoderma* (p. 85)
4'   Upper lip without fold .... Little white-shouldered bat, *Ametrida* (p. 86)
5   Forehead above eyes with large fleshy horizontal fold that presses down on noseleaf, ears
      without horizontal lobe extending over brow, no ladderlike pattern in wing ....
      .... Visored bat, *Sphaeronycteris* (p. 86)
5'   Whole face covered with complex folds, no united noseleaf, horizonal lobes from ears
      cover brow, ladderlike pattern in wing .... Wrinkle-faced bat, *Centurio* (p. 87)
6   Central upper incisors bilobed, with broad, parallel tips, only slightly longer than outer
      incisors 7
6'   Central upper incisors bilobed or single-pointed, more than twice as long or longer than
      outer incisors, tips may be narrow and convergent 8
7   Midback and facial stripes always present .... Tent-making bats, *Uroderma* (p. 79)
7'   No midback stripes, facial stripes present or absent, upper facial stripes usually end at
      forward edge of ear, rarely to middle of ear .... Fruit-eating bats, *Artibeus* (p. 84)
8   Fur pale at base 9
8'   Fur with dark band at base, sometimes faint, like shadow 10
9   No facial or midback stripes, tail membrane naked, without hairy fringe on rear edge,
      forequarters white, hindparts grayish, Central America only ....
      .... Honduran white bat, *Ectophylla* (p. 83)
9'   Pale facial stripes present, tail membrane with hairy fringe on edge 12
10   Large (FA ≥ 38), soft depression detectable by pressing behind noseleaf, with or without
      stripes, central upper incisors long, narrow, with single points, outer incisors tiny, cal-
      car about three-fourths of length of foot .... Big-eyed bats, *Chiroderma* (p. 81)
10'   Small (FA ≤ 39), rostrum behind noseleaf hard, with or without midback stripe, central
      upper incisors much longer than outer, bilobed, calcar about half length of foot ... 11
11   Facial stripes present, midback stripe present only if bat large, if tiny, then without back
      stripe and tail membrane edge with slight hairy fringe, two molars ....
      .... Yellow-eared bats, *Vampyressa* (p. 82)
11'   Size tiny, no facial or midback stripes, tail membrane naked, without hairy fringe .....
      .... Macconnell's bat, *Mesophylla* (p. 83)
12   Size large (FA > 45; WT > 25 g), color warm cinnamon, back and facial stripes bright
      and sharp, posterior edge of facial stripes reaches rear of ear, sides of noseleaf and base
      of ears yellow .... Great stripe-faced bat, *Vampyrodes* (p. 81)
12'   If large (FA > 45), color blackish brown, warm brown only if smaller; sides of noseleaf
      and base of ears cream or gray; three molars ....
      .... White-lined fruit bats, *Platyrrhinus* (p. 80)

Genera of Vampire Bats (Desmodontinae)

1   Tail membrane almost absent, reduced to a narrow, very hairy band down side of leg, fur
      soft and dense .... Hairy-legged vampire bat, *Diphylla* (p. 89)
1'   Tail membrane well developed, forms a complete band joining legs, fur short, thin, and
      straight 2
2   Lower central incisors with deep notch ...... Common vampire bats, *Desmodus* (p. 88)
2'   Lower central incisors without deep notch ....
      .... White-winged vampire bat, *Diaemus* (p. 89)

Genera of Vespertilionid Bats (Vespertilionidae)

1   Ears enormous (> 26 mm), short stub of tail tip extends free beyond membrane, South
      America only .... Big-eared brown bats, *Histiotus* (p. 95)

1' Ears enormous, tail ends at point within membrane, Central America only . . . .
. . . . Central American long-eared bat, *Bauerus* (p. 97)
1" Ears not enormous 2
2   Tail membrane densely furred for at least half its length, ears short and rounded . . . .
. . . . Hairy-tailed bats, *Lasiurus* (p. 96)
2' Not as above 3
3   Pale yellow frosted brown above, pale yellow below, fur pale at base, two upper incisors,
size tiny . . . . Little yellow bats, *Rhogeessa* (p. 94)
3' Black, brown, or red, fur dark at base, four upper incisors 4
4   Tragus rounded, first tooth behind upper canine large, no gap between canine and the first
large tooth, FA = 37–54 . . . . Big brown bats, *Eptesicus* (p. 93)
4' Tragus narrow and pointed, first tooth behind upper canine tiny, leaving an apparent gap
in front of first large tooth, FA = 29–44 . . . . Little brown bats, *Myotis* (p. 93)

Genera of Free-tailed Bats (Molossidae)

1   Upper lip with deep vertical wrinkles 2
1' Upper lip smooth 3
2   Ears joined broadly at midline of brow, four lower incisors . . . .
. . . . Broad-eared free-tailed bats, *Nyctinomops* (p. 100)
2' Ears almost meet but are not joined on midline of brow, six lower incisors . . . .
. . . . Free-tailed bats, *Tadarida* (p. 100)
3   Midline of muzzle between eye and nose raised in a ridge 4
3' Midline of muzzle flat, not raised in a ridge 5
4   Hair on crown longer than on neck, forming a slight crest, four lower incisors, tail usually
> 60% of head and body length, palate between upper tooth rows deeply concave . . .
. . . . Crested mastiff bats, *Promops* (p. 102)
4' Hair on crown not longer than on neck, two lower incisors, tail usually 50–60% of head
and body length, palate between tooth rows flat . . . . . Mastiff bats, *Molossus* (p. 102)
5   Ears joined broadly at midline of brow, reach to nose when flattened forward . . . .
. . . . Bonneted bats, *Eumops* (p. 101)
5' Ears may meet or be separate but are not joined, reach midway between nose and eye
when flattened 6
6   Forearm skin above sprinkled with bumps, head and body extremely flattened, size tiny
. . . . Flat-headed bat, *Neoplatymops* (p. 99)
6' Not as above 7
7   Ears with pointed tips, widely separated, slight accordion-like fold where ear meets
crown, two lower incisors, hair of crown diminishes gradually down face, tail close to
50% of head and body length . . . . Dog-faced bats, *Molossops* (p. 98)
7' Ears rounded, no fold at junction with crown, hair of crown ends abruptly in a line, four
lower incisors, tail usually 40–50% of head and body length . . . .
. . . . Doglike bats, *Cynomops* (p. 99)

Families and Genera of Monkeys (Primates)

1   Size tiny to small (WT < 700 g), claws, not nails, on hands and feet . . . .
. . . . Callitrichidae, Callimiconidae . . . 2
1' Size small to large (WT > 600 g), nails, not claws, on hands and feet . . . . Cebidae . . . 6
2   Size tiny (WT < 150 g), west and central Amazon Basin only . . . .
. . . . Pygmy marmoset, *Cebuella* (p. 105)
2' Size larger (WT > 200 g) 3

3 Lower incisors narrow at tips, tightly clustered to form a scoop, lower canine little differentiated from incisors and clustered with them, ears often with tufts, south of the Amazon and east of the Madeira including SE Brazil . . . . . Marmosets, *Callithrix* (p. 106)

3' Lower incisors broad at tips, not clustered into a scoop, lower canine distinctly larger than incisors, may have mane but not ear tufts, whole region 4

4 Bright gold or gold and black with prominent, centrally parted mane on head, E Brazil only . . . . Lion tamarins, *Leontopithecus* (p. 118)

4' Mane, if present, not centrally parted, color not gold or gold and black, or found outside SE Brazil only 5

5 Color black, face entirely black, hair on head forms two tiers: a shorter, even cap bordered on nape and sides by longer ruff . . . . Goeldi's monkey, *Callimico* (p. 120)

5' Black or colored; if black, then muzzle usually grizzled or patterned with white, hair on head not two-tiered . . . . Tamarins, *Saguinus* (p. 110)

6 Tail nonprehensile, not carried coiled at tip, size small to medium (WT = 0.6 – 4 kg) . . . 7

6' Tail prehensile, often coiled at tip, size medium to large (WT = 2 – 15 kg) 12

7 Eyes very large, crown with three black stripes, nocturnal . . . . Night monkeys, *Aotus* (p. 121)

7' Eyes normal size, crown not striped, diurnal 8

8 Fur very short, tail slender, white mask around eyes, small, slender, and agile . . . . Squirrel monkeys, *Saimiri* (p. 125)

8' Fur medium to long, tail not slender, no white mask around eyes, if small (≈ 1 kg), then body not slender 9

9 Size small (WT = 0.9 – 1.7 kg), tail thickly furred but not bushy . . . . Titi monkeys, *Callicebus* (p. 123)

9' Size larger (WT = 1.6 – 4 kg), tail distinctly bushy 10

10 Tail much shorter than head and body . . . . Uakari monkeys, *Cacajao* (p. 135)

10' Tail longer than head and body 11

11 Fur on tail, head, and body long and shaggy, tail carried low . . . . Saki monkeys, *Pithecia* (p. 131)

11' Fur on body not long and shaggy, bushy tail often carried in high arc over back . . . . Bearded saki monkeys, *Chiropotes* (p. 133)

12 Size medium (WT = 1.2 – 4.5 kg), top of head with black or brown cap or wedge, underside of tail tip hairy . . . . Capuchin monkeys, *Cebus* (p. 127)

12' Size large (WT = 3.5 – 15 kg), crown without contrasting dark cap, underside of tail tip naked 13

13 Head large, chin with prominent beard, throat swollen . . . . Howler monkeys, *Alouatta* (p. 136)

13' Head and throat not large, chin without beard 14

14 Limbs and tail robust, not thin and greatly elongated, thumbs well developed, does not usually travel by swinging by arms or tail below branches . . . . Woolly monkeys, *Lagothrix* (p. 140)

14' Limbs and tail long and thin, thumb almost absent, often swings below branches 15

15 Ears almost naked, inconspicuous, does not occur in SE Brazil . . . . Spider monkeys, *Ateles* (p. 142)

15' Ears thickly furred, stand out from side of head, SE Brazil only . . . . Muriqui, *Brachyteles* (p. 144)

Families and Genera of Carnivores (Carnivora)

1 All feet with four weight-bearing toes, fifth toe if present is a dewclaw that does not touch the ground 2

1' Feet with five weight-bearing toes 8

2 Claws of forefeet sharp and retractable, fifth toe present, tail not bushy . . . . Cats (Felidae) . . . 3

2′   Claws of forefeet blunt and nonretractable, fifth toe absent in wild species, tail bushy . . .
. . . . Dogs (Canidae) . . . 6
3   Body heavily spotted 4
3′   Body unspotted 5
4   Size large (HB > 1,000 mm), body, including neck, spotted, or black with shadows of
     spots . . . . Jaguar, *Panthera* (p. 168)
4′   Size small or medium, neck striped . . . . Small cats, *Leopardus* (p. 164)
5   Size large (HB > 800 mm) . . . . Puma, *Puma* (p. 167)
5′   Size small (HB < 700 mm) . . . . Jaguarundi, *Herpailurus* (p. 166)
6   Tail short and stumpy, does not reach ground, legs and neck short . . . .
     . . . . Bush dog, *Speothos* (p. 148)
6′   Tail long, to ground or nearly so when hanging 7
7   Body color black . . . . Small-eared dog, *Atelocynus* (p. 146)
7′   Body brownish gray . . . . Crab-eating fox, *Cerdocyon* (p. 147)
8   Large webs between all toes, tail very thick at base, tapering to tip . . . .
     . . . . Otters (Mustelidae) . . . 9
8′   No webs or at most partial webs between toes, tail not much thicker at base than near tip
     10
9   Size very large (HB > 1,000 mm), tail strongly flattened dorsoventrally, throat usually
     spotted, belly dark . . . . Giant otter, *Pteronura* (p. 162)
9′   Size smaller (HB < 900 mm), tail not strongly flattened, throat not spotted, belly pale or
     dark . . . . Neotropical otter, *Lontra* (p. 161)
10   Tail minuscule, less than 10% of head and body length . . . .
     . . . . Andean bear, *Tremarctos* (p. 150)
10′ Tail short, 60% of head and body length or less 11
10″ Tail long, 80% of head and body length or more . . . . Procyonidae . . . 15
11   Large (HB > 400 mm), face with black mask around eyes, tail with sharp black and pale
     rings . . . . Raccoons (Procyonidae), *Procyon* (p. 151)
11′ Large or small, tail never with dark and pale rings . . . . Mustelidae . . . 12
12   Sharply bicolored black or dark brown with pure white hood over top of head and neck
     and down back as stripe, tail pure white except at base . . . .
     . . . . Hog-nosed skunk, *Conepatus* (p. 161)
12′ Not sharply bicolored, tail not pure white 13
13   Head sharply patterned, tricolored: crown gray, white stripe over eye down side of neck,
     black muzzle, cheek, and throat . . . . Grisons, *Galictis* (p. 159)
13′ Head not tricolored, if white stripe or spot present, crown not gray 14
14   Large (HB > 400 mm), doglike, legs quite long, throat usually with pale patch, under-
     parts blackish . . . . Tayra, *Eira* (p. 159)
14′ Small (HB < 400 mm), not doglike, legs extremely short, underparts pale . . . .
     . . . . Weasels, *Mustela* (p. 157)
15   Tail prehensile . . . . Kinkajou, *Potos* (p. 155)
15′ Tail nonprehensile 16
16   Snout and front claws greatly elongated . . . . Coatis, *Nasua* (p. 153)
16′ Snout and front claws not elongated 17
17   Rings on tail indistinct or absent, no black and white eye rings, Central and South America
     . . . . Olingos, *Bassaricyon* (p. 154)
17′ Rings on tail sharp, black and pale, prominent black and white eye rings, Central Amer-
     ica only . . . . Cacomistles, *Bassariscus* (p. 156)

Rodents

1   Tail always long, bushy, completely covered above with straight, multibanded hairs
     > 10 mm long that stand out perpendicular to backbone, four to five cheek teeth on
     each side of each jaw . . . . Squirrels (Sciuridae)

1' Tail long or short, if completely covered with long hair above, then hair not multibanded, never more than four cheek teeth 2
2 Has large, externally opening cheek pouches . . . . Heteromyidae, Geomyidae
2' Without cheek pouches 3

3	Hystricognathi	Muridae
Four cheek teeth on each side of each jaw	+	−
Three or fewer cheek teeth	−	+
May have < 5 hindtoes	+	−
May have lateral mammae	+	−
May have HB < 150 mm	−	+
May have HB > 300 mm	+	−
May have HB > 150 mm and pelage of spines or bristles	+	−
May have HB < 150 mm and pelage of spines or bristles	−	+
May have fringe of down-curling hair on side of foot	−	+
May have HB > 200 mm and a hairy tail	+	−

Genera of Squirrels (Sciuridae)

1 Size tiny (HB < 115 mm, WT < 50 g) . . . Neotropical pygmy squirrel, *Sciurillus* (p. 194)
1' Size larger (HB > 115, WT > 50 g) 2
2 Ears long, standing well up above crown . . . . Common squirrels, *Sciurus* (p. 185)
2' Ears short, not standing high above crown 3
3 Tail hairs dense and bushy, incisors grooved; Talamanca highlands Panama and Costa Rica only . . . . Montane squirrel, *Syntheosciurus* (p. 195)
3' Tail relatively slender, incisors not grooved; Central and South America . . . .
. . . . Neotropical dwarf squirrels, *Microsciurus* (p. 193)

Families and Genera of Heteromyid, Geomyid, and Murid Rodents

Many of the small rodents are externally similar, and although those familiar with them can usually tell the genera and species apart in the field by differences in size, shape, fur texture, and behavior, these features are difficult to express in a key that has to work for a huge array of species over an enormous geographic range. Some characters are difficult to interpret, and in one case below the key is designed to bring out a genus from different possible interpretations of the same character. An individual species is given by name where one, but not other, species in a genus terminates in the key. This key may not always result in success.

1 Large externally opening cheek pouches present, body size large (HB > 170 mm), fur not spiny . . . . Pocket gophers, *Orthogeomys* (p. 195)
1' Large externally opening cheek pouches present, body small (HB < 150 mm), fur spiny . . . . Spiny pocket mice (Heteromyidae), *Heteromys* (p. 197)
1" No cheek pouches present, spiny or not spiny . . . . Muridae . . . 2
2 Dorsal fur includes stiff bristles or spines, not apparent visually, detectable when fur is rubbed backward 3
2' Dorsal fur without spines 4
3 Tail longer than head and body, size medium (HB > 120 mm) . . . .
. . . . Abrawaya's spiny rat, *Abrawayaomys* (p. 208)
3' Tail at least as long as head and body, size small ((HB < 110 mm), color brown with orange sides, underparts white and/or orange . . . . Spiny mice, *Neacomys* (p. 201)
3" Tail much shorter than head and body, size small (HB < 110 mm), color gray, underparts gray . . . . Gray spiny mice, *Scolomys* (p. 202)
4 Tail conspicuously hairy, with or without tuft at tip 5
4' Tail looks naked, may have fine hairs that protrude very slightly beyond tip 12

5   Tail entirely covered with flat-lying hairs that completely cover scales, tip without thicker tuft of hairs, muzzle blunt, eyes and ears small 6

5'   Tail hairs not adpressed (flat-lying), or hairs not completely hiding scales, and/or tail tip with conspicuously larger tuft, muzzle pointed, ears and eyes normal 8

6   Underparts dark gray or brown not sharply contrasting with sides, hindfeet not strongly paddle-shaped, hindtoes not partially webbed ....
     .... South American water mice, *Neusticomys* (p. 207)

6'   Underparts white or silver sharply contrasting with sides, hindfeet paddle-shaped, hindtoes partially webbed at base 7

7   Forefeet with four pads on palm, Central America only ....
     .... Central American water mice, *Rheomys* (p. 206)

7'   Forefeet with five pads on palm, Central and South America ....
     .... Crab-eating rats, *Ichthyomys* (p. 206)

8   Underparts (and upperparts) entirely rich orange-rufous, SE Brazil only ....
     .... Brazilian arboreal mouse, *Rhagomys* (p. 204)

8'   Underparts white, gray, yellow, or faintly orange 9

9   Color of back and/or sides bright orange or red, belly white 10

9'   Color dull red, brown, or tawny yellowish 11

10   Tail much longer than head and body, foot with broad white patch across top, SE Brazil only .... Rio de Janeiro arboreal rat, *Phaenomys* (p. 203)

10' Tail slightly longer than head and body, foot dusky with white toes, eye very large and ringed with black, Central America only .... Vesper rat, *Nyctomys* (p. 209)

11   Tail thickly or thinly haired, with distinct slight or prominent tuft of denser hair at tip, whiskers coarse and long, several reach shoulder, hair base dark gray, females with six mammae .... Climbing rats, *Rhipidomys* (p. 204)

11' Tail finely haired, hairs protruding slightly beyond tail tip, but not forming a distinctly thicker tuft, whiskers dense but fine, reach ear tip but few or none reach shoulder, hair base pale gray, females with eight mammae .... Arboreal rice rats, *Oecomys* (p. 200)

12   Hindfoot broad and paddle-shaped, with fringe of down-curling silver hairs along side, hindtoes partly webbed at base, whiskers fine and short, ears small, body size large, fur with gloss or sheen, tail hairs often longer and denser below than above 13

12' Hindfeet not paddle-shaped, without fringe or webs, whiskers long or short, body size large or small, fur glossy or dull, tail hairs similar above and below 14

13   Tail slightly longer than head and body, ears naked at tips, sole of hindfoot to heel covered with roundish scales .... Water rats, *Nectomys* (p. 202)

13' Tail usually shorter than head and body, ears hairy to tips, sole of heel without scales ...
     .... Marsh rats, *Holochilus* (p. 212)

13" Tail much longer than head and body, > 120%; ears hairy to tips, heel without scales ...
     .... Rice rats, *Oryzomys ratticeps, O. buccinatus* (p. 198)

14   Tail much shorter than head and body, 66% or less 15

14' Tail 66% to longer than head and body 18

15   Muzzle blunt, eyes and ears tiny, lost in fur, tail < 50% of head and body, SE Brazil only .... Brazilian shrew mouse, *Blarinomys* (p. 207)

15' Muzzle pointed, eyes and ears normal, tail 50–80% of head and body 16

16   Muzzle greatly elongated, claws greatly elongated ....
     .... Long-nosed mice, *Oxymycterus* (p. 211)

16' Nose pointed but not elongated, claws slightly elongated 17

17   Color dark brown to olivaceous, South America only ..... Grass mice, *Akodon* (p. 210)

17' Color blackish with reddish tints, Central America, W coast of Colombia and Ecuador, N Venezuela only .... Rice rat, *Oryzomys caliginosus* (p. 198)

18   Tail tip white, sharply divided from dark base, size large (HB > 180; WT > 150 g) ....
     .... Naked-tailed climbing rats, *Tylomys* (p. 208)

18' Tail tip not white, or size smaller 19

19   Size tiny (< 110; adult WT < 40 g, usually < 30 g) 20

19' Size larger (adult WT > 40 g) 23

20   Hindfeet broad and short, with pink soles, color uniform reddish brown, with pure white
     underparts, females with eight mammae . . . .
     . . . . Bicolored arboreal rice rat, *Oecomys bicolor* (p. 200)
20′ Hindfeet long and narrow, soles dark, color tawny, gray, or brown 21
21   Tail about the same length as head and body, hindfoot ≤ 20 mm, underparts gray or
     grayish not sharply contrasting with sides, females with ten mammae . . . .
     . . . . House mouse, *Mus musculus* (p. 215)
21′ Tail about the same length as head and body, hindfoot ≥ 22 mm, females with eight
     mammae . . . . Pygmy rice rat, *Oligoryzomys microtis* (p. 199)
21″ Tail much longer than head and body, underparts white or gray contrasting with sides,
     females with fewer than ten mammae 22
22   Incisor teeth grooved, sides of head, neck, or body often orange or cinnamon, females
     with six mammae . . . . Harvest mice, *Reithrodontomys* (p. 213)
22′ Incisors not grooved, sides of head often grayish, females with eight mammae . . . .
     . . . . Pygmy rice rats, *Oligoryzomys* (p. 199)
23   Large and robust, fur sparse and coarse, with very long guard hairs, tail naked, slightly
     bristly, with coarse, large scales, females with ten or more mammae . . . .
     . . . . Brown and black rats, *Rattus* (p. 214)
23′ Small, fur soft and dense, tail either slender, with fine, inconspicuous scales, or dark and
     shiny, with rings of scales, females with fewer than ten mammae 24
24   Ears greatly enlarged, tail completely naked, dark and shiny as if varnished, with con-
     spicuous rings of scales, Central America only . . . .
     . . . . Big-eared climbing rat, *Ototylomys* (p. 209)
24′ Ears not greatly enlarged, tail not shiny, without broad rings of scales 25
25   Tail longer than head and body, whiskers long, reaching shoulder, females with six mam-
     mae, Central America only 26
25′ Tail shorter than head and body, or if not, whiskers not reaching shoulder and females
     with eight mammae, Central and South America 27
26   Brown to tawny, tail slightly longer than head and body . . . .
     . . . . Mexican deer mouse, *Peromyscus mexicanus* (p. 212)
26′ Cinnamon to orange, tail much longer than head and body . . . .
     . . . . Isthmus rats, *Isthmomys* (p. 213)
27   Hindfeet short and broad, soles entirely pink, tail finely haired, hairs protruding slightly
     beyond tip in a tiny tuft, ears relatively short, hair base pale gray . . . .
     . . . . Arboreal rice rats, *Oecomys* (p. 200)
27′ Hindfeet long and narrow, soles dark, ears large, hair base dark gray 28
28   Hind toepads pink, midback sometimes with black stripe, SE Brazil and NE Argentina
     only, females with six or eight mammae . . . . . Atlantic forest mice, *Delomys* (p. 205)
28′ Hind toepads black, midback not striped, females always with eight mammae, entire
     region . . . . Rice rats, *Oryzomys* (p. 198)
NOTE: The final genera in (28) may not be distinguishable in the hand.

Families and Genera of Hystricognath Rodents

1   Body size large (WT > 0.8 kg), body shape and tail not ratlike or squirrel-like, some-
     times piglike, tail long or short 2
1′   Body size smaller (WT < 0.8 kg), body and tail ratlike or squirrel-like, tail always natu-
     rally long but may be broken off . . . . Spiny rats (Echimyidae) . . . 8
2   Clothed with stout, sharp spines (these may be partially hidden by woolly hair), tail over
     one-third of head and body length, thick, tapered toward tip, prehensile 3
2′   Not spiny, tail almost absent, or if present, nonprehensile 4
3   Spines on hindquarters not barbed, flexible and dry, like broom straws, SE Brazil only . . .
     . . . . Bristle-spined porcupine, *Chaetomys* (p. 222)

3′   Spines on hindquarters stiff and barbed, throughout rainforest region . . . .
. . . . Neotropical porcupines (Erethizontidae), *Coendou* (p. 217)
4   Spotted, always large (WT > 5 kg) 5
4′   Not spotted, large to small 6
5   Tail large, about 30–50% of head and body length, color black with rows of white spots
from shoulder to rump . . . . Pacarana (Dinomyidae), *Dinomys* (p. 225)
5′   Tail minuscule, 5% of head and body length, color brown or red with rows of white spots
from neck to rump . . . . Pacas (Agoutidae), *Agouti* (p. 224)
6   Size very large (> 20 kg), piglike, with stout legs, webs between toes . . . .
. . . . Capybara (Hydrochaeridae), *Hydrochaeris* (p. 223)
6′   Smaller (WT < 6 kg), with slender legs and high rump, toes not webbed . . . .
. . . . Dasyproctidae . . . 7
7   Large (WT = 2–6 kg), tail a tiny, naked, inconspicuous blackish stump . . . .
. . . . Agoutis, *Dasyprocta* (p. 226)
7′   Smaller (WT = 0.8–1.5 kg), tail short, hairy, but conspicuous, pencil-thin with a white tip
. . . . Acouchys, *Myoprocta* (p. 230)
8   Fingers of hands elongated, with nails, not claws, middle fingers separated by a gap . . . 9
8′   Fingers not elongated or separated by a gap, with claws, not nails 10
9   Tail naked, scaly, from Amazon Basin (or tail hairy and from Andean montane habitat in
S Peru or N Bolivia) . . . . Amazon bamboo rats, *Dactylomys* (p. 244)
9′   Tail hairy, from SE Brazil and N Argentina . . . .
. . . . Southern bamboo rat, *Kannabateomys* (p. 245)
10   Muzzle long, ears distinctly longer than broad, protruding above crown, hindfeet long
and narrow, tail slender, usually naked, and ratlike (may be missing) 11
10′   Muzzle short and blunt, ears short, not much if at all longer than wide, do not protrude
above crown, hindfeet short and broad, tail robust, naked, tufted, or hairy 12
11   Lower back with stiff spines standing upward in a geometrically spaced pattern, Central
America and Pacific coastal Ecuador and Colombia only . . . .
. . . . Armored rat, *Hoplomys* (p. 233)
11′   If spiny, spines lie flat and are inconspicuous, except in SE Brazil; Central and South
America . . . . Spiny rats, *Proechimys* (p. 232)
12   Fur soft, not bristly when brushed backward, tail fully furred and bushy 13
12′   Fur with stiff bristles or flexible spines 14
13   Color olivaceous, tail often orange and/or dusky; Amazon Basin only . . . .
. . . . Brush-tailed rats, *Isothrix* (p. 241)
13′   Color rusty, or rusty especially on rump, Central America, Pacific Coast, and mountains
of N Colombia and Venezuela only . . . . Rufous tree rats, *Diplomys* (p. 242)
13″   Color dirty white with striking black markings, SE Brazil only . . . .
. . . . Painted tree rat, *Echimys pictus* (p. 238)
14   Densely covered with flat, flexible spines, size medium to small (HB = 155–220), tail
hairy, Amazon Basin and the Guianas 15
14′   If densely spined, then either large (HB > 230), or tail naked-looking and scaly, or from
Atlantic coastal forests of Brazil from Ceará south 16
15   Size small (HB < 210), brown, tail entirely sparsely covered with red-brown hairs that
form a small tuft at tip (but tail sometimes missing) . . . .
. . . . Spiny tree rats, *Mesomys* (p. 234)
15′   Slightly larger (> 220), gray-brown, tail scaly except for long, coarse hairs in prominent
tuft at tip, Brazil, S of Amazon and E of Madeira only . . . .
. . . . Tuft-tailed spiny tree rat, *Lonchothrix emiliae* (p. 235)
16   From the Amazon Basin, Colombia, Venezuela, and the Guianas 17
16′   From Brazil southeast of the Amazon Basin . . . . Atlantic tree rats, *Nelomys* (p. 236)
17   Color blackish or gray; tail with white tip . . . . Tree rats, *Echimys* (p. 235)
17′   Color brown with reddish nose, or blackish, tail without white tip . . . .
. . . . Red-nosed tree rats, *Makalata* (p. 236)

Appendix C

Classification, Study, Biogeography, and Conservation of Neotropical Rainforest Mammals

The System of Classification of Organisms

You will see in this book that a Venezuelan cannot talk to a Peruvian about a paca and be understood: each uses a different name for the same animal. Scientific names are a necessity for unambiguous communication, even on the same continent among people who speak the same language. Only because the scientific names are given for each species can anyone in the world read this book and know exactly what animal is meant—now, or a century hence.

Organisms are scientifically classified according to a universal system. It is a practical system devised before evolution was discovered, originally based on how people perceive the world, and very similar to the way peoples worldwide classify living things in their own native languages. Humans naturally tend to group and name everything they see according to visible relationships or similarities, such as trees and oak trees, birds and sparrows. Science has refined the natural way we classify things into a precise system. The system is hierarchical: organisms are grouped into a series of nested, increasingly inclusive categories. Recent classifications often have the underlying principle, or goal, that the pattern of the classification reflect true evolutionary relationships: all the animals grouped under one name at a given level of classification should be more closely re-

lated to each other than to any animals under another name at the same level. For example, all members of the cat family (Felidae) are more closely related to each other (share a more recent common ancestor) than to any member of the dog family (Canidae); within the Canidae, all members of the dog genus *Canis* are more closely related to each other than they are to any members of the fox genus *Dusicyon*.

The level of classification that a name refers to is sometimes shown by the ending of the name. The classification of a rice rat *(Oryzomys capito),* including the categories of classification used in this book, is shown in Table 1. There are other levels, such as superfamily or subtribe, that are not used here. Subcategories (subfamily, subgenus, subspecies, etc.) have been defined in only some cases; some taxa do not have any. A taxon (plural: taxa) is the group of organisms under a single name at any of the levels below (e.g., family, genus, species, or subspecies).

The binomial or two-part name (genus and species, e.g., *Oryzomys capito*) is always used to designate a species. According to the rules, every generic (genus) name has to be different, but the species names do not (as long as they apply to different genera); therefore, the species name alone does not unambiguously designate a species. The generic name is always capitalized; the species (and subspecies) name is always in lowercase; both names are always italicized. The

Table 1

Kingdom: Animalia	animals
Class: Mammalia	mammals
Order: Rodentia	rodents
Suborder: Sciurognathi	rodents with squirrel-like jaws
Family: Mur**idae**	rats and mice
Subfamily: Sigmodont**inae**	New World rats and mice
Tribe: Oryzomy**ini**	rice rat group
Genus: *Oryzomys*	true rice rats
Subgenus *Oryzomys*	typical rice rats
Species: *capito*	common Amazon rice rat
Subspecies: *velutinus*	the populations in the Guiana region

subgenus name is placed in parentheses following the genus. The names are usually abbreviated as follows, for example, from the full name, *Oryzomys (Oryzomys) capito velutinus: O. capito* to designate the species; *O. c. velutinus* to designate the subspecies; *O. (Oryzomys)* to designate the subgenus. The singular abbreviation sp., meaning a species, is used after the generic name when it is not clear what the specific name is, or in the case of a new, unnamed species (usually written sp. nov.); the plural spp. is used after the generic name to designate all of the species in that genus.

On Species, Speciation, Why We Need Museums, and Why Scientists Can't Agree About Names

Species. If you opened this book, you are probably curious about what kinds of mammals live in American rainforests. You perhaps couldn't care less about that scientific name attached to each mammal, especially since even the experts often cannot agree on what it should be. However, neither this nor any other guidebook to living organisms in the world could have been written without the sciences that concern themselves with what those names should be: systematics, taxonomy, and evolutionary biology. This book describes kinds of animals. The kinds of animals are species. To write a book like this, we need to know what species there are, where they are, and how they differ from each other so that we can tell them apart.

A species is a real thing although it can be defined in various ways. A common definition is "groups of actually or potentially interbreeding organisms separated from other such groups"; in other words, a bunch of related animals that can breed among themselves so that genes are exchanged among them and not with other such groups. There are other definitions of species, but it would take several chapters to discuss them. Species are composed of populations of interbreeding individuals; if geographically separated populations or groups of populations of a species have evolved sets of different characteristics (such as different color patterns) so that by looking at an individual

it is possible to tell which group of populations it came from, then the different populations are often named as subspecies. Subspecies are arbitrary units that are defined for the sake of convenience; two neighboring subspecies may blend into each other with no clear dividing line between them, while two others are sharply defined and geographically separated.

Speciation. The great discovery of Charles Darwin and Alfred Russel Wallace was that individual animals and plants vary, that their variations can be inherited by their offspring, and that by the process of natural selection, some variations make the individuals that carry them more successful than other individuals in producing offspring. Through time, the perpetual process of variation of individuals, the weeding out of some individuals and selective success in producing successful offspring of others, causes populations to change, or evolve, into sets of individuals with characteristics quite different from those of their ancestors. At some point in the process, a population can become so different from its ancestors that it becomes (has evolved into) a new species. We now know (which Darwin and Wallace did not) that heritable characteristics are encoded in genes on molecules in reproductive cells, and that changes in the genes (mutations) create the genetic (heritable) variations among individuals.

If a population of a species becomes physically separated from others of its kind, such as by living on a separate island or on different sides of a mountain range too high to cross, then the evolutionary changes that occur within it can no longer be freely shared by interbreeding with members of other populations (there is no "gene flow" between them). An isolated population can gradually become more and more different from its nearest relatives, perhaps adapting to a different environment, until it is so changed that the two populations have become different species: their members look different, act differently, have different sets of genes, and if they come into contact with their former relatives again, either they cannot interbreed, or if they do, the "hybrid" offspring are less successful in reproducing than are offspring of members of either parent population. Speciation has occurred, and if the two species differ enough in their ecological roles, they could now share the same geographic range as noninterbreeding, separately evolving populations. It is not surprising that both Darwin and Wallace discovered natural selection during voyages among islands, where steps in the processes of the evolution of species (speciation) are most clearly laid out among isolated insular populations.

Why scientists disagree about species' names. If we could stand in one place in the Amazon rainforest and look at all the mammals that live together in that patch of forest—perhaps as many as 150 species—it would be relatively easy to sort them all out, count the number of species, describe them, and give each one a name. The individuals of each species in that forest interbreed constantly, mixing their genes with each generation, so that all members of that species in the same population look much alike. If we could then jump across the River Amazon—so wide in places that you cannot see across it—and again gather up all the mammals from one patch of forest on the other side, we could again easily sort them, separating the individual species. But then, if we could take the animals from both sides, line them up together, and again try to sort them, we would immediately have a problem. Some kinds of animals would occur on only one side of the river, with none at all like them on the other side; but for many of the remaining animals we would find that each side of the river had animals of a very similar kind (clearly members of the same genus), but not exactly alike. Similar animals from different sides of the river might have slightly or very different colors; they might be different sizes or have tails of different lengths. We could tell by looking at them which side of the river they came from. These differences between the populations are maintained because the physical barrier between them (the river) prevents interbreeding. Are they different species or not?

What we are seeing is geographic variation, often an indication of slight genetic isolation among populations. Geographic variation across a species' range may represent stages in the process of the separation of lineages that can lead to formation of new species. South America is a continent with exceptionally large numbers of barriers that animals and plants cannot easily cross. The rainforest region is split into many pieces:

the Andes divide the west coast and Central America from the interior of the continent; the Amazon splits the region across the middle from east to west; on each side a series of giant tributaries cuts the forest into long strips often 300 km wide and over 1,000 km long; and on the north and southeast of the forest, wide areas of grasslands or dry scrub divide the Amazon Basin rainforest from rainforests in wet coastal regions of Venezuela and eastern Brazil (see maps 1, 2, 3). It is quite natural then, and even to be expected, that the populations of animals we find on different sides of these barriers do not look the same. How do we decide which of these slightly or distinctly different-looking populations of animals are species and which are simple variations (subspecies)?

The best way to go about it is to gather together, side by side, representatives of every different population of that kind of animal, making sure we include representatives of all the kinds of individuals that occur within the same unbroken population, as well as from all the "separated" populations. For example, for the animals found between the Amazon, the Madeira, and the Purus rivers, we might want to look at individuals from the south bank of the Amazon, from 200 km upstream on each side of each tributary, from 400 km upstream, and from 1,000 km up in the headwaters. We would then have before us what we need to make a study of the variation that occurs in those populations. We might see immediately that although animals from one side of the Madeira at its mouth near the Amazon are small and pale brown and those from the other side are large and black, that as we go farther and farther southwest toward the river source, the animals become more and more similar until they merge into a single medium-sized, dark brown population 1,200 km away in Peru and Bolivia around the headwaters. In this case, we could infer that we were looking at two populations of the same variable species. On the other hand, if we find that everywhere we look there are either small, pale brown animals or large black ones, with no intermediate-looking individuals, and that at the headwaters there is even a small zone where both kinds of animals are found together in the same forest but do not interbreed, then we can be confident that we are dealing with two species.

To find out what species there are, and to study the evolutionary relationships among them, we therefore need to compare individuals from different places in as much detail as possible: not only their size or color, but every feature of their anatomy that we find to vary—each bone of the skull, the cusps on the teeth, the soles of the feet, the anatomy of the reproductive tract, and increasingly, the very structure of their molecules. This is done with museum specimens. You will soon notice if you leaf through the "variation" headings in the accounts above that, for a very large number of genera, scientists do not yet agree on how many species there are: whether different-looking populations represent true species or just geographically separated varieties of the same species. One of the main reasons for such disagreements is that we simply do not have enough specimens in museums, from enough places in the animals' ranges, to be able to do the simple study outlined in the previous paragraph. We may have individuals from a few locations along the Amazon but none at all from any headwaters; we often do not even know where the geographic limits of a species lie. Often the museum specimens do exist, scattered throughout many museums in Europe, the United States, and South America, but no one has yet undertaken the job of comparing them all. In such cases we simply have to make educated guesses about how to divide the populations into species. In other cases we know that the geographic ranges of the different varieties do not overlap, and thus we cannot ever observe whether they might interbreed in nature. In this case also, whether we call them a single variable species or several species is our best educated opinion. When there is not enough information available to truly resolve the question, scientists' opinions on species vary depending on their views about different processes of evolution and differing concepts of what degrees of difference species are likely to show. Some prefer to "lump" varieties under one species name until they have been proved separate, while others tend to "split" varieties into as many individual units as are recognizable, and call them all species. There are also different schools of thought about how the concept of a species should be defined. Despite these philosophical differences, the study of the nature and boundaries of species in natural populations presents major scientific questions of great theoretical and practical importance, as well as describing the world we live in. In summary, it is the natural variations among organisms, and lack of sufficient knowledge about them, that causes the most problems when we try to classify them into species.

If two species are so similar looking that we can barely tell them apart, who cares whether we give them the same or different scientific names? Why worry about it? Species that look morphologically alike can have strikingly different ecologies and behaviors (sometimes it is these differences that have led to the discovery that more than one species was lumped under the same name). Closely related species can have drastically different effects in agroeconomics or in the epidemiological cycles of human diseases. The most similar species are also the best ones for studies of the finest levels of the workings of ecology and evolution.

Most field guides sweep the messy unresolved problems of taxonomy under the rug of authority and present to the reader a static world where all species are sharply defined and ordered. But this is not true for any set of organisms anywhere in the world. We have decided to present the *taxonomie réalité,* not to exasperate and confuse readers and convince them that scientists are nitwits, but to impress upon readers how little we know, how much still needs to be done before we can write that perfect field guide, and most of all, that tropical America is a living example of the changeable nature of organisms and mechanisms of evolution, where mammals are as diverse and diversifying as anywhere on earth.

Why we need museums. Museums of scientific specimens are libraries where organisms are carefully preserved so that they can be studied in the future. Many of the specimens collected by great scientific explorers such as Darwin, Audubon, von Humboldt, and Wallace are still preserved today. We can directly compare them now, side by side with species discovered decades or centuries later. As we saw in the previous section, we cannot study species, or evolution, or the distribution of organisms without museum collections. The differences between species of mammals can be subtle and easily over-looked: the lobes on a tooth, the number of nipples, or the position of a whisker can be a distinguishing feature. We may have to look again and again over the years before we discover important differences between two taxa. We must be able to re-examine an animal when a new method is discovered by which we can study the relationships between organisms, or when a new machine allows us to look at what we have not been able to see before. Variation is characteristic of all the forms of life on earth and part of their essential nature, but a single specimen can tell us little or nothing about it.

If we want to understand much about species, we need to preserve enough individuals from different populations and geographic locations to encompass their ranges of variation. This means collecting animals—not one, but a whole series. Many people object strongly to killing wild animals, for sentimental reasons. This is certainly an understandable viewpoint. But collections of animals are also vital for the preservation of living ones.

Most of us, including those who object to all killing of animals, wish to conserve the other species of life on earth. Man has already overrun the earth, and it seems likely that by the end of our own lifetimes nothing will remain of the richest natural ecosystems on earth, such as the rainforests, except those places deliberately chosen for preservation. We now face the irrevocable choices: to condemn this forest and all its inhabitants to death and replace it with crops to feed more people; or to save that one for the sake of keeping a few species. Could you choose which group of forest preserves would save the most species?

To make a wise choice, you would need to know what kinds of species there were in each forest and whether those same species were found in many other forests, or in none. For the majority of species in the world, that information is derived solely and directly from museum collections. If the information cannot be found for the forests one wants to save or exploit, then new collections of specimens need to be made there. These constitute the only credible and lasting record and proof of where organisms are, or, increasingly, used to be. The collecting of mammals

for museums is on far too small a scale to harm any of the species or individual populations concerned. Unlike hunters for meat or commerce, museum collectors usually visit an area for only a short period, take a limited number of specimens, and then do not return to the same area again for years, if at all. Unlike the situation with butterflies or birds' eggs, mammal specimens have no commercial value, and specimens are not taken for private collections.

This book was written based on museum specimens. The range maps were drawn by compiling where they came from. The illustrations were painted by examining series of specimens and choosing typical individuals as models of the color patterns. Museum collections are the origin of the lists of species that are our only way of actually measuring the species richness or diversity of a habitat. Their vital importance for both science and conservation should be understood by everyone interested in the inhabitants of our planet.

The Biogeography of Neotropical Mammals

The geographic distribution of species, or biogeography, is the combined result of the geological history of the earth and the biological history of the evolution of organisms. The collection of animals and plants that is present at any one place and time always tells a fascinating story of events that we cannot see, but that can be unravelled like a mystery. The study of biogeography asks simple questions: (1) What group of species (fauna or flora) is found in a region (such as the Neotropics)? (2) How does the fauna change from place to place within the region? (3) Where did the different species come from (what is the geographic history of the elements of the fauna)? and (4) What can we infer about the events and mechanisms that led to the distribution of species on the landscape at a particular time? Answering these questions is not simple.

There are only two ways for a species to occur at a given geographic place: either it came there by immigration (or transport) from somewhere else, or it evolved in that place by speciation from ancestors that were already there. In the second case, the ancestors could themselves have reached that place by immigration at any time; between a remote era, say 100 million years ago, when a primitive marsupial may have traveled, to just yesterday, say about 10,000 years ago, when the rainforest grew back over formerly arid areas and a rainforest mammal could move with it.

Whether a species can spread from one region to another essentially depends on the kinds of barriers there are to prevent such spread. Barriers can be of many kinds: physical, such as wide oceans, rivers, mountain ranges, or inhospitable habitat types; or biological, such as a disease or a predator that might prevent a species from surviving, or another species that is already established and that uses the same resources.

Speciation itself, as well as the assembly of lineages and species into regional faunas, occurs on geological time scales: the fauna of the Neotropics today is the accumulated product of millions of years of events, some very ancient, some recent, some happening now. Several major events are responsible for the main features of the Neotropical fauna.

The most momentous factor that has determined what mammals live in South America has been its history of connections with other continents. The earliest ancestral mammals evolved and spread on the landmass of Pangaea before continental drift broke the continents apart and separated them with oceans, beginning in the early Jurassic period about 190 million years before the present (MYBP). By about 135 MYBP, in the earliest Cretaceous, Pangaea had separated into the southern continent of Gondwanaland and the northern continent of Laurasia; by the end of the Cretaceous to the early Paleocene, about 65–55 MYBP, as the dinosaurs died out, Gondwanaland itself had split, and South America had become an island. It remained so for about the next 50 million years.

This period was the dawn of the time when, finally free of the dominance of the dinosaurs, mammals exploded in an exuberance of evolution and filled the earth's ecosystems with creatures as strange and wonderful-seeming to us now as were the dinosaurs themselves. In South America, several endemic orders of mammals radiated into diverse assemblages containing many bizarre-looking animals entirely unlike those that were evolving on other continents, much as Australia today is full of beasts unlike those anywhere else. Most of these orders are now extinct, but the last remnants of Xenarthra (anteaters, sloths, armadillos) are still with us. The splitting up of the continents was a slow process, and all the while mammals were evolving. Some early members of orders also found on other continents somehow managed to immigrate to South America before the continents got very far apart, but once there, they were isolated and evolved into endemic families of species. These included the marsupials, primates, and hystricognath rodents, many of which still dominate the rainforest mammal fauna today.

About 7 MYBP an event took place that dramatically changed the history of South America: the Isthmus of Panama rose above sea level, and for the first time in about 130 million years animals could walk on land between North and South America. The "great faunal interchange" then began. Xenarthra, marsupials, primates, and caviomorph rodents moved north, and carnivores, deer, peccaries, squirrels, and murid rodents went south. This interchange completed the major groups of the fauna as they are today. Other major geological events, such as the uplifting of the Andes, had profound effects on temperate-climate and high-elevation species, but relatively little effect on lowland rainforest genera, apart from their roles in influencing the weather and presenting a barrier to migration.

The present fauna of the Neotropics is thus a mixture of families that evolved in the Northern and Southern hemispheres. We still do not know much about the history of some important families of mammals in South America, such as the leaf-nosed bats and murid rodents.

If we look in general at the faunas of the New World, some interesting facts about the rainforest appear. It is common knowledge that there are more species of mammals in rainforest than in other environments, but what kinds of species are these? Of the approximately 500 rainforest species, 60% belong to endemic South American families. Among these are all of the living sloths and platyrrhine monkeys and most of the species of leaf-nosed bats, echimyid rodents, anteaters, and marsupials, so that the rainforest is now a refuge for the last or only representatives of some of the old endemic radiations. If the rainforest is lost, a large and perhaps final chapter in the evolution of mammals in South America will be closed forever.

More recent events shaped the fauna at the species level. The extent of the Neotropical rainforest was not always as we see it today. During the turbulent Pleistocene, when glaciers advanced and retreated many times in the Northern Hemisphere, the Neotropics (and many other tropical areas) experienced a series of alternating wetter and drier periods. There is evidence that during the drier periods the rainforest shrank and became fragmented, perhaps like a series of forest islands (called refugia) with drier savannas or scrub between. The rainforest animals and plants were then isolated in these refugia. With the exception of a few, mostly large species, most rainforest mammals can live only in rainforest climates and habitats. A savanna or desert is as great a barrier to their movement as a wide river. As we have seen, this scenario of isolation of populations on islands is a situation likely to allow animals to evolve differences that can result in the formation of new species.

As described in the earlier section on speciation, the Neotropical rainforest region is also today divided into many pieces by mountains, rivers, grasslands, and arid zones. Certain barriers are large and almost complete (the Andes, the arid regions between Amazonia and southeastern Brazil); others allow animals to cross from time to time (rivers). Many scientists think that this dynamic history of past and present fragmentation explains the high species richness of the Neotropical rainforest today, which has more species in one place than do the world's other rainforests.

When we look at the present distribution of species in the Neotropics, we see that we can divide the rainforest region into large sections that have series of endemic species within them. They have their own faunas of birds, mammals, lizards, butterflies, and so forth. The faunas of adjacent sections are often very similar, with adjacent sections having what one might call "sister" species: species that are clearly nearest relatives and that may have the same ecological roles (also known as parapatric species).

There are four main faunal areas for rainforest mammals (see maps 1 and 2): (1) Central America and the Pacific lowlands of Colombia, Ecuador, and northern Peru; (2) the great block of the Amazon Basin; (3) the Atlantic rainforests of southeastern Brazil; and (4) the Caribbean coastal rainforests of Colombia and Venezuela. Each rainforest block is separated from the others by a combination of barriers, including mountains, rivers, and dry habitats. As one might expect, the vast forests of the Amazon harbor by far the most species. The Central American rainforest fauna is relatively small, with a number of endemic species that are close relatives of those in Amazonia, but few endemic rainforest genera. The northern Caribbean coastal forests have even fewer species, a number of them also endemic. The barriers between the latter two regions are leaky, and a number of exclusive species and even some endemic genera (e.g., *Diplomys*) are shared between them. In contrast, the rainforest fauna of the Atlantic forest of southeastern Brazil is now, and has for a long time been, totally isolated from other regions, and is extraordinarily different from that of the Amazon Basin, with many endemic genera and species, and unique animals with no close living relatives, such as the bristle-spined porcupine.

These four large regions by no means have homogeneous faunas: they can once again be divided into areas that each have a number of their own endemic species. For example, the "Guiana" region, which includes all the rainforest east of the Rio Negro and north of the Amazon, and south of the Amazon east of the Xingu, has a series of species of its own, such as the Guianan saki monkey, brown bearded saki, and Brazilian bare-face and golden-handed tamarins; but north of the river it also lacks many species and genera found elsewhere in Amazonia, such as red squirrels, night monkeys, titi monkeys, and olingos.

In numbers of mammal species present, the richest part of the Neotropical rainforest is the area along the base of the Andes in Colombia, Peru, and Ecuador, and the poorest is the Guiana region. But it is essential to reemphasize that even this relatively "poor" region is very rich, and includes its own endemic species that are not found elsewhere.

Conservation of Neotropical Rainforest Mammals

Conservation is the protection of species against the actions of man. Many factors determine what makes a given species more or less vulnerable to extinction or serious damage by human action. The main ones are outlined briefly below. As stated in the epigraph of this book, we have now become responsible for the fate of a major portion of the other species on earth. If we are to succeed in saving many of them, we must understand how human action affects the organisms we wish to preserve.

We exterminate organisms in three basic ways: (1) by singling them out and destroying them individually for human consumption (in the broad sense), as we have done for Steller's sea cows, passenger pigeons, and dodos (among others); (2) by destroying resources they need in order to survive, such as habitat or food supply, as is happening to giant pandas and golden lion tamarins; and (3) by introducing exotic species, including diseases of domestic animals, that kill or outcompete the native flora or fauna. Two or all three factors may be involved in specific cases, as in the decline of California condors and muriqui monkeys. The first factor was the primary one historically, when human populations were low; the second factor is becoming and will in the future be by far the most important one as human populations reach catastrophic levels. The third is of lesser importance in tropical rainforests, but is one of the major causes of extinctions on islands and in aquatic habitats, and is threatening increasing numbers of ecosystems.

For the rainforest mammals described here, we can define which characteristics make a species more or less vulnerable to extinction from each cause. As can be seen repeatedly throughout this book, hunting of individual species for food, skins, or commerce has affected populations of many species throughout the Neotropics. Several features of a species' biology determine how intensely it is hunted, how well it can withstand hunting pressure, and for how long. The first is size: animals are shot for meat only if there is enough meat on them to be worth the price of a shotgun shell (a relatively large sum for subsistence hunters). The cutoff point is 0.5–1 kg: the size of a large squirrel or small monkey, although these are not often hunted. A hunter will tend not to risk scaring away larger game by shooting at smaller animals until he is at the end of a day's fruitless hunt. The amount of meat on an animal determines its value, which in turn determines the amount of special effort a hunter will give to finding it. Because of their weight, the largest species, such as manatees, tapirs, spider monkeys, and woolly monkeys, are relentlessly pursued and have been locally extirpated in many inhabited areas.

A second important characteristic determining vulnerability is conspicuousness; this determines whether hunters will be able to find and kill every individual. Animals that live in noisy groups are more conspicuous than solitary ones, and large animals that live in the canopy are much easier to see and follow than large terrestrial species. Secretive animals that favor dense thickets, such as deer or agoutis, are often able to escape detection or flee out of range and sight before a hunter can shoot, and some individuals always escape.

A third factor is whether an animal's habitat is easily reached by hunters. There are few roads in the Amazon Basin, and rivers are the main access routes into the forest; any place that a motor canoe can reach is accessible to hunters. The only species that appear genuinely endangered in the Amazon Basin today are those restricted to riverine habitats: giant otters, manatees, black caiman, crocodiles, giant river turtles, giant osteoglossine fish, and so forth. Giant otters have every characteristic to make them vulnerable: valuable skins, large size, large, noisy groups, and riverine habits; they are now extinct from much of their former range. The Amazon Basin is so large and thinly inhabited that most terrestrial species can still find remote refuges regardless of how much they are worth.

The fourth major characteristic that affects a species' ability to withstand hunting pressure is its reproductive rate. This determines how quickly animals that are shot can be replaced by new individuals in the population and how well the population can withstand steady removal of individuals. Deer, in which females can first breed as yearlings and can have one or twin young every year thereafter, have a strong ability to maintain their populations under some hunting pressure, especially when this is coupled with their solitary, cryptic habits. Spider monkeys, in which females may not breed until they are five or six years old and have only one young every three years thereafter, have almost no ability to replace the individuals shot under even light hunting pressure, especially since their low reproductive rate is coupled with large body size and conspicuous arboreal behavior that allows hunters to find almost every individual.

Susceptibility to the other cause of extinction—loss of resources—mainly involves two characteristics of species: (1) the size of their geographic range and (2) their ecological resilience or flexibility. The first of these is fairly self-evident: the smaller an area a species inhabits, the more likely that a single catastrophic event or human project can wipe it out. The second determines how well a species can adapt (as a living animal, not in the sense of adaptation as evolutionary change) to human-caused changes in its habitat. For example, a species that can live in secondary forest and hedgerows may not be threatened by cutting of the primary forest, but a species that can live only in primary forest will be. An example is the near-extinction of lion tamarins while in the same region, many populations of tufted-ear marmosets (which can live in secondary scrub) are succeeding well.

A final characteristic of species affects their susceptibility to extinction by any means: their natural population size and density of individuals. Population density is a result of many factors; while there is not enough space to discuss them here, suffice it to say that the fewer individuals there are, both in total and per square kilometer, the more easily a species seems to be eliminated by habitat destruction and fragmentation or by hunting.

As populations of animals become fragmented and restricted to isolated parks, they are increasingly susceptible to extirpation due to epidemic disease. Introduced diseases, including canine distemper, have recently decimated wild species. Keeping all domestic animals out of parks, and never reintroducing to the wild any animal that has been captive, may prove to be more important for long-term species survival than is the prevention of poaching.

The intact rainforest could produce an indefinitely sustainable yield of its myriad resources of plants and animals useful to man. For it to do so, each exploited species needs to be regulated with respect to its individual ability to support cropping. If rainforests can be managed for production of products, then many more of them may be saved. Almost all large terrestrial game species except tapirs (including deer, peccaries, common armadillos, and large rodents) have all the characteristics that make them able to withstand considerable cropping: high reproductive rates, ability to avoid detection, high densities, and large geographic ranges. Game laws should allow subsistence hunters to take all of these, while hunting of large primates should be curtailed, and hunting of small primates strictly limited.

In conservation planning, the biogeography of a region must be considered. One piece of the Amazon rainforest may have a different set of species in it than a superficially similar piece somewhere else: to save all the species, areas throughout the region must be preserved. Certain regions of Neotropical rainforest are currently far more threatened than others. Because of high human populations, deforestation, and relatively small geographic size, forests of southeastern Brazil, northern Colombia, the west coasts of Peru, Ecuador, and Colombia, much of Central America, and the mid- and high-elevation Andean slopes are now at a critical stage where only immediate action seems likely to save many endemic species from extinction. Because so many of its species are endemic, and so little of it is left, the Atlantic forest of Brazil is probably the most severely endangered of all Neotropical rainforests. The Amazon Basin is not in much danger relative to the peripheral regions mentioned above, except for its riverine fauna. However, both the Amazon Basin and Guianas, at little risk when the first edition of this book was written, are now under new threats: massive commercial logging on the northern coast and agroindustry in the drier South are consuming the edges of Amazonia, while within it, resources are being rapidly wasted by uncontrolled local exploitation.

Although the rainforest is of great economic importance to man, both for its global effects on climate and water supply and for its many individually valuable species, the chief reason to save it is a philosophical one: for the species themselves.

Appendix D
Tracks of Large Mammals

The tracks of large mammals in rainforest habitats are far more often seen than are the animals themselves. The presence of tracks can be the only sign that a species is present in an area. Tapirs, for instance, are shy, silent, largely nocturnal, and like to lurk in dense, inaccessible swamps, but they are great walkers, and their weight sinks their tracks deep in the mud. Such tracks can remain easily visible for days or weeks. A single tapir can thus leave a deceptively large number of tracks throughout its haunts.

The feet of most mammals are flexible, dynamic structures that both mold to the terrain and adjust to the speed of travel. The amount of a foot that touches the ground or is impressed in a track varies both with the gait and with the slope and softness of the ground: in hard ground peccary tracks show only the tips of the toes, in softer ground the entire outline of the main hooves will show, and in deep mud the tiny side toes will appear. Tracks of the same animal are thus highly variable. Because tracks are so variable, they contain a great deal of informa-

tion: sometimes the species, age, sex, activity, and time the tracks were made can be read from a single series of footprints.

The best places to look for tracks are on stream and river banks, on beaches near the vegetation line, on the fringes of swamps, and on trails and roads after rain. A cheap, easy, and portable method of making and keeping records of tracks is simply to lay a sheet of clear, stiff plastic (such as thin Mylar or overhead projector transparency sheets) over the track and trace it with a fine-point permanent marking pen (nonpermanent markers will rub off). The date, place, and any notes can be written directly on the same sheet. The tracks shown on the following pages (figs. 14–19) were traced by this method in Manu National Park, Madre de Dios Department, Peru, and in southern Bolivia. An excellent reference is M. J. M. Aranda, *Rastros de los mamíferos silvestres de México* (Xalapa: Instituto Nacional de Investigaciones sobre Recursos Bioticos, 1981).

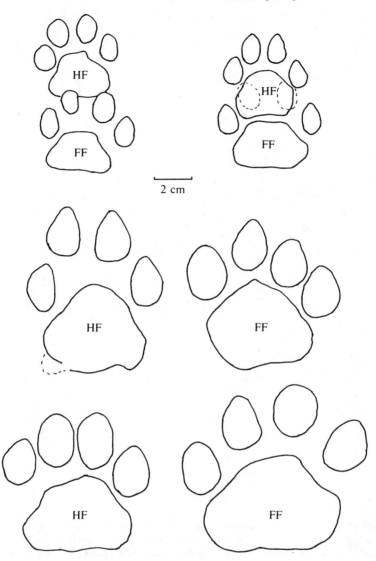

2 cm

Figure 14. *(top)* **Ocelot,** *Leopardus pardalis;* left, female; right, male. Track positions as found in nature, with hindfoot forward of forefoot during walk; left feet. Note forefoot wider than hindfoot; in **margay,** *L. wiedii,* hindfoot and forefoot are about the same size. *(center)* **Puma,** *Puma concolor;* male, hindfoot on left; note slightly pointed toes, length of foot slightly greater than width (both measured at maximum dimension), large gap between center hindtoes and pad. *(bottom)* **Jaguar,** *Panthera onca;* male, hindfoot on left; note round toes, foot wider than long.

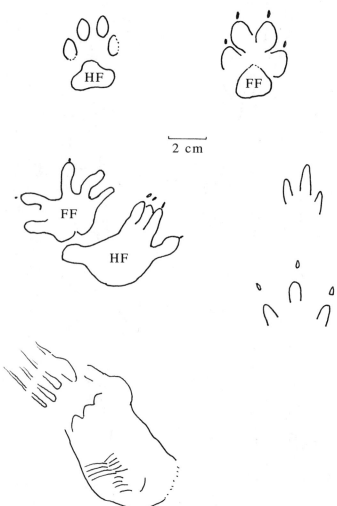

Figure 15. *(top)* Left, **jaguarundi,** *Herpailurus yaguarondi,* note small, narrow toe pads. Right, **crab-eating fox,** *Cerdocyon thous. (center)* Right, **common opossum,** *Didelphis marsupialis,* front followed by hindfoot in natural position, right feet. Right, **nine-banded armadillo,** *Dasypus novemcinctus,* one side toe on front foot and two on hindfoot not visible, as is usual on hard ground. *(bottom)* **Porcupine,** *Coendou prehensilis,* right hindfoot. Forefoot track would appear immediately behind and parallel to hindfoot. Note striations on sole.

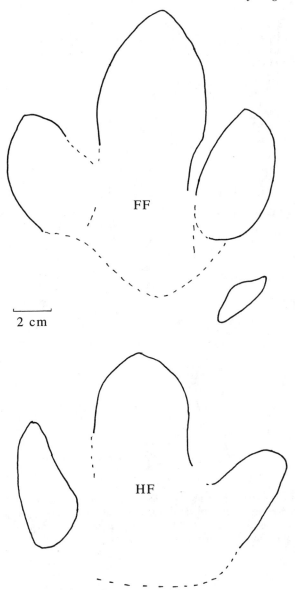

FF

2 cm

HF

Figure 16. Brazilian tapir, *Tapiris terrestris,* right feet. The imprint of the fourth toe of the front foot does not often appear.

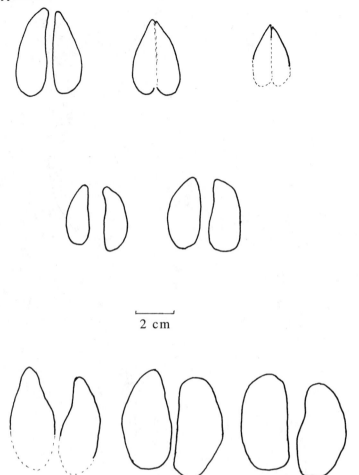

Figure 17. *(top)* Right and middle, **red brocket deer,** *Mazama americana,* on soft and hard ground. Left, **gray brocket deer,** *M. gouazoubira.* Note that the rear of the foot shows prominently in red brocket and faintly or not at all in gray brocket. All left feet. *(center)* **Collared peccary,** *Tayassu tajacu,* hindfoot on left. *(bottom)* **White-lipped peccary,** *T. pecari,* hindfoot (left) and two front feet, all from different individuals.

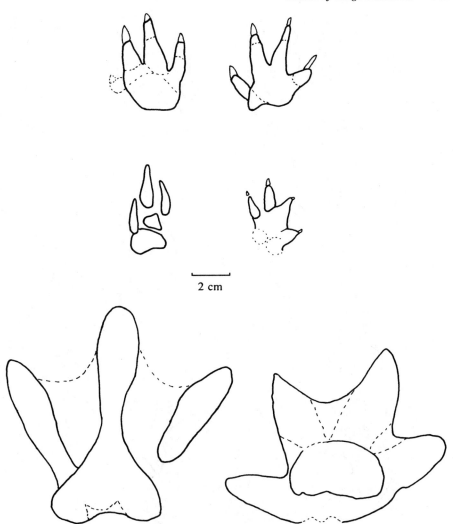

Figure 18. *(top)* **Paca,** *Agouti paca;* hindfoot on left; left feet. Note that the outer digit often does not appear in tracks (dotted on hindfoot). *(center)* **Agouti,** *Dasyprocta variegata;* hindfoot on left; left feet. Tracks of **acouchys,** *Myoprocta,* are similar but much smaller. *(bottom)* **Capybara,** *Hydrochaeris hydrochaeris;* hindfoot on left, left feet.

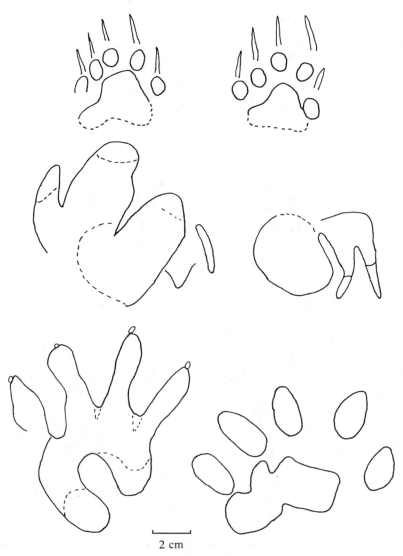

2 cm

Figure 19. *(top)* **Coati,** *Nasua nasua;* hindfoot on left; left feet. *(center)* **Giant anteater,** *Myrmeco-phaga tridactyla;* two left forefoot tracks from same series on different ground (and perhaps different gait). *(bottom)* **Giant otter,** *Pteroneura brasiliensis;* hindfoot on left lacks the impression of one toe (the foot has five toes). The large webs between the toes to their tips did not show in either of two sets of tracks traced, but may show on other substrates. The tracks are associated with drag marks of the tail.

Appendix E
Checklist and Index of Scientific Names

Species of the lowland rainforest fauna are shown in **boldface.** Species or genera from other habitats but treated or mentioned in the text are in regular type. Plate numbers are shown in **boldface** following page numbers.

Marsupialia
Didelphidae, 11–37
Opossum Family
Large Opossums, 11–22
Caluromys derbianus   Central American woolly, 13, **1**
Caluromys lanatus   western woolly, 11, **1**
Caluromys philander   bare-tailed woolly, 12, **1**
Caluromysiops irrupta   black-shouldered, 13, **1**
Chironectes minimus   water, 19, **1**
Didelphis albiventris   white-eared, 16, **1**
Didelphis aurita   southeastern common, 16
Didelphis marsupialis   common, 14, **1**
Didelphis virginiana   Virginia, 17, **1**
Glironia venusta   bushy-tailed, 14, **1**
Lutreolina crassicaudata   lutrine, 20
Metachirops (see _Philander_)
Metachirus nudicaudatus   brown four-eyed, 21, **1**
Philander andersoni   Anderson's gray four-eyed, 18, **1**
Philander frenata, 18
Philander mcilhennyi   Mcilhenny's four-eyed, 19
Philander opossum   common gray four-eyed, 17, **1**
Mouse Opossums, 22–33
Gracilinanus aceramarcae
Gracilinanus agilis   agile gracile, 33
Gracilinanus dryas   wood sprite, 31
Gracilinanus emiliae   Emilia's gracile, 32
Gracilinanus kalinowskii   Kalinowski's gracile, 32
Gracilinanus marica   northern gracile, 31
Gracilinanus microtarsus   Brazilian gracile, 33, **2**
Marmosa lepida   little rufous, 25, **2**
Marmosa mexicana   Mexican, 26
Marmosa murina   murine, 24, **2**
Marmosa robinsoni   Robinson's, 26, **2**
Marmosa rubra   red, 25, **2**
Marmosops cracens, 28
Marmosops dorothea   Dorothy's slender, 29
Marmosops fuscatus   gray-bellied slender, 27
Marmosops impavidus   Andean slender, 28
Marmosops incanus   gray slender, 30, **2**
Marmosops invictus   slaty slender, 31
Marmosops neblina   Cerro Neblina slender, 29
Marmosops noctivagus   white-bellied slender, 26, **2**
Marmosops parvidens   delicate slender, 28, **2**
Marmosops paulensis   São Paulo slender, 30
Micoureus alstoni   Alston's woolly, 24
Micoureus constantiae   pale-bellied woolly, 23

Micoureus demerarae   long-furred woolly, 22, **2**
Micoureus phaea   little woolly, 23
Micoureus regina   short-furred woolly, 23
Short-tailed Opossums, 33–37
    Monodelphis adusta   sepia, 34, **3**
    *Monodelphis americana*ʼ three-striped, 36, **3**
    Monodelphis brevicaudata   red-legged, 33, **3**
    Monodelphis dimidiata, 36
    Monodelphis domestica   gray, 36, **3**
    Monodelphis emiliae   Emiliaʼs, 34, **3**
    Monodelphis iheringi   Iheringʼs, 36
    Monodelphis kunsi   pygmy, 35
    Monodelphis maraxina   Marajó, 36
    Monodelphis osgoodi   Osgoodʼs, 35
    Monodelphis rubida   chestnut-striped, 37
    Monodelphis scalops   long-nosed, 35
    Monodelphis sorex   shrewish, 35
    Monodelphis theresa   Theresaʼs, 36
    Monodelphis unistriata   one-striped, 37

    Xenarthra
Myrmecophagidae
Anteater Family, 38–42
    Cyclopes didactylus   silky or pygmy, 41, **4**
    Myrmecophaga tridactyla   giant, 38, **4**
    Tamandua mexicana   northern tamandua, 41
    Tamandua tetradactyla   southern tamandua, 39, **4**

    Bradypodidae
Three-toed Sloth Family, 42–44
    Bradypus torquatus   maned, 44, **4**
    Bradypus tridactylus   pale-throated, 43, **4**
    Bradypus variegatus   brown-throated, 43, **4**

    Megalonychidae
Two-toed Sloth Family, 44–45
    Choloepus didactylus   southern, 45, **4**
    Choloepus hoffmanni   Hoffmannʼs, 44, **4**

    Dasypodidae
Armadillo Family, 45–51
    Cabassous centralis   northern naked-tailed, 47
    Cabassous unicinctus   southern naked-tailed, 46, **A**
    Dasypus kappleri   great long-nosed, 50, **A**
    Dasypus novemcinctus   nine-banded long-nosed, 49, **A**
    Dasypus septemcinctus   seven-banded long-nosed, 50, **A**
    Euphractus sexcinctus   yellow, 46, **A**
    Priodontes maximus   giant, 48, **A**

    Chiroptera
Emballonuridae
Sheath-tailed Bat Family, 52–58
    Balantiopteryx   least sac-winged, 56, **B**
    Balantiopteryx infusca
    Balantiopteryx io

Balantiopteryx plicata
Centronycteris maximiliani   shaggy, 54, **B**
Cormura brevirostris   chestnut sac-winged, 56, **B**
Cyttarops alecto   smoky, 58
Depanycteris (see *Diclidurus*)
Diclidurus   ghost, 57, **5, B**
Diclidurus albus
Diclidurus ingens
Diclidurus isabellus
Diclidurus scutatus
Peronymus (see *Peropteryx*)
Peropteryx   doglike sac-winged, 55, B
Peropteryx kappleri
Peropteryx leucoptera
Peropteryx macrotis
Rhynchonycteris naso   long-nosed, 53, **5, B**
Saccopteryx   white-lined sac-winged, 53, **5, B**
Saccopteryx bilineata
Saccopteryx canescens
Saccopteryx gymnura
Saccopteryx leptura

Noctilionidae
Bulldog Bat Family, 58–60
Noctilio   fishing or bulldog, 59, **5, B**
Noctilio albiventris
Noctilio leporinus

Mormoopidae
Leaf-chinned and Mustached Bat Family, 60–61
Mormoops blainvillei, 61
Mormoops megalophylla   leaf-chinned, 61, **B**
Pteronotus   mustached or naked-backed, 60, **5, B**
Pteronotus davyi
Pteronotus gymnonotus
Pteronotus parnellii
Pteronotus personatus

Phyllostomidae
Leaf-nosed Bat Family, 62–90

Phyllostominae
Spear-nosed Bat Subfamily, 62–70
Anthorhina (see *Mimon*)
Bartionycteris (see *Micronycteris*)
Chrotopterus auritus   woolly false vampire, 69, **C**
Lonchorhina   sword-nosed bats, 63, **C**
Lonchorhina aurita
Lonchorhina fernandezi
Lonchorhina marinkellei
Lonchorhina orinocensis
Macrophyllum macrophyllum   long-legged, 64, **C**
Macrotis, 63
Micronycteris   little big-eared, 62, **C**
Micronycteris brachyotis

Micronycteris daviesi
Micronycteris hirsuta
Micronycteris megalotis
Micronycteris microtis
Micronycteris minuta
Micronycteris nicefori
Micronycteris pusilla
Micronycteris sanborni
Micronycteris schmidtorum
Micronycteris sylvestris
Mimon   hairy-nosed, 66, **5, C**
Mimon bennettii
Mimon cozumelae
Mimon crenulatum
Phylloderma stenops   pale-faced spear-nosed, 67, **C**
Phyllostomus   spear-nosed, 66, **C**
Phyllostomus discolor
Phyllostomus elongatus
Phyllostomus hastatus
Phyllostomus latifolius
Tonatia   round-eared bats, 65, **C**
Tonatia bidens
Tonatia brasiliensis
Tonatia carrikeri
Tonatia evotis
Tonatia minuta
Tonatia nicaraguae
Tonatia saurophila
Tonatia schulzi
Tonatia silvicola
Trachops cirrhosus   fringe-lipped, 68, **5, C**
Vampyrum spectrum   false vampire, 69, **C**

Lonchophyllinae
Spear-nosed Long-tongued Bat Subfamily, 70–72
   Lionycteris spurrelli   chestnut, 70, **D**
   Lonchophylla   spear-nosed, 71, **D**
   Lonchophylla bokermanni
   Lonchophylla handleyi
   Lonchophylla hesperia
   Lonchophylla latidens
   Lonchophylla mordax
   Lonchophylla robusta
   Lonchophylla thomasi

Glossophaginae
Long-tongued Bat Subfamily, 72–76
   Anoura   hairy-legged, 73, **D**
   Anoura caudifera
   Anoura cultrata
   Anoura geoffroyi
   Anoura latidens
   Choeroniscus   long-nosed, 74, **D**
   Choeroniscus godmani
   Choeroniscus intermedius

Choeroniscus minor
Choeroniscus periosus
Glossophaga   common, 72, **6, D**
Glossophaga commissarisi
Glossophaga leachii
Glossophaga longirostris
Glossophaga morenoi
Glossophaga soricina
Hylonycteris underwoodi   Underwood's, 75
Leptonycteris, 73
Lichonycteris   dark, 74, **D**
Lichonycteris degener
Lichonycteris obscura
Scleronycteris ega   Ega, 74

Carolliinae
Short-tailed and Little Fruit Bat Subfamily, 76–78
   Carollia   short-tailed, 76, **6, E**
   Carollia brevicauda
   Carollia castanea
   Carollia perspicillata
   Carollia subrufa
   Rhinophylla   little fruit, 77, **E**
   Rhinophylla alethina
   Rhinophylla fischerae
   Rhinophylla pumilio

Stenodermatinae
Fruit Bat Subfamily, 78–87
   Ametrida centurio   little white-shouldered, 86
   Artibeus   large fruit-eating, 84, **6, E**
   Artibeus amplus
   Artibeus fimbriatus
   Artibeus fraterculus
   Artibeus hirsutus
   Artibeus incomitatus
   Artibeus inopinatus
   Artibeus intermedius
   Artibeus jamaicencis
   Artibeus lituratus
   Artibeus obscurus
   Artibeus planirostris
   Artibeus (Dermanura)   dwarf fruit-eating, 85
   Artibeus (Dermanura) anderseni
   Artibeus (Dermanura) aztecus
   Artibeus (Dermanura) bogotensis
   Artibeus (Dermanura) cinereus
   Artibeus (Dermanura) concolor
   Artibeus (Dermanura) glaucus
   Artibeus (Dermanura) gnomus
   Artibeus (Dermanura) hirsutus
   Artibeus (Dermanura) phaeotis
   Artibeus (Dermanura) toltecus
   Artibeus (Dermanura) watsoni
   Artibeus (Enchisthenes) hartii

Centurio senex   wrinkle-faced, 87, **E**
Chiroderma   big-eyed, 81, **E**
Chiroderma doriae
Chiroderma improvisum
Chiroderma salvini
Chiroderma trinitatum
Chiroderma villosum
Ectophylla alba   Honduran white, 83, **6**
Enchisthenes (see *Artibeus*)
Mesophylla macconnelli   Macconnell's, 83
Platyrrhinus   white-lined fruit, 80, **E**
Platyrrhinus aurarius
Platyrrhinus brachycephalus
Platyrrhinus dorsalis
Platyrrhinus helleri
Platyrrhinus infuscus
Platyrrhinus lineatus
Platyrrhinus recifinus
Platyrrhinus umbratus
Platyrrhinus vittatus
Pygoderma bilabiatum   Ipanema or double-lipped, 85, **E**
Sphaeronycteris toxophyllum   visored, 86, **E**
Sturnira   yellow-shouldered or hairy-legged, 78, **E**
Sturnira aratathomasi
Sturnira bidens
Sturnira bogotensis
Sturnira erythromos
Sturnira lilium
Sturnira ludovici
Sturnira luisi
Sturnira magna
Sturnira mordax
Sturnira nana
Sturnira oporaphilum
Sturnira tildae
Uroderma   tent-making, 79, **6, E**
Uroderma bilobatum
Uroderma magnirostrum
Vampyressa   yellow-eared, 82
Vampyressa bidens
Vampyressa brocki
Vampyressa melissa
Vampyressa nymphaea
Vampyressa pusilla
Vampyrodes   great stripe-faced, 81, **E**
Vampyrodes caraccioli
Vampyrodes major
Vampyrops (see *Platyrrhinus*)

Desmodontinae
Vampire Bat Subfamily, 88–90
Desmodus rotundus   common, 88, **6, F**
Diaemus youngi   white-winged, 89
Diphylla ecaudata   hairy-legged, 89, **F**

Cynomops greenhalli
Cynomops planirostris
Eumops bonneted, 101, **G**
Eumops auripendulus
Eumops bonariensis
Eumops glaucinus
Eumops hansae
Eumops maurus
Eumops perotis
Eumops trumbulli
Molossops dog-faced, 98, **G**
Molossops neglectus
Molossops temminckii
Molossus mastiff, 102, **G**
Molossus bondae
Molossus coibensis
Molossus macdougalli
Molossus molossus
Molossus pretiosus
Molossus rufus
Molossus sinaloae
Neoplatymops mattogrossensis flat-headed, 99, **G**
Nyctinomops broad-eared, 100, **G**
Nyctinomops aurispinosus
Nyctinomops laticaudatus
Nyctinomops macrotis
Promops crested mastiff, 102, **G**
Promops centralis
Promops nasutus
Tadarida brasiliensis Brazilian, 100, **7**, **G**

Primates
Callitrichidae
Marmoset and Tamarin Family, 104–120
Callithrix argentata silvery marmoset, 106, **8**
Callithrix emiliae Snethlage's marmoset, 107
Callithrix humeralifera tassel-ear marmoset, 107, **8**
Callithrix jacchus tufted-ear marmoset, 108, **8**
Callithrix mauesi Rio Maués marmoset, 108
Cebuella pygmaea pygmy marmoset, 105, **9**
Leontopithecus chrysomelas golden-headed lion tamarin, 119, **8**
Leontopithecus chrysopygus golden-rumped lion tamarin, 119
Leontopithecus rosalia golden lion tamarin, 118, **8**
Saguinus bicolor Brazilian bare-face tamarin, 115, **10**
Saguinus fuscicollis saddleback tamarin, 110, **9**
Saguinus geoffroyi Geoffroy's tamarin, 118, **10**
Saguinus imperator emperor tamarin, 115, **10**
Saguinus inustus mottled-face tamarin, 117, **10**
Saguinus labiatus red-chested mustached tamarin, 113, **9**
Saguinus leucopus silvery-brown bare-face tamarin, 116, **10**
Saguinus midas golden-handed or midas tamarin, 112, **9**
Saguinus mystax black-chested mustached tamarin, 114, **9**
Saguinus nigricollis black-mantle tamarin, 113, **9**
Saguinus oedipus cotton-top tamarin, 117, **10**
Saguinus tripartitus golden-mantle tamarin, 112, **9**

Index of Genera and Common Names